Cosmic Messengers

Martin Harwit, author of the influential book *Cosmic Discovery*, asks key questions about the scope of observational astronomy. Humans have long sought to understand the world we inhabit. Recent realization of how our unruly Universe distorts information before it ever reaches us reveals distinct limits on how well we will ultimately understand the Cosmos. Even the best instruments we might conceive will inevitably be thwarted by ever-more complex distortions and will never untangle the data completely. Observational astronomy, and the cost of pursuing it, will then have reached an inherent end. Only some totally different lines of approach, as yet unknown and potentially far more costly, might then need to emerge if we wish to learn more. This accessible book is written for all astronomers, astrophysicists, and those curious about how well we will ever understand the Universe and the potential costs of pushing those limits.

Martin Harwit is Professor Emeritus of Astronomy at Cornell University, New York. For many years he also served as Director of the National Air and Space Museum in Washington, DC. For much of his astrophysical career he built instruments and made pioneering observations in infrared astronomy. His advanced textbook, *Astrophysical Concepts* (1973), has taught several generations of astronomers through its four editions. Harwit has had an abiding interest in questions first raised in *Cosmic Discovery* on how science advances or is constrained by factors beyond the control of scientists. His subsequent book, *In Search of the True Universe* (Cambridge University Press, 2014), explores how philosophical outlook, historical precedents, industrial progress, economic factors, and national priorities have affected our understanding of the Cosmos. This new book rounds out his informal trilogy on the themes of cosmic discovery. Harwit is a recipient of the Astronomical Society of the Pacific's highest honor, the Bruce Medal, which commends "his original ideas, scholarship, and thoughtful advocacy."

Cosmic Messengers

The Limits of Astronomy in an Unruly Universe

MARTIN HARWIT

Cornell University, New York

CAMBRIDGE
UNIVERSITY PRESS

CAMBRIDGE
UNIVERSITY PRESS

University Printing House, Cambridge CB2 8BS, United Kingdom

One Liberty Plaza, 20th Floor, New York, NY 10006, USA

477 Williamstown Road, Port Melbourne, VIC 3207, Australia

314–321, 3rd Floor, Plot 3, Splendor Forum, Jasola District Centre, New Delhi – 110025, India

79 Anson Road, #06–04/06, Singapore 079906

Cambridge University Press is part of the University of Cambridge.

It furthers the University's mission by disseminating knowledge in the pursuit of education, learning, and research at the highest international levels of excellence.

www.cambridge.org
Information on this title: www.cambridge.org/9781108842440
DOI: 10.1017/9781108903318

First published 2021

Printed in the United Kingdom by TJ Books Limited, Padstow Cornwall

A catalogue record for this publication is available from the British Library.

ISBN 978-1-108-84244-0 Hardback

Contents

Preface

Astronomy is largely a curiosity-driven science. We wonder how our home, the Universe, came into being; how myriad galaxies like our Milky Way may have originated; and how the billions of stars in those galaxies, many of them resembling our Sun, were formed. How many stars have planets similar to those in our Solar System? How many of them harbor life? Does life quite different from any found on Earth exist elsewhere? And could it be intelligent?

Answers to such questions come at considerable cost with no guarantees that our curiosity will ever be fully satisfied.

Certainly, astronomy is not solely driven by curiosity. Astronomers take pride in recalling that our forebears produced charts of the heavens to help sailors navigate the high seas, explore our planet Earth, and set up new trade routes. Today, we are similarly charting the Solar System to map the orbits of kilometer-sized asteroids, piles of rock traversing the spaces between the planets. We need to know whether any asteroid might foreseeably collide with Earth any time soon so we might avoid calamities threatening the survival of life on our planet.

Such practical concerns, nevertheless, are not the main reason many of us chose to become astronomers. Rather, it is curiosity about the origin of the world we inhabit and hope that we might someday discover a guiding principle driving the workings of the Cosmos so we could convince ourselves that "Yes, all this now makes good sense!"

By itself, this wish for clarity is not all we should consider. The public supporting our efforts through the taxes it pays has every right to ask, "Could your curiosity-led search turn out to be endless? Will you then endlessly ask for further support for your quest? Will you astronomers ever be satisfied you have gone as far as possible even though you fall short of meeting your ultimate goals?"

Such questions deserve answers, not only because our fellow citizens have every right to know, but because the clarity we'd gain in formulating a reply could also help us find better ways to pursue our quest.

My purpose in writing this book thus was partly to answer these questions, at least to my own satisfaction, though when I started I didn't know whether I'd succeed: I found myself straying from traditional questions about the origins and workings of the world of planets, stars, supernovae, quasars, and ultimately the entire Cosmos.

Cosmic Discovery, my first attempt to provide a broader look, had described how entirely new astronomical phenomena had frequently been unveiled by unanticipated novel instruments. The invention of telescopes, and the improved designs Galileo had introduced, had revealed moons orbiting Jupiter, mountains broaching the lunar landscape, and a Milky Way resolving into countless individual stars. Similarly, the introduction of radio techniques, three to four centuries later, had led to a string of baffling discoveries of quasars, pulsars, gamma-ray bursts, and other phenomena initially merely given descriptive names because the basic mechanisms at work in them were not yet understood.[1]

In Search of the True Universe, my second effort devoted to similar questions, investigated the extent to which novel approaches of theoretical physics had led to major new astronomical insights. Hans Bethe at Cornell University in Ithaca, New York, had shown how nuclear physics could explain the enormous amounts of energy ordinary stars like the Sun radiate in their lifetimes. And Albert Einstein's general relativity had predicted the potential existence of gravitational waves first indirectly detected more than twenty years after his death.[2]

Cosmic Messengers, the concluding volume in this trilogy, now investigates how information reaches us from across the Cosmos. It focuses on how the messengers – the radio waves, X-rays, visible light, infrared radiation, neutrinos, gravitational waves, cosmic rays, and occasionally odd mixtures of atomic isotopes or stray dust grains – bear witness to remote astronomical events, often after surviving long journeys across vast regions of space.

At times any of these messengers may arrive unscathed; at others irrevocably transformed. Physical processes by now well understood determine which messengers can be trusted to reach us intact, and which cannot. Most messengers can convey information reliably only within limited energy ranges. The information they transmit is necessarily bounded and finite.

Because of these finite bounds, a finite range of instruments may suffice to teach us all we will ever be able to learn through remote observation of the Cosmos. Even though instruments able to detect wider ranges of messenger traits could in principle be constructed, they might serve few astrophysical purposes if the information the messengers transmitted could not be trusted.

We still are uncertain about the full set of messengers the Universe actually conveys. Successive generations of astronomers will surely explore this question further and reassess how much additional information on the Cosmos such messengers could reliably yield.

Two trends have surfaced in the past few years. The first has shown technological advances steadily introducing powerful new messengers conveying new insights on the remote Universe. X-rays and ever-more energetic gamma rays are providing new panoramas of supermassive galactic black holes; gravitational waves and equally elusive neutrinos document vast distant explosions; and the most energetic cosmic-ray particles the Cosmos may ever generate have by now been captured, their energy bounds clearly revealed. A new way of studying the world has arisen, a *multi-messenger science*, in which each of these powerful new cosmic tracers complements and reinforces the messages delivered by all the others. If traditional ways of studying the Universe fail us, any of the newer means of probing the world around us should find paths to break through.

Less well known is another realization just sinking in. It is the recognition that not only massive stars can readily deflect light beams from more remote regions, as Albert Einstein's general relativity first predicted. In a universe like ours, now seen to be filled with masses large and small, light reaching us from great distances has to slalom through crowded fields of intervening stars and galaxies before ever arriving at Earth. After so many gravitational twists and turns, how can we be sure about just where that light originated? How can we be certain about the sequence in which light reaching us today was originally emitted? This too depends on the winding path followed. Passing a massive body close by slows the light beam more than passage at greater distances; and often a light beam splits and does both.

Within our own Milky Way galaxy, we now realize, most stars no longer appear isolated. Surrounded by planets and asteroids similar to ours, they too provide more complex gravitational terrains that passing radiation has to navigate to reach us. Making matters even more complex, planets and asteroids are now found wandering on their own throughout the Galaxy, having broken away from their parent stars. They emit virtually no light of their own; we only know they are there because they occasionally act like gravitational lenses, bending the trajectories and magnifying the light from more distant stars as its rays pass close by them.

With the increasingly powerful telescopes we have by now built, we can detect ever-finer deflections such small bodies induce. But rather than yielding more accurate information on the location and timing of remote astronomical events, we are beginning to realize that we are, instead, more desperately trying to untangle just where those sources might be located and when the light reaching us was emitted, and finding that unseen intervening masses, ranging from dark matter to planets and asteroids, are interfering with that quest.

These are not problems that multi-messenger science will resolve. Einstein's work taught us that not just light, but all other messengers as well, follow trajectories identically bent and delayed by gravitation. If we cannot trace the trajectories gravity induces on light, we will not find complementing paths followed by other messengers. They all are captives of the same gravity.

The next few years and their multi-messenger approaches are thus likely to teach us a variety of new ways of studying the Universe, but only until the telescopes now being built for detecting these newer messengers will reach the limiting capabilities already being approached by optical and radio astronomy – capabilities ultimately hostage to gravitational effects that have already become evident. As we approach the bounds at which the reliability of any and all the messengers reaching us wanes, further observations with instruments already in hand could continue to yield occasional new insights; but building more powerful instruments and observatories with superior sensitivity or resolving powers is likely to yield diminishing returns. A finite set of observational tools could then suffice to take us as far as conceivable instrumentation ever will.

Added insight might then be gained only through extensive searches for events occurring so seldom that their very existence had never been suspected. Searches for these rare occurrences could potentially require century-long surveys conducted with combinations of optical telescopes, radio arrays, cosmic-ray observatories and other major facilities. To make these affordable they would have to be thoroughly automated. For cost will inevitably determine how well we may eventually come to understand the Universe!

Ultimately, we may find that we already possess whatever instruments could usefully teach us more about the Universe. We may then reach the realization that major cosmic phenomena could exist that simply do not generate messengers able to reach us unscathed and intact, and therefore might never be directly observed. Circumstantial evidence might still convince us that some of these, among them a little-understood *dark matter*, or an equally obscure *dark energy* do indeed generate messengers. But doubt of their utility could long linger.

Two considerations, utility and cost, will then force themselves on us. "Will we astronomers at this stage still be asking for further support for our curiosity-driven cosmic quest?" and "How long will it take us to be satisfied that we have pursued our searches to a fruitful conclusion?"

The simple answer to these questions is that astronomy as conducted at that stage will no longer be particularly interesting or worth pursuing purely by means of remote observations.

If society, for whatever pressing reason, wished astronomers to pursue their search further than observations from afar permit, the alternative of launching exploratory voyages beyond the bounds of the Solar System might need to be seriously considered, though recognizing that the expenditures entailed in such ventures would, at least as far as we can judge today, vastly exceed those ever spent on observational astronomy.

A third alternative may then still prevail. An unanticipated class of messengers has recently emerged, delivering information previously thought to be well beyond observational reach. They are the asteroids, comets, or similar bodies ejected from exoplanetary systems. Two such interplanetary stragglers have recently been discovered crossing the inner Solar System along distinct trajectories, making it likely that countless similar intruders may follow.

If we can intercept these stragglers with spacecraft to probe their nature up close, sampling their chemical and structural, or even biological properties before they wander on, we could gather valuable evidence, over time, on planetary processes occurring elsewhere in our Galaxy, the Milky Way. This could avoid costly voyages lasting centuries if not millennia because even the nearest stars are so distant and the spacecraft we could readily build are likely to travel at speeds far below the speed of light.

We should, by all means, derive as much information as possible by studying these intruders, though recognizing that the knowledge potentially gained will almost certainly be restricted solely to an improved understanding of nearby exoplanetary systems. Extragalactic astronomy would, most likely, not be significantly advanced.

Given where we now stand today, the present book aims to explore the range of information accessible to us by purely observational means from within the Solar System, and to compile a list of instruments we may ultimately require to observationally derive most of, if not all, the useful information the Universe can convey. For want of a better name, we may call this assembly of instruments the *Cosmic Toolkit*.

Five considerations may illustrate the usefulness of envisioning such a toolkit.

- First, and foremost: We understand the physical processes governing the transmission of information across the Universe rather better, today, than we grasp most of the complex astrophysical phenomena we have by now observed. This includes the formation of stars and planetary systems, the explosions of supernovae, the mergers of galaxies, or any number of other readily observed events. Without our reliance on the properties of transmitted messengers, we would not even be able to speak with confidence about the nature of those more intricate astrophysical processes.

- Second: Defining the size and complexity of the cosmic toolkit will provide us with estimates of the efforts, time, and funds we may need to invest on designing and assembling a toolkit containing at least fully working prototypes of all required tools. Experience shows that once a satisfactory prototype tool is in hand copies of it may generally be fabricated at appreciably lower cost.

- Third: Over the past 75 years, ever since the end of World War II, new astronomical tools have been adopted with surprising speed. Given our recognition of the primary messengers transmitting information our way, and knowing how rapidly we have assembled the fraction of the prototypical toolkit already in hand today, it seems quite likely that, at current annual expenditures on astronomy world-wide, the entire toolkit could be in place within another century or two. We may then have to provide many clones of this kit for use in observatories across the globe to discover all we may ultimately come to know. The magnitude of those added expenditures would then depend on how rapidly we want to complete our search.

- Fourth: In the course of investigations to date, we have found that major astronomical discoveries generally follow within less than a century of the first inkling or prediction that some new messenger might exist. On this basis it might take no longer, at current expenditure rates on astronomy internationally, than the one or two centuries just cited, before we have all the tools at hand to complete the toolkit and make the major discoveries they would enable. If so, it would resolve one of the two questions this book seeks to address: "How much will it cost and how long will it take to discover the major astronomical phenomena available cosmic messengers may ultimately reveal?"

- Finally: Once the complete toolkit is firmly in place, the future of observational astronomy, as currently practiced, will gradually phase out, as we successively reach saturation levels on information gathered by available means.

The tools required to capture all reliably available information will define the scope of the entire observational approach. Embarking on cosmic voyages to gather additional information directly would then almost certainly escalate costs unaffordably. Such voyages might conceivably be justified if life on our planet were seriously threatened and finding a new habitat for future generations was the only hope for survival of life as we know it.

A new realization then arises. With the finite number and scope of the messengers anticipated to complete our quest, we should expect that the number of major cosmic phenomena we may ultimately discover observationally will remain finite as well – as anticipated in my earlier book *Cosmic Discovery*.[3] The larger body of observational data available today permits an update not only of the list of major cosmic phenomena recognized by now, four decades after *Cosmic Discovery*'s original compilation. It also enables a new estimate of the total number we may ultimately discover, which still appears to be of the order of 100. Other phenomena might then still exist, but the finite scope of the messengers reliably transmitted might not suffice to reveal them.

Some readers may be troubled by the thought that the scope of astronomy may ever become bounded – that someday we may find ourself blocked from learning more. We have become so used to taking for granted that new vistas will always be revealed by ever-improved instruments, that it now seems difficult to accept that technological advances will no longer serve a purpose once further information transmitted by cosmic messengers arrives corrupted.

With these introductions on the ways in which the Universe transmits information, the costs entailed in the capture and reading of that information, and the ultimate limits these two factors impose on how much we may learn about the Universe, we now can outline four questions the book addresses in its four distinct parts.

Part I *Instruments, Messengers, and Cosmic Messages* sketches the role that instrumentation and observations have played in advancing the scope of astrophysics and cosmochemistry.

Part II *The Bounded Energies of Nature's Messengers* dwells on the nature of the particles and radiations conveying all the information reaching us today. It explains how basic physical processes inevitably restrict the energies of different types of messengers to distinct, bounded energy ranges.

Part III *Parameters Specifying Individual Messengers* historically traces instrumental achievements to date, and lists fundamental limits on the information that any known messengers will ever be able to transmit.

Finally, Part IV *The Pace of Progress* lists the most powerful astronomical gains to date, and cites the expense of implementing the observational means and the sociological changes inevitably entailed. It envisions projections into the future based on the history of discoveries to date.

Astronomy as pursued today has been enabled by fundamentally changed social structures, largely reflected in rapidly expanding research groups and the progressive compartmentalization of skills. In turn these changes have raised economic and societal concerns we still have not satisfactorily resolved, but urgently should. They directly affect the lives of the very people on whose efforts the future of astronomy will ultimately depend.

If our research has been sound, it will provide alternatives for society to consider once we arrive at the edge of knowledge, look out into our surroundings as far as observations alone will ultimately permit, and humanity's isolation and confinement to our Solar System begins to fully sink in.

Martin Harwit

Notes

1   *Cosmic Discovery: The Search, Scope and Heritage of Astronomy*, M. Harwit, Basic Books, New York, 1981; reissued by Cambridge University Press, 2019

2   *In Search of the True Universe: The Tools, Shaping, and Cost of Cosmological Thought*, M. Harwit, Cambridge University Press, 2013

3   Ibid., Harwit, 1981

Acknowledgments

I am indebted to my fellow astrophysicists, Alain Omont at the Institut d'Astrophysique in Paris, Ira Wasserman at Cornell University in Ithaca, NY, and Tom Wilson of the Max Planck Institute for Radio Astronomy in Bonn, for their willingness to read and critique the first draft of this book. Each brought a different perspective to the task, as I had hoped; stressed points on which he disagreed with my approach; and recommended inclusion of topics I had not given enough consideration. I thank all three for their willingness to do so much.

James R. Johnson, a dedicated amateur scientist and author of a recent book *Comprehending the Cosmos: An Analysis*, envisions the world around us almost entirely in terms of visual displays – symmetries, diagrams, groupings, flow charts – as well as thought-provoking epithets. I greatly appreciated his unique perspective and recommendations.

Three anonymous referees provided useful assessments solicited by Cambridge University Press. I am indebted to them for their forthright advice.

Rachel Ivie at the American Institute of Physics pointed out the existence of a substantive literature on the prospects of young researchers currently entering astronomy and astrophysics as graduate students or as post-docs. And Drew Brisbin, a former student at Cornell and now a member of the post-doc community, commented on the topic from a personal perspective.

Help with finding some of the early photographs appearing in the book, and possibly finding who might own their copyrights, was provided by a number of colleagues. Jean E. Mueller provided advice on the holdings of the Palomar and Caltech archives, from which the portrait of Fritz Zwicky in Figure 5.1 was obtained. Ruth Isaacson of the Genetics Society of America secured permission for me to publish the likeness of Ronald Aylmer Fisher, in Figure 8.3. Thijs de Graauw, now working on a space project with colleagues in Russia, indefatigably searched for the copyright ownership of the expressive portrait of Yakov Borisovich Zel'dovich reproduced in Figure 2.3. Haruyuki Okuda and Takenori Nakano in Japan similarly were of great help in documenting Chushiro Hayashi's portrait dating back to the early 1940s, and now appearing as Figure 2.4. Nakano, who had carefully preserved this photo since his own student days at that time, later worked closely with Hayashi on a theory of how stars first form. I am deeply indebted to all of these colleagues for taking the time to help me locate the ownership of these sources.

I thank Dietrich Müller at the University of Chicago for referring me to a comprehensive review he had published on cosmic-ray instrumentation and physics.

Anonymous members of the Cornell University Library staff provided access to the countless references to articles listed in the bibliographies at the end of each chapter. I literally could not have written this book without their efficient help and support over the years.

Members of Cambridge University Press, publishers of this book made a wonderful team. Vince Higgs, editor of two previous books I had published with Cambridge University Press, once again provided imaginative advice. Esther Migueliz Obanos, now for a second time, successfully managed a book I had submitted, through all the successive stages it has to transit before finally appearing in print. Margaret Patterson was key in this process as copy editor, instilling cohesion and clarity throughout. The Press artists patiently sought ways of producing black and white drawings that could satisfactorily capture the complexities I wished to display in Figures such as 6.6, 6.7, 6.8, and others.

Our children, Alex Harwit, Eric Harwit, and Emily Harwit, provided unique perspectives, largely shaped by their respective professional insights. Alex is a physicist and engineer in the US aerospace industry. Eric has written about recent industrial revolutions in Asia, primarily in China. Emily has worked on aid to post-war nations and other fragile economies. In one way or another all these communities are seeking to adjust to an ever-evolving modern technological world – as is the astrophysics research community world-wide.

Finally, I want to thank my wife Marianne who, over the years, was always first to read a new book I was drafting. She'd comment on style and substance, and often could sense uncertainties reflected in awkward sentences. Questions she had asked on reading earlier books seemed to her to have accumulated, largely unanswered in this new draft, though by now so obvious that they no longer should or could be ignored. If I was writing about the scope of astronomy and how far we might still have to go, would it not also make sense to ask who was going to pay the bills for all that? And why should they want to, given so many other urgent priorities crying out for support? Did astronomers have a right to ask for public support without at least attempting to assess how many more years, decades, centuries, perhaps even millennia, it might take to satisfactorily understand our Universe? And even if we could answer that question, would the cost of that enterprise demand ever greater public expenditures? And then, what could society reasonably expect to gain in return. Not everybody, after all, was interested in what was going on out there in the Universe, when school lunches for their children were a more immediate concern.

It seemed to me that perhaps, but only perhaps, answers to some of those questions might by now lie within reach, emerging from a perspective of astronomy gained not by immediately focusing on major phenomena, such as supernova explosions, newly discovered planetary systems, or the nature of the microwave background radiation, but by asking how the information gained on these phenomena was reaching us. Which were the messengers, the light, the radio waves, the cosmic rays, the occasional meteorites falling on Earth, conveying all this information? How much more information than we already had in hand by now, could we, in principle, extract from them? And if that information was finite, how could we best assess its scope and how much its extraction could cost?

Initially, I wasn't sure I could succeed in answering all these questions. I thought we still knew too little. Later, it occurred to me that we might know enough, by now, to at least partially assess what observational astronomy and astrophysics might eventually offer.

This book describes the perspective that largely guided me. Finding how earlier generations of astronomers advanced our understanding also inspired me throughout. Readers will find me often referring to their work as I provide my best estimate of how far astronomical observations will ultimately take us; where the end of the current journey may lie; how long it may take us to get there; what we may have learned by the time we arrive; and what all this will cost.

I hope readers will find the book useful.

How to Read This Book

I wrote this book for people interested in the Universe around us. This includes astronomers and astrophysicists, of course. But physicists, chemists, biologists, mathematicians, public officials, governmental representatives, and others may also wish to read it. Many among them will be unfamiliar with astronomical usage – definitions, symbols, various units of energy, time, length, energy, and how these are mutually related.

The book's Appendix provides much of this information, defining vocabulary terms, supplying lists of commonly used symbols, and tabulating units. It begins with an alphabetically ordered list of symbols and then provides a dictionary of commonly used astronomical and physical terms. A number of tables and sketches clarify relationships.

The Appendix is self-contained. It does not list pages in the text on which a given subject is further discussed.

The Index at the book's end is similarly self-contained. It excludes material listed in the Appendix, but serves as a guide to the main pages in Chapters 1 to 9, on which an indexed topic, or the contributions made by a listed scientist may be found. The Index and Appendix thus serve parallel purposes, the Appendix largely acting as a dictionary for non-astronomers, the index guiding readers to subjects of especial interest.

For astronomers, astrophysicists, and other scholars wishing to pursue presented topics to greater depths, the text interweaves references to original papers in which a subject being discussed may be further pursued. Within a chapter these references are marked by numbered superscripts and then spelled out in numerical order in the Notes ending the chapter. Because the text is intended to be self-contained, most readers may rarely need to turn to these added sources.

Occasional footnotes within the text provide supplementary information tagged by alphabetic superscripts. The supplement then appears in smaller print at the bottom of the tagged page.

Part I

Instruments, Messengers, and Cosmic Messages

1

Instruments, Messengers, Astrophysics, and Cosmochemistry

Well before the start of the twentieth century two major classes of instruments revolutionized our understanding of the Universe – the telescope and the spectroscope. A third, the photographic plate, the first means to establish a permanent record which any astronomer could independently examine, became fully embraced only as the twentieth century was about to dawn.

Historians have long celebrated Galileo's contributions to science through his introduction of the improved telescopes he constructed to study the Moon, the planets, and the stars. The revolution introduced by the spectrometer has not been publicized as widely but was of comparable significance. It led to the startling realization that many of the chemical elements that abound on Earth also exist in the outer layers of the Sun and stars.[2] [a; b]

1.1   Fraunhofer's Prisms

To understand the origins of astronomical spectroscopy we need to recall the craftsmanship of Joseph Fraunhofer. In 1815 this 28-year-old, largely self-taught optician was trying to perfect an achromatic lens – a lens to sharply

---

[a] This chapter could not have been written without extensive reference, throughout, to Klaus Hübner's insightful biography *Gustav Robert Kirchhoff: Das Gewöhnliche Leben eines außergewöhnlichen Mannes*, loosely translated as "Gustav Robert Kirchhoff: the ordinary life of an extraordinary man." Surprisingly, as Hübner himself points out, no other comparably complete biography of Kirchhoff surfaced until Hübner's own work appeared in 2010, 123 years after Kirchhoff's death in 1887.[1]

[b] Translations of German passages appearing as footnotes in this chapter are my own (MH).

Fig. 1.1 Portrait of Joseph Fraunhofer. (From Deutsches Museum München: Bild-Nr. 43952, with permission of the Deutsches Museum.)

focus all colors of light onto a single point. This wasn't easy, but for Fraunhofer nothing had ever been.

Born on March 6, 1787, and orphaned at age 11 with no means of support, the young Joseph was bonded through a six-year apprenticeship to a Munich glass grinder and mirror maker. Denied further schooling he remained weak in writing and arithmetic. In 1801, the second year of his apprenticeship, two houses in Munich suddenly collapsed. After four hours of strenuous efforts, Joseph was pulled out of the rubble as sole survivor. The Crown Prince of Bavaria, Maximilian Joseph, who had come to inspect the disaster was impressed by the youngster's intelligence and awarded him 18 Dukats, also promising him further support if needed.

This gesture enabled Fraunhofer to extricate himself from bondage, acquire his own glass grinding and polishing machinery, and buy several books to make up for lost schooling.

Years later, at a newly established institute for mathematical optics in Bavaria, Fraunhofer, shown in Figure 1.1, realized his most ambitious optical achievements, among them the design and construction of achromats fabricated from two mutually compensating types of glass, neither of which by itself could bring all the colors of light to a single focus.[3]

While testing one of these achromats, he hit on the idea of first testing its performance in several narrower monochromatic ranges. To obtain sufficiently bright beams Fraunhofer used sunlight and a prism to isolate the various colors the Sun emits. On closer examination he was surprised to discover that sunlight, ranging from blue at one end of the spectrum to red at the other, was interrupted by hundreds of narrow dark streaks interspersed along his prism's display.[4]

He carefully mapped 324 of these streaks, designating the most prominent among them by the letters of the alphabet by which we still know them today.[5] The dark streak designated D was double. He found it present not only in the spectrum of the Sun, but also in the spectra of the stars Betelgeuse, Capella, Pollux, and Procyon; and its position in the Sun's spectrum coincided precisely with that of a bright yellow feature he had already identified in the light emitted by terrestrial flames.[6]

There, Fraunhofer needed to cease this pursuit and return to his primary goal of designing and fabricating superior achromats. He wrote:

> In these experiments lack of time permitted me to consider solely matters of optical consequence, leaving the rest untouched or insufficiently pursued. Since the path set by these physical-optical experiments appears to promise interesting results, it would be highly desirable for trained scientists to award them attention.[7] [c]

Fraunhofer's contemporaries did not follow this advice. Prisms had long been known to divide light into the colors of the rainbow but appeared to serve no other recognized purposes.

Fraunhofer died in 1826 at age 39. Thirty-three years would have to pass before his efforts once again bore fruit.

1.2   Bunsen's Burner and Kirchhoff's Spectroscope

In the fall of 1859, two colleagues at the University of Heidelberg, the 48-year-old professor of chemistry Robert Wilhelm Bunsen, and his younger colleague, the 35-year-old professor of physics Gustav Robert Kirchhoff, both shown in Figure 1.4, embarked on a collaboration that was to change forever our understanding of the chemical composition and elemental transformation of the Universe![8]

---

[c] Bei allen meinen Versuchen durfte ich, aus Mangel der Zeit, hauptsächlich nur auf das Rücksicht nehmen, was auf praktische Optik Bezug zu haben schien, und das übrige entweder gar nicht berühren oder nicht weit verfolgen. Da der hier mit physisch-optischen Versuchen eingeschlagene Weg zu interessanten Resultaten führen zu können scheint, so wäre sehr zu wünschen daß ihm geübte Naturforscher Aufmerksamkeit schenken möchten.

Four years earlier, Bunsen had invented a burner that now bears his name. It burnt an adjustable mix of air and lighting gas – a mixture of methane, hydrogen, and carbon monoxide extracted from coal.

Bunsen found that, by throttling back the air fed to the burner, he could obtain a low-luminosity flame that exhibited strikingly bright colors when heating the salts of different metals. He wondered whether the colors of these flames, by themselves, might reliably identify the chemical composition of the various salts he was heating. Although his attempts were partly successful, the bright yellow emission of sodium present even in the slightest traces of airborne dust was a major nuisance. Seeking to eliminate this shortcoming, Bunsen considered various ways to filter out the yellow sodium light.

That same year, 1859, Kirchhoff had just constructed a spectroscope for determining the refractive indices of the birefringent crystal aragonite.[9] To help out Bunsen, he used this apparatus to examine substances that strongly absorbed yellow light. Two possibilities appeared suitable. Both cobalt glass and the dye indigo proved opaque to sodium's yellow flame emission and transmitted light at both shorter and longer wavelengths.

Using these findings, Bunsen was able to enhance his method of chemical analysis by flame color, by viewing the flames through a glass-walled prism filled with indigo dye. With this he could filter out sodium's yellow emission by viewing a flame through shorter or longer light paths through the prism, depending on the strength of the yellow emission he needed to suppress. Bunsen published these findings in September 1859.[10]

As Bunsen later recalled, he and Kirchhoff were discussing their respective findings, that September, when Kirchhoff proposed that Bunsen's chemical analyses by color might prove more incisive if he were to obtain the flames' spectra rather than merely their color judged by eye. He thought it would not take too much added effort to modify the spectroscope with which he had been studying the refraction of aragonite: To denote the wavelengths at which he was conducting his measurements Kirchhoff had been passing sunlight through the crystal, using Fraunhofer's dark sunlight features as fiducial wavelength markers.

The two friends agreed to give this method a try.[11]

1.3   Curiosity's Drive

Later, Kirchhoff would write a colleague, "The apparatus we used was assembled in haste from parts, a majority of which we already possessed; it is thence incomplete in some aspects."[12] [d]

---

[d] "Der von uns benutzte Apparat ist in aller Eile aus Theilen, die wir zum größten Theile besaßen, zusammengesetzt; er ist daher in mancher Beziehung unvollkommen."

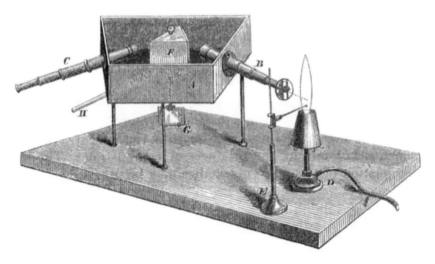

Fig. 1.2 Woodcut of the Spectroscopic Apparatus of Gustav Kirchhoff and Robert
Bunsen. A trapezoidal box A, blackened on the inside, houses a prism F. Mounted on
the sides of the box are two small telescopes B and C. The ocular lenses of telescope B
have been removed and replaced by a plate in which a slit formed by two knife edges
is placed in the focal plane of the telescope's objective. The flame of Bunsen's burner
D is aligned with the axis of telescope B. The end of a fine platinum wire supported
by a small stand E is bent into a hook and placed near the bottom of the flame. On
this hook is a melted globule of metal chloride to be inserted in the flame. Between
the objective lenses of B and C, a hollow prism F, having a refracting angle of 60°, is
filled with carbon disulfide. The prism can be rotated about a vertical axle carrying
a mirror G, above which a handle H simultaneously turns both the prism and the
mirror. A small auxiliary telescope, not shown but placed at some distance, views
the mirror and, through it, the reflection of a horizontal scale indicating the prism's
rotational angle. The telescope C has a fine vertical wire, the wavelength of light
falling on which is indicated by the scale viewed by the auxiliary telescope. A small
light placed near the wire can illuminate it to make it more visible. (This figure was
published in the *Philosophical Magazine* series 4, 20, 88–109, August 1860 with an
English translation of Kirchhoff and Bunsen's original paper. The figure caption
presented here is somewhat shorter than that reproduced in the translated article.)

Complete or not, the insights Kirchhoff and Bunsen gained by means of this
apparatus shown in Figure 1.2 were momentous!

Whereas before 1859 nothing was known about the composition of the Sun
or the myriad stars populating the Galaxy, a year later in 1860 it was clear that
the Sun, and most likely also the stars, contained many if not all of the same
elements known on Earth – and possibly consisted entirely of these elements.
The way to find out was now open!

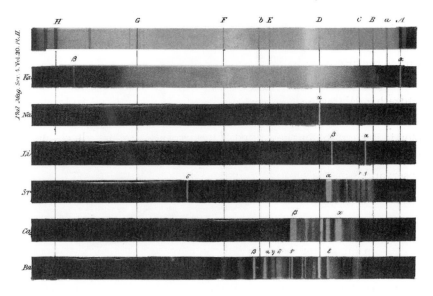

Fig. 1.3 This set of spectra of different elements purified in the laboratory, with their wavelengths calibrated against the Solar spectrum shown at top, was published in the English translation of Kirchhoff and Bunsen's original paper by Henry Enfield Roscoe. At the time the authors identified potassium by the letters *Ka* for *Kalium*, at left, in place of today's accepted abbreviation *K*. The laboratory emission lines for potassium and sodium clearly match the corresponding absorption lines marked *A* and *D* in the Sun. For the remaining elements the correspondences are not as apparent in this reproduction; but the distinct spectrum of every highly purified element does appear to uniquely identify each. (This plate was reproduced in the *Philosophical Magazine* series 4, 20, 88–109, August 1860.)

Kirchhoff and Bunsen first wanted to make sure that the characterizing flame spectra their alkali salts emitted when heated in their flames were due entirely to the alkali component of the salt, and remained constant, whether the alkali salt in question happened to be a chloride, bromide, nitrate, sulfate, etc. After exhaustive tests of hundreds of such highly purified salts, they found each alkali salt's flame spectrum to invariably emit the alkali's characteristic spectrum regardless of the salt the alkali had formed. This assured them that the salts of sodium, potassium, lithium, etc. in their laboratory spectra also implied the existence of these same sodium, potassium, lithium, etc. metals in the Sun.[13] In a second paper, published a year later, Kirchhoff and Bunsen conjectured that this unique spectral identification of the alkali metals their salts contained might be that, at the extreme temperatures of their flames, the salts dissociated to expose the free alkalis.[14]

Although Bunsen and Kirchhoff often mentioned the possibility of determining the chemical composition of the stars by these means, neither of them ever undertook these astronomical measurements.

For another year, Kirchhoff continued his studies of the Solar spectrum and the physical identification of many of its spectral features. The prolonged exposure to the bright sunlight entering his spectroscope's slits began taking its toll on his eyes. To save his eyesight, he eventually abandoned this line of research altogether. With the significance of Kirchhoff and Bunsen's work quickly sinking in, a number of other astronomers began to follow their lead. By 1864, Lewis Morris Rutherford in the United States, Angelo Secchi in Italy, and William Huggins and William Allen Miller in England, respectively, had published their observations on the spectra of Jupiter, Mars, the Moon, and various stars.[15]

1.4   The Discovery of Two New Chemical Elements

In the meantime, Kirchhoff and Bunsen began to concentrate on even more challenging matters. The first was a physical explanation for a ubiquitous reversal of the spectral emission features they found in their laboratory flames into corresponding absorption features in the Sun's spectrum. The second was the discovery and identification of two previously unknown chemical elements giving rise to absorption features they had detected in their spectra of the Sun and in faint emission features they had occasionally encountered in laboratory flames.

To this end, it is worth identifying the distinct roles that Kirchhoff and Bunsen played in reaching their conclusions about the chemical composition of the Sun. On November 13, 1859, Bunsen wrote Henry Roscoe, shown with Kirchhoff and Bunsen in Figure 1.4:

> At the moment I and Kirchhoff are collaborating on work leaving us sleepless. Kirchhoff has made a beautiful, totally unexpected discovery, in determining the origin of the dark lines in the Solar spectrum, and in artificially amplifying these in line-less flame spectra, specifically at the identical position as the Fraunhofer lines. Thereby the way is now clear for determining the chemical composition of the Sun and stars with the same assurance as we can determine strontium chloride, etc., in our chemical reagents. On Earth one can differentiate by this method between materials with the same assurance as on the Sun ... If you have a mix of lithium, potassium, sodium, barium, strontium, calcium, you need only to bring a milligram of this into our apparatus in order to immediately read off all these constituents simply by observing them through a telescope.[16] [e]

---

[e] Im Augenblick bin ich und Kirchhoff mit einer gemeinschaftlichen Arbeit beschäftigt, die us nicht schlafen läßt. Kirchhoff hat nämlich eine wunderschöne, unerwartete Entdeckung gemacht, indem er die Ursache der dunklen Linien im Sonnenspektrum

Fig. 1.4 Photograph of Gustav Robert Kirchhoff at left and Robert Wilhelm Bunsen seated, taken during a joint visit to England in 1862. Their host, Henry Enfield Roscoe at right, a former student of Bunsen and by then professor of chemistry at Owens College in Manchester, had regularly spent his summers in Heidelberg working with Bunsen. By March 1861 Roscoe had lectured on the work of Kirchhoff and Bunsen at the Royal Society in London, and was working on an English translation of their entire first paper on chemical spectroscopy and the nature of the Solar atmosphere, later published in the *Philosophical Magazine*. Through the near-simultaneous publication of their series of papers, in Germany as well as in England, Kirchhoff and Bunsen's astounding spectroscopic work found rapid and enthusiastic reception throughout science. (This photograph taken in 1862 is reproduced here with the kind permission of the Heidelberg University Archives, where it is identified by shelf-mark: Universitätsarchiv Heidelberg, BA Pos I 388.)

Bunsen's words "read off" indicate that he and Kirchhoff already were audaciously interpreting the Solar spectrum's dark Fraunhofer features as a

aufgefunden und diese Linien künstlich im Sonnenspektrum ferstärkt und in Linienlosen Flammenspektren hervorgebracht hat, und zwar der Lage nach mit den Fraunhoferschen identische Linien. Dadurch ist der Weg gegeben, die stoffliche Zusammensetzung der Sonne und der Fixsterne mit der selben Sicherheit

coded inventory of the Sun's chemical composition. If you could identify a metal by its bright emission lines in laboratory flames, you could be sure that Fraunhofer's corresponding dark absorption features in the Solar spectrum meant that the same metals were also present on the Sun.[f]

What about the reverse argument? Must a similar match of spectral features in the Sun and in laboratory flames inevitably imply that one and the same chemical was responsible for both? If so, perhaps the chemical identification of substances imprinting distinct but as-yet-unidentified spectral features in their laboratory flames might imply the existence of as-yet-unidentified chemical elements both here on Earth and in the Sun!

The two friends set to work to find out. Here Bunsen's chemical virtuosity shone through. Although the two articles they jointly published did not identify which of them had been responsible for taking the lead in different parts of the investigation, wherever the work of one of them dominated, the article would only point out that "one of us"[g] had been responsible. Given the vast efforts involved in extracting the tiny amounts of metal responsible for the Solar spectral features they could faintly identify in some of their laboratory flames, Bunsen was clearly the "one" responsible for the vast chemical task of first extracting, and then identifying two previously unknown chemical elements on Earth, and by inference then also in the Sun.

The first of the two elements they found, they named *caesium*, for the two closely spaced blue lines emitted in their flame spectra, reminding them of the Latin term for the blue sky. The second they named *rubidium* after the two closely spaced deep red lines their flame spectra exhibited.[17]

These were colossal achievements in chemical identification and purification. They involved extraction of a few grams of pure caesium salts from 44 tons of brine from the spa Bad Dürkheim in southwest Germany. From 150 kilograms

---

nachzuweisen, wie wir Strontiumchlorid usw. durch unsere Reagenten bestimmen. Auf der Erde lassen sich die Stoffe nach dieser Methode mit der selben Schärfe unterscheiden und nachweisen wie auf der Sonne ... Haben Sie ein Gemenge von Li, Ka, Na, Ba, Sr, Ca, so brauchen Sie nur ein Milligramm davon in unseren Apparat zu bringen, um dann unmittelbar durch ein Fernrohr alle diese Gemengeteile durch bloße Beobachtung abzulesen.

[f] More than a century before bar-coded packages were to appear in twentieth-century grocery stores to identify the contents of a package a customer was buying, Bunsen already understood that Fraunhofer's dark-lined Solar spectrum bar-coded the chemical contents of the Sun's atmosphere!

[g] Einer von uns.

of the mineral lepidolite they similarly extracted small amounts of the pure rubidium salts giving rise to their flame spectra.[18]

The audacity of figuring that the Solar absorption spectrum might well be no more than an encoding of chemical elements present in the Solar atmosphere, and deciding on that basis that flame spectra exhibiting matching emission features would mean that these elements must also exist on Earth, *and should be sought here*, is truly astounding! The breathtaking scope of this work, and speed with which Bunsen's chemical expertise apparently overcame an all-but-overwhelming list of challenges, makes for thrilling reading, even today!

1.5   Inventing Blackbodies and Radiative Transfer

During one phase of their ongoing experiments, Kirchhoff passed a beam of sunlight through a flame containing vaporized sodium salts. To his surprise, instead of filling the dark solar line D with compensating light emitted by the flame, the dark solar feature actually deepened. The light emanating from the Sun appeared to be absorbed by the flame. Another scientist might have set this problem aside but Kirchhoff's curiosity would not let him rest. He needed to explain this unanticipated darkening at least to himself.

When nothing is known about a problem, a useful guide may be offered by basic conservation principles. Kirchhoff's primary guide now became the conservation of energy.

By 1860, most physical scientists, and certainly Kirchhoff, believed in the principle of conservation of energy that the German physician Julius Robert Mayer had proposed in 1842.[19] Kirchhoff argued that, in thermal equilibrium with its surroundings, any substance should emit precisely as much radiation as it absorbed. That way, the amount of energy entering and leaving the substance would be conserved. Here, the substance involved was the flame-vaporized sodium salt. If, by some means, the vapor in his flame were to absorb sunlight at just those wavelengths at which the flame emitted most strongly, the flame would be heated to a higher temperature. It would emit more radiation at other wavelengths, but would weaken the light coming from the Sun at just those wavelengths emitted by the sodium salts in his flame. The reason the flame was absorbing the sunlight, rather than adding to it, was simply that the flame's temperature was far lower than the temperature of the Solar atmosphere where the sunlight originated.[20]

Although this simple explanation is correct, Kirchhoff wanted to be certain he had not overlooked any essential factors: His flame observations dealt solely with a narrow wavelength range of light traversing a highly confined path through his spectroscope. Could his argument falter if the actual events were more complex?

He undertook to simulate the absorption of sunlight by shining limelight through a flame emitting the characteristic yellow light of heated sodium salts. Limelight is a brilliant featureless white light emitted by lime at extremely high temperatures. On passing through the flame's evaporated sodium salt the limelight was absorbed at precisely the same wavelength as the previously observed emission feature from the flame. This left no doubt in Kirchhoff's mind about the validity of his theoretical approach.[21]

He noted that he and Bunsen had likewise reversed the brighter lines of the spectra of lithium, potassium, calcium, strontium, and barium, by heating chlorates of the respective metals during passage of the Sun's rays.[22] He wrote,

> What has been stated concerning sodium is equally true of every other substance which, when placed in a flame of any sort, produces bright lines in its spectrum. If these lines coincide with the dark lines of the solar spectrum, the presence in the Sun's atmosphere of the substances which produce them must be concluded, provided always that the lines in question cannot have their origin in the atmosphere of the earth. In this way means are afforded of determining the chemical constitution of the sun's atmosphere; and the same method even promises some information concerning the constitution of the brighter fixed stars.[23]

Kirchhoff thus reached the important conclusion that the Sun must be a hot glowing body emitting a continuum spectrum. The emitted sunlight, however, escapes the Sun's hot interior through an absorbing atmosphere of cooler gases, which imprint the myriad Fraunhofer absorption features on the Solar spectrum.

One more note on Kirchhoff's theoretical paper on the absorption and emission of flames needs to be emphasized because of its wide-ranging consequences: To test whether he might have overlooked anything, Kirchhoff broadened his investigations to cover all conditions that conceivably could be relevant to his observations. He appeared keenly aware that if he did not consider all eventualities, other scientists would quickly raise objections and lose interest in his arguments.

This was a daunting undertaking. It forced him to conjecture hypothetical bodies that perfectly absorbed and perfectly emitted light at each and every conceivable wavelength. He called these *blackbodies* – a name by which every physicist still knows them today. At low temperatures, such bodies would emit no visible light. As their temperatures increased, they would start emitting red light. At progressively higher temperatures their glow would turn white as the body emitted over an ever-wider range of wavelengths. The intensity of this *blackbody radiation* measured at any wavelength, Kirchhoff concluded, should depend on two, and only two, parameters – the temperature of the radiating body, and the wavelength at which the radiation's intensity was measured.[24]

Further than this broad characterization of the blackbody spectrum Kirchhoff was unable to go. But his realization that blackbody radiation was fully characterized by merely two parameters, its temperature and wavelength, made a deep impression, four decades later, on Max Planck, as Planck was struggling to find a way to interpret the *blackbody spectrum* emitted by heated bodies – a process that would lead him to develop a novel, quantized theory of radiation.[25]

1.6   The Twin-Advent of Astrophysics and Cosmochemistry

The great achievement of Kirchhoff and Bunsen was their decoding the chemical message imprinted in the Solar spectrum, and their demonstration that processes in the Sun and presumably the stars could be understood through the application of trusted physical and chemical methods.

To substantiate this broad claim they meticulously documented each step in their chain of reasoning. They designed and fabricated a spectrometer suitable for both laboratory experiments and Solar observations. They worried whether some of the Sun's dark spectral lines might actually be produced on passage through Earth's atmosphere, but were able to reject the notion.[h] Kirchhoff developed a novel theory of *radiative transfer*, the absorption and emission of radiation of gases as a function of temperature. He tested this theory through experiments on the absorption of limelight in laboratory flames heating well-understood salts. On the basis of spectroscopic signatures registered both in the Solar spectrum and weakly also in some of their laboratory flames, they predicted the existence of two new chemical elements, minute traces of which should be found in previously unknown salts. Through their subsequent discovery and isolation of caesium and rubidium by arduous chemical means they established that the Solar spectrum could indeed be reliably read as a register of the chemical composition of the Solar atmosphere and, by implication, also more generally elsewhere in nature.

With these comprehensive steps, Kirchhoff and Bunsen succeeded in documenting their overarching conclusion: The laws of physics and chemistry applying on Earth also appear to largely hold in the heavens!

---

[h] Presumably they noticed that the dark features did not appreciably change between observations of the Sun at zenith or near the horizon.

In so doing, they simultaneously gave birth to two new research fields – cosmochemistry and astrophysics.

All this they accomplished in two miraculous years between 1859 and 1861!

I have dwelled on the work of Kirchhoff and Bunsen at such length because it so remarkably deepened our appreciation of the means by which the Universe transmits information through naturally generated messengers. Anyone properly equipped could intercept these to interpret the messages they conveyed. For Kirchhoff and Bunsen the messengers were photons arrayed in narrow, well-defined spectral frequency ranges.

The questions the observation of these groupings of photons posed were resolved when the two friends found that many of the narrow absorption features the Solar spectrum exhibits occurred precisely at those wavelengths at which the various alkali salts heated in their flames radiated most strongly.

Though most astronomers would have been satisfied to stop at this juncture and publish their findings, Kirchhoff could not rest until he could also provide an explanation – a theory – of why the Solar features were dark, whereas the laboratory flame features were luminous.

Similarly, Bunsen would not rest until two of the Solar features also found in laboratory flames could be identified as produced by salts of metals which until then had been completely unknown. Figuring out the environments in which such terrestrial elements might be found, and then actually extracting minute quantities of them from tons of mined soils suspected of harboring them, required insight as well as a will to show that not only some, but all of the Solar absorption features were to be associated with well-defined elements – thus confirming that the Fraunhofer lines provided a trustworthy compilation of elements found in the Solar atmosphere.

It was this completeness with which Kirchhoff and Bunsen clarified the physical and chemical information the Solar spectrum conveyed that was so remarkable and unique – impressing the scientific community that not just one, but two new scientific disciplines had just been born, astrophysics and cosmochemistry!

1.7   The Versatility But Also the Ambiguity of Messengers

In retrospect we still need to ask whether the chemical data Kirchhoff and Bunsen had derived from their spectra constituted all the information that astronomical spectroscopy would ultimately yield. Or might additional information be found buried in these spectra that had no connection to chemistry and could ultimately lead to ambiguities and confusion.

The spectrometer Kirchhoff and Bunsen had constructed was a low-resolution instrument. As long as no telescopes were in use to gather increasing amounts of starlight and the human eye was the sole available sensor of radiation, the construction of higher-spectral-resolution instruments would most likely have yielded signals too faint for the human eye to discern. Later, as larger telescopes came into general use, higher spectral resolving powers initiated deeper analyses.

A gift to astronomy that spectroscopy of the late nineteenth century offered was that, besides identifying the chemical composition of stars, it also provided a way for measuring those stars' velocities along an observer's line of sight.

Working in Prague in 1842, the Austrian mathematician and physicist Christian Andreas Doppler had first predicted that both sound and light emitted by moving bodies should arrive at an observer systematically shifted to higher or lower frequencies, depending only on whether the radiating source was approaching or receding from an observer.[26] For sound waves the shift in frequency emitted by sirens could readily be confirmed. Nearly another half century had to pass before an instrument developed by Hermann Carl Vogel, at the Potsdam Observatory near Berlin, could be constructed to reliably measure the small shift in the spectra of stars that line-of-sight velocities as low as a few kilometers per second generally imposed.

By 1888, Vogel and Potsdam's Julius Scheiner were able to photographically record and measure the Doppler shifts and corresponding line-of-sight velocities of several stars.[27]

Photography in use by then made the process much more reliable than observations by eye, and provided a permanent record that could be shared with others. The following year, Vogel and Scheiner also determined that the spectrum of the apparently variable star Algol showed the existence of two sets of superposed spectra in periodic relative motion, indicating Algol to be an eclipsing binary – a pair of stars whose orbital plane happens to be seen edge-on from Earth.

The mass and size of each of the stars now also became apparent. The line-of-sight velocities yielded orbital dimensions and accelerations along a radial direction. Application of Newton's laws of motion, together with the duration of the observed eclipses, then provided corresponding transverse velocities as well as stellar diameters and masses.[28; 29] The nature of such *spectroscopic binaries*, whose velocities as well as variations in magnitude could thus be simultaneously measured, introduced a powerful new method for assessing the masses and sizes of different types of stars. Later, Algol was revealed to actually consist of three stars orbiting a common center of mass.

The *radial velocities* of stars, toward or away from Earth, make themselves apparent in a uniform shift of a star's entire spectrum of chemical features. If the star is approaching, its spectrum is shifted to higher electromagnetic frequencies, shorter wavelengths. Visible features are displaced toward the blue end of the spectrum. If the star is receding, its visible features are displaced toward lower frequencies at the red end of the spectrum – redshifted. The radial velocities of galaxies are measured by these same redshifts.

Spectroscopic methods thus ultimately yielded not only the chemical composition of the Universe, but also the masses of stars and galaxies, as well as the galaxies' internal velocities and mass distributions. And were it not for spectroscopy we would still not know today that increasingly remote galaxies are progressively more highly redshifted – that the Universe is expanding!

By the early twentieth century, spectroscopy of electromagnetic radiation thus had become indispensable not only for identifying chemical elements in stars and galaxies but simultaneously also for measuring their radial velocities.

As if this were not enough, Einstein's theory of general relativity soon also predicted that strong gravitational fields would impose an additional spectroscopic redshift, permitting a determination of the gravitational potential of the region from which a photon had been launched.

Spectroscopy thus not only revealed the chemical constituents of astronomical bodies but also a mix of other physical properties, some of which could be mutually disentangled only through further analysis.

1.8   The Range of Electromagnetic Radiations

With the successes astronomy had garnered though observations conducted in the visible wavelength domain, a few astronomers of the late nineteenth and early twentieth centuries began to search for thermal radiation, heat, emitted by stars and planets in the infrared wavelength domain. By 1933, Karl Jansky, working for the Bell Telephone Laboratories in Holmdel, New Jersey, had also detected faint radio emission appearing to reach us from around the center of our Galaxy, the Milky Way.[30]

Great strides in advancing both infrared and radio observations, however, had to await the introduction into astronomy of novel, more sensitive instrumentation developed for military purposes in World War II. Later, during the Cold War that soon followed, X-ray and gamma-ray instruments also found their way into astronomy through contributions of technologies originally developed for military purposes.[31] By the end of the twentieth century, astronomical signals carried by photons were reaching us and readily detected from across a broad but bounded energy range. Arriving low-energy photons have generally survived exposure to a succession of environments that may have left their imprint on these messengers, deflecting their initial trajectories, potentially delaying their time of arrival, or distorting their original state of polarization. A sizeable fraction of these messengers may even be annihilated before ever reaching us. To correctly interpret the information an arriving photon might convey, the history of its journey to reach us needs to be sorted out.[i] As Figure 1.5 indicates, cosmic photons with wavelengths exceeding a few kilometers cannot reach us even from nearby stars.

If we begin reading Table 1.1 which presents information on radiation starting at the right end of Figure 1.5 and progress toward the left, we quickly note that photons in quite narrow, mutually adjoining spectral ranges often convey information on distinctly different cosmic traits.

This is well illustrated by the relatively narrow ranges in which the dispersion measure of interstellar or intergalactic regions can be detected by virtue of the progressively delayed arrival of radiation at increasing wavelengths λ or equivalently lower frequencies ω. The group velocity U – the velocity at which radio waves at frequency ω propagate – can be shown to be

$$U = c[1 - \omega_p^2/\omega^2]^{1/2},$$

making clear that significant changes in velocity over narrow frequency ranges can only occur close to the plasma frequency ω_p. Similarly narrow ranges over which given types of emission occur also affect other emission mechanisms. The frequencies at which rotating molecules in interstellar space radiate generally are significantly lower than the vibrational frequencies emitted by these same molecules, or the emissions their atomic constituents would usually radiate. As a result, photons emitted by these distinct mechanisms generally fall into different spectral ranges as one proceeds from low to higher energy photons.

In recent decades we have also learned to read and interpret messages conveyed by cosmic rays, highly energetic charged particles – atomic nuclei, electrons, and positrons – some reaching us from the Sun, others generated in violent cosmic explosions within the Galaxy and sometimes in remote galaxies across the Universe.[j]

---

[i] Astronomers observing the Universe in different parts of the electromagnetic spectrum have developed their own designations for units of energy and their equivalent wavelengths and radiation frequencies. Tables A.1 and A.2 of the Appendix explain and cross-reference these various designations.

[j] Astronomers usually refer to the Milky Way, an aggregate of roughly a hundred billion stars, as "the Galaxy," spelled with a capital "G", and refer to all other galaxies with a lower-case "g."

TABLE 1.1

Transmission of Electromagnetic Radiation Toward Earth

- At wavelengths longer than $\sim$3 kilometers, corresponding to radio frequencies lower than 100 kHz or 10^5 cycles per second, the Galaxy's interstellar medium is steadfastly opaque. Though observations of the Sun, planets, and asteroids are still possible, particularly with instruments borne beyond Earth's ionosphere aboard rockets or satellites, transmission of information on more remote astronomical sources is blocked.[32]

- At wavelengths $\leq$ 100 meters information about the Galaxy and the ambient Universe progressively emerges. Interference from Earth's ionosphere also diminishes and some ground-based radio observations can be carried out.

- At wavelengths as short as $\sim$10 meters, corresponding to frequencies of 30 MHz and energies of $\sim 10^{-6}$ electronvolts (eV), transmission through the interstellar medium clears almost entirely, and investigations of pulsars and other Galactic radio sources become possible even with ground-based telescopes.[33]

- Wavelengths in the 20-cm range led to the discovery of fast radio bursts, FRBs, transient events lasting only a few milliseconds, often never to be detected again. Only a few FRBs are repeating sources. To date, FRBs have been detected solely in distant galaxies. As we will later see, low-frequency waves travel more slowly through space than waves at higher frequencies. This effect, called *dispersion*, takes place only in regions populated by freely floating electrons. The observed time delay in a given radio frequency range is directly proportional to the source distance and electron number density along the radio wave's trajectory. Currently FRBs provide the best direct estimate for the number densities of extragalactic electrons, $n_e \sim 300$ pc cm^{-3}. When the FRB's radiation is linearly polarized, with the direction of polarization rotating as a function of radio frequency, this *Faraday* rotation can provide the mean magnetic field along the photons' trajectory.[34]

- As shorter wavelengths in the centimeter range are reached, radio spectroscopy reveals absorption and/or emission by interstellar molecules.

- Thereafter, substantial parts of the millimeter and infrared wavelengths, down to a few micrometers and into the optical range, permit direct spectroscopic detection of interstellar atomic ions, small molecules, interstellar dust, and the spectra of stars. The submillimeter and infrared work is best carried out from above the atmosphere. Much of the optical work can be conducted using ground-based observations.

- Once the ultraviolet regime is broached, Earth's atmosphere again becomes opaque and remains opaque to X-rays and gamma rays. Observations in these ranges are successfully carried out with instrumentation aboard spacecraft. In the gamma-ray energy range from $\sim 4 \times 10^7$ eV to $\sim 2.5 \times 10^{11}$ eV, instruments aboard spacecraft such as NASA's Fermi space observatory regularly provide almost instantaneous data on unanticipated cosmic outbursts.

- Although observation of individual high-energy gamma photons remains beyond reach, Earth's upper atmosphere interacts effectively with these energetic radiations, as well as with comparably energetic charged particles, to generate atmospheric air showers through Cherenkov radiation, providing clear evidence for highly energetic Galactic gamma-ray sources.[a]

---

[a]Cherenkov radiation will be described in Section 3.6.

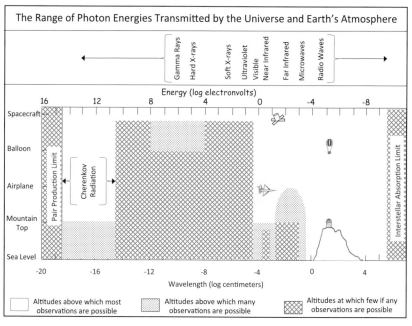

Fig. 1.5 The Wavelengths and Photon Energy Ranges of Electromagnetic Radiation Transmitted by the Universe and our Planet's Atmosphere. The horizontal logarithmic scale at the bottom identifies the wavelengths of electromagnetic radiations reaching Earth from above the atmosphere. The vertical scale shows how deeply radiation at each wavelength can penetrate the atmosphere without being absorbed. Cross-hatched regions of the diagram indicate wavelengths at which no energy is transmitted. The cross-hatching at the extreme right refers to radiation absorbed by the Galaxy's interstellar gases; at the extreme left it refers to annihilation through collisions with the ubiquitous cosmic microwave background. Both of these extreme radiations are absorbed long before ever reaching the Solar System, let alone the upper regions of our atmosphere. Between these two bounds, Earth's atmosphere unevenly absorbs photons at distinct wavelengths. To obtain a clear view of the Universe, astronomical observatories may then need to be placed on mountain tops, high-flying aircraft, high-altitude balloons, or spacecraft well beyond the atmosphere's reach. The region labeled "Cherenkov Radiation," however, also shows benefits the atmosphere offers for detecting highly energetic cosmic-ray particles and radiation. The scales at top and bottom of this figure are logarithmic, meaning that an energy identified by the number 10 is 10^{10} times higher than an energy denoted by 0, because $10^0 = 1$. Table 1.1 explains some of the features shown here in greater depth.

In 1987 we became aware that *neutrinos* could provide astronomical information well beyond the reach of any other messengers. Neutrinos carry no electric charge, and are far less massive than atomic nuclei or even the electrons orbiting those nuclei in atoms and molecules. A single spike of neutrinos reached us that year from a supernova explosion in a small companion galaxy to our Milky Way, the Large Magellanic Cloud at a distance of 170,000 light years. A steady flow of neutrinos soon was also convincingly shown to be arriving from deep inside the Sun.

Neutrinos and antineutrinos from the Sun or generated in supernova or hypernova explosions are the messengers most difficult to detect, but have recently also become central to much of our understanding. They are the messengers best suited to inform us on nuclear processes prevailing deep in the interior of stars. In astronomy, neutrinos play a role analogous to that of X-rays in medicine. They reveal internal processes otherwise hidden from view.

Gravitational waves similarly provide their own unique perspectives on cosmic events. In late 2015, we directly detected a first pulse of gravitational waves. Although Albert Einstein's general theory of relativity had predicted the existence of these waves, we had not known for nearly a century whether cosmic events can ever generate them with sufficient power for unambiguous detection.[35]

We expect that these waves may be emitted in a variety of ways. By now we already know that mutually attracting black holes or neutron stars abundantly radiate these waves as they spiral toward each other and plunge into a final merger, generating a colossal flash of gravitational radiation. The staggering amount of radiation they release in less than their final second before merging well exceeds the total energy the Sun will ever radiate in the course of the billions of years from birth to eventual insignificance.

The first directly observed gravitational waves, veritable tsunamis, provided clear evidence that they had been generated in the merger of a pair of stars each of which had earlier collapsed to form a black hole. Each of these stars was roughly 30 times more massive than the Sun and so compact that not even light could escape its gravitational grip. Whether such pairs exist in nature had previously not been known. Now it seemed they could be fairly common.[36; 37]

In 2017, the merger of two *neutron stars* was also detected. Each was the remnant of a star collapsed into a body resembling a gigantic atomic nucleus consisting predominantly of *neutrons*. This merger distinguished itself through its emission of both gravitational waves and electromagnetic radiation.[38] And, just as Einstein had predicted a century earlier in 1918, both the light and the gravitational waves reached us almost simultaneously. After crossing vast stretches of the Universe for a hundred million years to reach Earth, the emitted gravitational waves and the accompanying gamma rays had arrived at Earth within less than two seconds of each other. We now knew we could rely on gravitational waves to move at the same speed as light.

1.9   Cosmic Messengers and Their Inherent Bounds

Having looked back now at how astrophysics and cosmochemistry gained their start, primarily through information conveyed by electromagnetic radiations, we may begin to examine how the same levels of dedication, carried out on a far broader front, will play out in years ahead until all the messengers that can profitably be observed become accessible with available instruments. To this end we need to be clear about the distinction between *messengers* and the *messages* they may convey.

A photon, a high-energy cosmic-ray particle, a neutrino, all can act as individual messengers, telling us – if we are using the required instruments – the time and the direction from which the messenger arrived. Arrival time and direction, however, are not necessarily the sole *message* conveyed. If the arriving messenger also is passed through a spectrometer, we can determine its energy as well. Here, the arrival time, the arrival direction, and the detected energy can jointly convey *information* we previously lacked. When further interpreted in terms of observations obtained with other messengers from one and the same source, a more complex message may well emerge.

One example already provided for this was clear from the Solar observations of Kirchhoff and Bunsen. Their messengers were photons throughout. The message they uncovered was that the the dark lines in the Sun's Fraunhofer spectrum were a bar-code spelling out the elemental composition of the Solar atmosphere. Like any other coded message it required insight, here provided by laboratory spectra of highly purified elements. But the message came through loud and clear once the data had been correctly interpreted. Ever since, astronomers have conducted chemical studies of stars, planetary atmospheres, interstellar clouds, and whole classes of other sources by spectroscopic means.

At times the messenger and the message it conveys may be difficult to separate. If the direction from which a photon of visible light reaches us points back at the position on the sky where a supernova recently exploded, at least one message it may convey is that this supernova emits visible photons. If we then note that no neutrinos were observed from that same direction, the absence of neutrinos could mean that this supernova did not emit a sufficient flux of neutrinos for our instruments to detect. This absence of a potential set of messengers may in itself provide valuable information – a characteristic of this particular supernova we previously had not noted.

Messengers thus are the conveyors of *messages* – factual information. At times they may not tell us much more than that a certain class of messenger is reaching us from a part of the sky from which no similar messengers had previously arrived. The implied message could then be that a previously unknown source of such messengers has flared up and now does emit them.

1.10   The Central Role of Astronomical Instruments

Even as late as the third part of the twentieth century, most astronomical information garnered had been conveyed by electromagnetic radiation – originally just visible light, later radio waves, and increasingly also ultraviolet photons. By the late 1960s, X-ray and gamma-ray detections, and a rising tide of far-infrared observations added further access to information transmitted by electromagnetic waves.[k]

Some of the observations just described might entail instruments with high spectral resolving power, as a search for the gravitational redshift from low-mass sources might require. Alternatively, they might involve the use of high-spatial-resolution images of an astronomical region to distinguish between emission from compact bodies like stars, as contrasted to nearby interstellar dust clouds. Others again could suggest the use of extremely high time resolution to capture the periodic bursts of radio emission from a pulsar. And, finally, the ability to measure the polarization of a radiating source might lead to insights that otherwise would be difficult to extract.

Assembling the full toolkit for satisfying all these observational capabilities, for electromagnetic radiation, or for the detection of any and all other astronomical messengers including gravitational waves, neutrinos, high-energy cosmic rays, and other carriers of information, can thus be seen to be an essential long-term aim of astronomy.

This might seem a tall order but, as we will see in later chapters, all messengers reaching Earth from regions beyond the confines of the Solar System exhibit finite energy bounds, meaning that the instruments required to detect them also need to respond only to messengers in those energy ranges. These ranges are generally distinct but inevitably finite. For electromagnetic radiation the messengers – the photons – are confined to an energy span from 10^{-10} to $\leq 10^{15}$ eV, as Figure 1.5 shows.

As we will later make clear, the messengers' upper energy bounds can be traced back to collisions among the various types of radiation traversing all

---

[k] Photons are quanta of electromagnetic radiation – radio waves, infrared and visible light, X-rays and gamma rays. Cosmic rays are highly energetic electrically charged particles, atomic nuclei, electrons, and positrons, though highest-energy gamma rays are often also referred to as cosmic rays. Neutrinos, fundamental particles with the lowest known rest-masses, propagate at almost the speed of light. I will refer to the hypothetical particles constituting gravitational waves as *gravitons* throughout the book. Like photons, they have no rest-mass, move at the speed of light and have energies $h\nu$, where h is Planck's constant and ν is the frequency of a transmitted gravitational wave. However, it is worth noting that in various quantum theories of gravitation the designation graviton has quite different connotations. Specifics on each of these messengers may be found in the Appendix.

space, and to the mutual destruction they suffer when they collide. The lower limits sometimes can dip down to zero energy.

Knowing the bounds confining messenger capabilities, we become able to also define the number of different instruments that will be required to satisfactorily capture any messages the Universe faithfully transmits.

1.11   A Cosmic Toolkit

Significant cosmic evolution occurs on time scales of billions of years, far too slow to show up noticeably over a human life span or even a millennium. An ideally complete astronomical toolkit we might hope to assemble as soon as practicable to learn almost everything about the current state of the Cosmos could thus serve those same purposes for countless years ahead without the need for inventing and implementing further tools.

We may call this a *prototypical astronomical toolkit*.

Defining this prototype will not concern itself with the number of copies of each tool in the kit that might ultimately be needed. For some astronomical purposes a whole battalion of identically constructed tools will certainly prove worthwhile. The Atacama Large Millimeter/Submillimeter Array, ALMA, at an elevation of 5 km on the Chajnantor plateau in Chile's Atacama desert, comprises sixty-six 12- and 7-meter aperture telescopes working in combination.

Constructing identical copies of such telescopes costs far less than designing and constructing a first, well-working prototype whose many potential failure modes have been sorted out and avoided in designing a final archetype. Once a first fully operational tool has been constructed and adequately tested, constructing clones is easier and less costly. Further expanding such a network to increase its capabilities also becomes easier.

Disregarding such duplicates or networks, then, how many distinct tools would our basic cosmic toolkit require? To answer this question, our guideline should come, as far as we are able to judge today, from a roster of every conceivable type of information the Universe is capable of transmitting.

For millennia, all we knew about the heavens was conveyed by light visible to the unaided eye and by occasional meteorites – chunks of iron or stone falling from the sky, their origins unknown.

Today, we recognize many more messengers conveying information. They comprise a wide range of electromagnetic and gravitational radiations: highly energetic cosmic-ray particles; electrons and positrons; and neutrinos and antineutrinos, some thought to be arriving from distant galaxies. From our more immediate surroundings, interstellar dust grains, interstellar atomic matter, and rare radioactive isotopes are swept up by the Solar System on its trek through the Galaxy. Some eventually rain down on Earth to convey a rich, variegated vista of the Cosmos.

1.12   The Tools Already in Hand

In several respects, we have already reached instrumental bounds we will never be able to surmount, and encountered astrophysical bounds the Cosmos itself cannot transgress. For now, I will only provide a somewhat random sample; but it will be important for readers to keep in mind that:

• Figure 1.5 shows that receivers for detecting radio waves at the longest conceivable wavelengths are futile for most astronomical studies. The interstellar absorption limit on the right warns us, for example, that highly redshifted long-wavelength radio waves emitted by some of the earliest-formed galaxies are strongly absorbed by gases in our Galaxy's interstellar spaces and will never reach us at Earth's location in the Milky Way. Instead, a better strategy would be to concentrate our instrumental efforts on improving apparatus for detecting short-wavelength radio waves, which are absorbed far less strongly.

• A *pulsar* is a highly compact, rapidly rotating *neutron star*. At high radio frequencies, a particular pulsar in the Crab Nebula remnant of a supernova explosion is known to occasionally emit brief bursts of individual radio pulses referred to as *nanoshots*. In two radio-frequency bands, 6 to 8.5 GHz and 8 to 10.5 GHz, such spikes, individually persisting no longer than $\Delta t \sim 0.4 \times 10^{-9}$ seconds, are evident.

These are the sharpest possible pulses that electromagnetic waves can ever convey within these narrow, $\Delta \nu = 2.5 \times 10^9$ Hz, frequency bands, because Fourier transform theory, and thus also Heisenberg's uncertainty principle, tell us that the product of uncertainty in arrival time and uncertainty in detected frequency, $\Delta t \Delta \nu$, can never be less than of order unity.[39] Existing radio techniques thus already appear to be in hand, or at least are within a factor of ~4 of resolving the sharpest possible radio pulses the Universe can transmit at these frequencies – pulses lasting an almost incredibly short billionth of a second.[40]

• The trajectories of low-energy, electrically charged cosmic rays are strongly deflected and scrambled by magnetic fields as they traverse turbulent interstellar clouds. The directions from which these particles arrive at Earth are almost random. Building instruments to determine their directions of arrival with exquisite directional resolution would thus be useless. Instead, our efforts may be better spent constructing instruments to precisely determine the directions of arriving X-rays and low-energy gamma rays, potentially radiated by these cosmic-ray emitters. Neither X-rays nor gamma rays are electrically charged.

So, neither radiation is deflected by magnetic fields as they journey through interstellar or intergalactic space.

• Later, we will discuss extragalactic gamma rays at energies above $\sim 10^{14}$ electronvolts (eV), hindered from reaching us over intergalactic distances because they are destroyed in collisions with low-energy photons constituting a pervasive *microwave background radiation* permeating the entire Universe. Gamma rays more energetic than $\sim 10^{13}$ eV thus generally cannot reach us from distances exceeding ~ 100 Mpc. Sources at distances of this order consistently lack photons in these high-energy ranges, even though gamma rays in the 10^{13} eV energy range and higher are clearly detected from several powerful sources within our own Galaxy. Chapter 3 will document how these extragalactic barriers arise.

Because many types of messengers are strongly deflected, if not entirely decimated along their intergalactic or even interstellar trek to reach Earth, we need to ask whether some maximum level of competence might suffice beyond which improved instrumental performance no longer increases the information careful observations could provide.

We knew about the potential limitations governing all these distinct astronomical messengers as long as half a century ago. But by now we also know how these messengers are generated, how they are deflected, absorbed or destroyed, and how – as we are beginning to realize – their energy ranges also appear to be bounded, sometimes both at a high- and at a low-energy end of the energy spectrum.

A question that naturally arises is, "Do circumstances exist that limit any and all information the Universe transmits? In short: Does this machine we call *the Universe* first launch and then successively deflect or destroy information, and thus harbor an innate uncertainty principle? If so, can it be explicitly quantified?"

Later chapters will show that such bounds do exist. They include the Heisenberg uncertainty principle of quantum mechanics, which nature imposes on any and all measurements; but we will also find other added constraints that often are more severe.

1.13   The Cost of the Cosmic Toolkit

We are beginning to reach bounds beyond which increasingly powerful instruments will no longer provide more informative data. Once there, our basic toolkit will be complete.

Although we do not know just how soon we may be able to assemble the complete cosmic toolkit, a useful step to take, even today, is to ask how much time, and how large an investment of funds it might require for us to establish the entire set of instrumental capabilities needed to fully capture all the information the Universe reliably provides.

Reaching this goal would at least help us estimate the magnitude of the efforts, the persistence required, and the expense to be met to reach a stage at which we are certain our instruments are tapping and resolving all the information the Universe faithfully transmits – and that we will not fail, through lack of attention or determination, to capture essential data.

Critics may wonder what difference it could make, today, for us to know the magnitude of such a complete toolkit: It is that astronomy by now requires high expenditures, making the assembly of a rationally conceived toolkit all the more important.

A list of tools that such a prototypical toolkit should contain will need to recognize that the full set of observational instruments already in hand is determined by the wide range of distinct messengers we recognize by now, each providing a mutually exclusive perspective on the Cosmos.

The cost of designing and constructing instruments for detecting new classes of messengers never detected before tends to be high. Building copies of instruments already vetted and calibrated is relatively inexpensive. An assessment of the cost of providing astronomers at least one prototypical instrument for any conceivable observation – while avoiding expenditure on capabilities decidedly exceeding useful bounds – should yield a first useful benchmark for assessing the cost of the entire set of tools.

Further costs that may later accrue as different combinations of the various instruments in the kit are duplicated or combined to solve particular problems will be considerably lower than those of entirely novel devices.

Combinations of such instruments may consist of large identical arrays designed to simultaneously observe ever-larger regions of the sky to search for rare events. Sparse arrays of telescopes can yield higher angular resolution and better images than individual telescopes. Arrays of identical telescopes, staring at one and the same small patch of sky, can yield higher sensitivities for detecting low-level signals than any individual telescope could.

Such combinations of instruments certainly will arise and prove useful; but their cost will be more predictable than those for detecting messengers which, for all we know, might not even exist because the Cosmos either is incapable of generating them or of faithfully transmitting them across distances required to reach us.

The list of instruments a complete toolkit will most likely need to comprise will emerge as we begin to document, in chapters ahead, how our reconstruction of the history of the Universe, at least as we understand it today, has advanced; how some of the most valuable insights were won; and how an evolving panorama has steadily been gaining credence over time, though governed by a logic that does not yet fit in readily with the laws of physics as we understand them today.

1.14   What Justifies an Estimate of the Required Toolkit Today?

Four inventories compiled in recent decades lend confidence that our understanding of the Universe by now suffices to at least assess the range of tools a comprehensive toolkit will ultimately have to include. As we learn more, this assessment undoubtedly will improve.

First, we have already assembled an increasingly comprehensive list of all potential messengers – the photons, cosmic-ray particles, neutrinos, gravitational waves, and atomic matter – informing us of events far out in the Universe. Demonstrating the very existence of neutrinos or gravitational waves reaching us from remote explosions required not only patience, ingenuity, and hard work. It also demanded confidence that the substantial funds invested in constructing the entirely novel observatories, never imagined before but essential for detecting these messengers, would eventually reap rewards.

Second, an inventory now in hand lists all the major energy sources at play in cosmic evolution. Originally published by Masataka Fukugita of the University of Tokyo and P. J. E. (Jim) Peebles at Princeton in 2004, and partially updated and abbreviated in Table 1.2, it represented a first comprehensive inventory of the major energy sources driving cosmic evolution today. This *Cosmic Energy Inventory* spelled out its authors' broad ambitions:[41; 42]

> We present an inventory of the cosmic mean densities of energy associated with all the known states of matter and radiation at the present epoch. The observational and theoretical bases for the inventory have become rich enough to allow estimates with observational support for the densities of energy in some 40 forms. The result is a global portrait of the effects of the physical processes of cosmic evolution.

In later chapters we will return to this inventory. For now it suffices to note how detailed our knowledge of the constituents of the Universe has become.

Together, these two inventories appear to require no more than a finite set of messengers respectively bounded by finite energy ranges, mutually evolving along lines governed by currently understood laws of physics, and in good agreement with an emerging cosmic history we will outline in Chapter 2.

Third, by now, we have a well-documented inventory delimiting the energy ranges across which individual messengers can usefully serve as carriers of information. As already indicated, interactions of distinct classes of messengers

TABLE 1.2

The Mass–Energy Density of Different Cosmic Constituents Averaged over the Entire Universe.[a]

Constituent	Fractional Density Parameter[b] $(8\pi G/3H^2)\rho$	Equivalent Mass-Density,[b] ρ in units of g cm^{-3}
Density Parameter, Ω_0	1.00 ± 0.006	9.7×10^{-30}
Dark Energy, Ω_Λ	0.70 ± 0.01	$(6.8 \pm 0.3) \times 10^{-30}$
All Matter, Ω_m	0.30 ± 0.01	$(2.9 \pm 0.3) \times 10^{-30}$
Cold Dark Matter, Ω_c	0.24 ± 0.01	$(2.3 \pm 0.1) \times 10^{-30}$
Cosmic Microwave Background Radiation	5×10^{-5}	4.64×10^{-34}
Cosmic Neutrino Background Component, Ω_ν	< 0.013	$< 1.3 \times 10^{-31}$
Total Baryonic Rest-Mass, Ω_B	0.046 ± 0.003	$(4.4 \pm 0.3) \times 10^{-31}$
Warm Intercluster Plasma	0.040 ± 0.003	$(3.9 \pm 0.3) \times 10^{-31}$
Intracluster Plasma	$(1.8 \pm 0.7) \times 10^{-3}$	$(1.8 \pm 0.7) \times 10^{-32}$
Main Sequence Stars in All Types of Galaxies	$(2.1 \pm 0.4) \times 10^{-3}$	$(2.1 \pm 0.4) \times 10^{-32}$
White Dwarfs	$(3.6 \pm 0.8) \times 10^{-4}$	$(3.5 \pm 0.7) \times 10^{-33}$
Neutron Stars	$(5 \pm 2) \times 10^{-5}$	$(5 \pm 2) \times 10^{-34}$
Black Holes	$(7 \pm 2) \times 10^{-5}$	$(7 \pm 2) \times 10^{-34}$
Substellar Objects	$(1.4 \pm 0.7) \times 10^{-4}$	$(1.4 \pm 0.7) \times 10^{-33}$
Planets	10^{-6}	10^{-35}
Gaseous Hydrogen and Helium Atoms	$(6.2 \pm 1) \times 10^{-4}$	$(6 \pm 1) \times 10^{-33}$
Molecular Gas	$(1.6 \pm 0.6) \times 10^{-4}$	$(1.6 \pm 0.6) \times 10^{-33}$
Dust	$(2.5 \pm 1.2) \times 10^{-6}$	$(2.4 \pm 1) \times 10^{-35}$
Baryons Sequestered in Massive Black Holes[c]	$4 \times 10^{-6}/(1-\epsilon)$	$4 \times 10^{-35}/(1-\epsilon)$
Radiant Energy Originating in Stars	2×10^{-6}	2×10^{-35}
Neutrinos from Stellar Core Collapse	3×10^{-6}	3×10^{-35}
Cosmic Rays and Magnetic Fields	$\sim 5 \times 10^{-9}$	$\sim 5 \times 10^{-38}$
Kinetic Energy in the Intergalactic Medium	$(1 \pm 0.5) \times 10^{-8}$	$(1 \pm 0.5) \times 10^{-37}$
Binding Energy from Primeval Nucleosynthesis	-8×10^{-5}	-8×10^{-34}
Gravitational Binding from Primeval Structure	$(-8 \pm 1) \times 10^{-7}$	$(-8 \pm 1) \times 10^{-36}$
Binding Energy from Gravitational Settling	-10^{-5}	-10^{34}
Binding Energy from Stellar Nucleosynthesis	-5×10^{-6}	-5×10^{-35}

[a] This table is largely based on the work of Masataka Fukugita and P. J. E. Peebles,[43] but with more recent updates by Shull, Smith, & Danforth;[44] E. L. Wright;[45] the WMAP collaboration;[46] and the Planck Collaboration.[47]
[b] Based on a Hubble constant $H_0 = 70$ km s^{-1} Mpc^{-1} and a Euclidean model of the Universe, whose density parameter is $\Omega_0 \equiv (8\pi G/3H^2)\rho = 1$.
[c] ϵ is the fractional mass–energy radiated away in the formation of the black hole.

lead to mutual annihilation or disfigurement. The highest messenger energies recognized today are attained by cosmic-ray particles, whose maximal energies appear to lie just above $\sim 10^{20}$ eV. The highest energies attained by all other messengers are orders of magnitude lower.

Fourth, an inventory is now in hand of the capabilities of all the known messengers for conveying information on the angular size of observed celestial sources; spectra of their emitted energies; detailed timing of observed sequences of events; or levels of polarization at distinct messenger energies. All these can be graphically displayed in terms of a multidimensional *phase space* we will encounter in Chapter 6, where currently available instrumental capabilities are clearly summarized. By inference these displays also identify instrumental capabilities not yet in hand though almost certainly attainable if the required investments were considered worthwhile.

Together these four inventories define the messengers on which all informative observations rely; the major sources of cosmic energy; the ranges of energies individual messengers can attain; and the characteristics of individual messengers to be mined to extract all the information they can convey.

1.15   The Pace of Progress and Its Dependence on Investments

Six centuries ago, it could have seemed as though we'd never know everything significant about our home planet Earth. We only had rather rough indications of Earth's surface area and did not know for certain how many oceans or continents there were. Yet, six hundred years later, we have maps of terrain identifying the size and height of features on scales of meters – or less. We'll never find an unknown new ocean on Earth, nor a mountain higher than Mount Everest. We've arrived at a point where the end of our geophysical search is coming into sight.

Why should we not expect our understanding of the Universe, our appreciation of its nature, to grow with comparable speed provided we can meet some simple prerequisites?

Our experience, to date, can be summarized fairly simply:

• We certainly will not comprehend the Universe fully until we have exhausted all available means to access the lode of available astronomical information. Can we not, at least, estimate how long this may take?

• It is easy to see that even this more modest task will certainly not be accomplished before we have assembled a complete toolkit enabling us to gather, at will, every form of information the Universe reliably transmits. If we lack this ability there may always be some crucial piece of information we will miss.

• From this point of view, it makes added sense to ask how long it may take and the efforts we will need to invest to assemble the complete astronomical toolkit.

• The history of astronomical discoveries, to date, has consistently recorded periods of rapid advances whenever new observational capabilities were introduced. Providing the tools for discovery thus constitutes the very least effort we should be prepared to expend before we can expect to comprehend the Universe as well as may ultimately become possible.

• The initial use of a tool generally is followed by a phase in which the early discoveries it enabled are fleshed out, but further major discoveries fail to materialize until some further tool, often constructed along quite different lines, becomes available and yields added insight.

• Acquiring the right tools usually is only a first step. The tools still have to be put to use if we are to learn anything. But the expense of constructing powerful new instruments has usually been as costly as the actual observational work that followed in the course of the tools' lifetimes.

1.16   An Inventory of Messengers and Their First Detections

The costs of detecting novel messengers with increasingly sensitive instruments operating reliably in dedicated observatories determine the speed at which astronomy advances and the expenses entailed.

Two of the newly noted messengers, listed immediately below, were introduced as the eighteenth century was waning and the nineteenth just starting. Further pursuit of novel methods continued into the late nineteenth and early twentieth centuries. A staccato of instrumental advances following World War II, without a doubt, provided the greatest impetus to date. So persistent was the cascade of advances in viewing the Cosmos, that only a summary of the myriad channels of information that suddenly opened and the stunningly rapid gains of astronomical insight that followed can fully convey the sheer rush of progress.

Meteorites In 1798, two Göttigen University students, H. W. Brandes and J. F. Benzenberg, both 22 years old, undertook measurements from two well-separated locations near Göttingen to trace the trajectories of meteors streaking across the night sky. By ordinary methods of triangulation, they reconstructed the heights at which the meteors were passing, and found these to be far greater than the atmosphere had been thought to extend. They also inferred the velocities at which the meteors were traveling. This soon made clear that the meteors must be messengers originating in space and impacting our planet. Deciphering just where they came from and what they might teach us would still require time.[48] [I] No delay occurred between these curiosity-driven observations, and the appreciation that meteors unquestionably had an extraterrestrial origin. Previously the origin of these falls had been quite uncertain.
[Elapsed time 0 years]

Infrared Radiation Two years later, in 1800, the German-born musician William Herschel, who had established new roots in England where he had developed an all-encompassing passion for probing the

---

[I]   Both Brandes and Benzenberg went on to become professors of physics – Brandes at Leipzig and Benzenberg at Düsseldorf.

heavens, discovered that the spectrum of radiation from the Sun extended well beyond the red end of the visible range. Our eyes could no longer detect this radiation, but the infrared stream of energy the Sun transmitted was readily sensed by a thermometer. Although serious infrared astronomical observations were undertaken within some decades following Herschel's investigations, the most significant breakthrough came 165 years later, in 1965 at Caltech, where Gerald Neugebauer, D. E. Martz, and Robert B. Leighton discovered a class of stars emitting radiation almost solely at infrared wavelengths.[49] Only five years later, in 1970, D. E. Kleinmann and Frank J. Low in the United States discovered galaxies radiating most of their energy at mid-infrared wavelengths.[50]
[Elapsed time 165–70 years]

Chemical Spectroscopy The introduction of spectroscopy for chemical analyses solved the mystery of the origin of the dark spectral lines in the Sun's spectrum, which Joseph Fraunhofer had discovered in 1815. Gustav Kirchhoff and Robert Bunsen at Heidelberg showed in 1860 that chemical elements could be uniquely identified in the Sun through their corresponding laboratory flame spectra.
[Elapsed time 45 years]

Radio Waves By 1865, the great Scottish physicist James Clerk Maxwell had extended Michael Faraday's work earlier in the nineteenth century to develop a theory showing electromagnetic radiation to be a transverse wave propagating at the speed of light; in 1888, the German physicist Heinrich Hertz explicitly showed the theory to hold for radio-frequency waves. By 1911, the Italian engineer Guglielmo Marconi was demonstrating that radio waves could be transmitted across the English Channel; and in 1933, Karl Jansky working at the Bell Telephone Company in the United States discovered radio emission from the Galactic Center.
[Elapsed time 22–68 years]

X-Rays Wilhelm Röntgen's discovery of X-rays at Würzburg in Germany in 1895 was followed in 1962 by the discovery of extrasolar X-rays by a group of scientists working with Riccardo Giacconi at the American Science and Engineering Corporation and Bruno Rossi at the Massachusetts Institute of Technology. The identification of the locations of these sources in 1964 by a group led by Herbert Friedman at the US Naval Research Laboratory in Washington then sealed the discovery.[51; 52; 53]
[Elapsed time 67–69 years]

Radioactivity Henri Becquerel's discovery of radioactivity in Paris in 1896, and Pierre and Marie Curie's discovery, also in Paris, of two new radioactive elements in 1898, were first shown in 1948, by the Austrian-born US scientist H. E. Suess, to have significant astronomical consequences.[54]
[Elapsed time 50 years]

Cosmic Rays In 1900 J. Elster and H. Geitel in Germany investigated the dissipation of electric charge from a body in air.[55] By 1933, the US physicist Thomas H. Johnson, carrying out observations in Mexico city at an elevation of 2250 m above sea level, had clearly identified an east–west asymmetry of cosmic radiation, indicating that primary particles appeared to be positively charged.[56] Six years later, through a series of balloon flights carried out in Panama, Johnson was able to establish that the positively charged primary cosmic-ray particles were protons.[57]
[Elapsed time 33–39 years]

Gamma Rays Also in 1900, the French chemical physicist Paul Villard discovered gamma rays emitted by radium. The first discovered cosmic gamma-ray sources were gamma-ray bursts, initially noticed through the US military's Vela project in the late 1960s but kept classified until 1973.
[Elapsed time ∼70 years]

Isotopes Between 1913 and 1917 Frederick Soddy in Aberdeen discovered that radioactive atoms can generate elements of less massive atoms, some of which exhibit distinct atomic weights but identical chemical properties. He referred to such atoms as mutual "isotopes" of an element.[58]

Positrons The potential existence of positrons, particles essentially identical to electrons but carrying a positive rather than negative charge, was postulated by the Cambridge theoretical physicist Paul A. M. Dirac in 1930.[59] At Caltech, Carl D. Anderson discovered positrons in his laboratory observations of cosmic rays in 1932.[60]
[Elapsed time 2 years]

Neutrinos The Austrian-born Swiss theoretical physicist Wolfgang Pauli predicted the existence of neutrinos in 1930.[61] The US experimentalists Clyde Cowan and Fred Reines discovered these neutrinos in 1960.[62] In 1987, astronomical neutrinos and antineutrinos were reliably detected, both in Japan and in the United States, when Supernova 1987A exploded in the relatively nearby Large Magellanic Cloud.[63; 64]
[Elapsed time 57 years]

Dark Matter In 1933 the Swiss astrophysicist Fritz Zwicky, working at Caltech in Pasadena, observed that individual galaxies in galaxy clusters exhibited velocities so high that the galaxies would escape the clusters unless some unseen matter were present whose gravitational forces kept them from escaping. He postulated the existence of what he called *dark matter* to keep the clusters of galaxies intact.[65]
By the late 1970s, Vera Rubin and W. Kent Ford at the Carnegie Institution in Washington, DC, had noted that the speeds at which stars and interstellar gases orbit the central regions of their galaxies were so high that the galaxies' gravitational attraction had to be produced by masses far exceeding the combined mass of all the stars and gas observed.[66]

Dark Energy In his first efforts to model the Universe by means of general relativity, in 1917, Einstein introduced his cosmological constant Λ, a mass–energy density designed to keep the Universe static – as he and most of his contemporaries thought it was. In 1998, two distinct supernova consortia determined that today's expansion rate of the Universe was considerably higher than anticipated based on the combined mass–energy densities of cosmic baryons and dark matter combined. They attributed this to a new dark energy that could be represented by a cosmological constant Λ based on the observed cosmic expansion rate, the *Hubble constant*.[67; 68]

Gravitational Waves Einstein's and Rosen's 1937 theoretical prediction of gravitational waves was tested through pulsar observations by the US physicists Hulse and Taylor in 1975, confirmed through observations by Taylor and Weisberg in 1982, and further refined by Weisberg and Taylor in 1984.[69; 70; 71; 72]
[Elapsed time 45 years]

The LIGO/Virgo international collaboration of roughly a thousand physicists established the existence of gravitational waves as information carriers by their detection of an event in late 2015 that clearly identified the merger of two stellar-mass black holes.[73]
[Elapsed time 79 years]

Table 1.3 summarizes the rates at which these new messengers and analytic instruments to detect them became available.

TABLE 1.3

Astronomical Messengers: From Experiment or Theory to Cosmic Observation

Information Carrying Messenger	Laboratory Measurement or Theoretical Prediction by Scientists and Dates, t_c	Initiators of First Astronomical Discoveries	Discovery Dates t_{np}	Elapsed Years Δt_e $t_{np} - t_c$
Light		Antiquity		
Meteorites		Benzenberg & Brandes	1798	0
Infrared Radiation	William Herschel, 1800	Neugebauer et al. / Low et al.	1965–70	165
Chemical Spectra	Fraunhofer, 1815	Kirchhoff & Bunsen	1860	45
Cosmic Rays	Elster & Geitel, 1900; Hess, 1912	T. H. Johnson	1933–39	27–39
Radio Waves	Maxwell, 1865; Hertz, 1888	K. Jansky / J. S. Hey	1931–46	66–81
X-rays	Röntgen, 1895	Friedman / Giacconi	1964	69
Radioactivity	Becquerel and the Curies, 1896–98	H. E. Suess	1948	50–52
Gamma Rays	Paul Villard, 1900	Vela Project Members	1973	73
Isotopes	F. Soddy, 1917	H. E. Suess	1948	31
Positrons	Paul Dirac, 1930	Carl D. Anderson	1932	2
Dark Matter (?)		Zwicky / Rubin & Ford	1933–79	(0?)
Neutrinos	Pauli, 1930	Kamiokande & IMB	1987	57
Dark Energy or Λ (?)	Einstein, 1917	Two Supernova Consortia	1998	(81?)
Gravitational Waves	Einstein & Rosen, 1937	LIGO / Virgo	2016	45–79

The table's first column lists the new classes of messengers. The second lists the scientists who introduced a potential messenger by either theoretical or experimental means, as well as the year, t_c, the concept was introduced. The third column lists the astronomers or consortia discovering a previously unknown phenomenon with the newly available messenger. The fourth records the year of the novel phenomenon's discovery, t_{np}. The final column shows the elapsed number of years, $\Delta t_e \equiv t_{np} - t_c$, from conception of a new messenger to the discovery of a novel astronomical phenomenon it enabled. The mean elapsed time, $\langle \Delta t_e \rangle$, is roughly 60 years, though Δt_e ranges from 0 to 165 years.

Many of the intervening periods, Δt_e, have been quite short. Complex capabilities, often pioneered at enormous expense for industrial or military purposes or for fundamental investigations in physics or chemistry, appear to have subsequently been imported and assimilated into astronomy with little delay and at relatively low cost.

Different astronomers and astrophysicists may disagree on the individuals or projects credited in columns 2 and 3 of Table 1.3, or the specific years listed in columns 3 and 4. But it is unlikely that their entries in the last column would differ by more than a few years, one way or another. The speed with which newly discovered scientific concepts have been assimilated into astronomy's armamentarium for discovery has been impressive. Provided this pace can be maintained, the effort to assemble the entire observational toolkit fully tapping all of the channels of information the Universe offers may take no more than another century or two – even if other potential carriers of information such as dark matter may some day need to be added to the list.

Will this application of new methods and construction of novel instruments continue forever to lead to further discoveries? Most likely it will not. Natural bounds on the energy of the available messengers will limit the different observations we can undertake. Increasing sensitivity will not necessarily lead to ever-more penetrating observations. Naturally occurring fluctuations; gravitational deflection along viewing directions; absorption, scattering, and diffraction of radiation by intervening gas and dust clouds; deflection of charged cosmic-ray particles by magnetic fields, and other factors will ultimately provide added restrictions to profitable angular, timing, and spectral resolving powers.

Further advances may then come only through simultaneous observations carried out by combined arrays of instruments already in hand rather than by improved apparatus. If so, the cost of astronomical observations should correspondingly decline, as we will mainly be making use of available instruments or readily fabricated duplicates. The task of completing the toolkit for observing all messengers the Universe transmits will then have been accomplished.

How will the conduct of our search thereafter advance before we find ourselves satisfied that we fully understand the leading cosmic processes that fashioned the evolution of the Universe? We may not yet be equipped to usefully answer this question. But we do have ways for making informed estimates of how long it may take for us to discover most of the major phenomena for which the Universe is able to transmit information. These are questions to which we will turn in Part IV of the book.

Notes

1 *Gustav Robert Kirchhoff: Das Gewöhnliche Leben eines außergewöhnlichen Mannes*, K. Hübner, Verlag Regionalkultur, Heidelberg, 2010

2 Bestimmung des Brechungs- und Farbenzersteuungsvermögens verschiedener Glasarten in Bezug auf die Vervollkommung achromatischer Fernröhre, J. Fraunhofer, München, 1817, pp. 193–226. Reprinted in Ostwald's *Klassiker der Exakten Wissenschaften #150*, published by Wilhelm Engelmann, Leipzig, 1905, pp. 1–34

3 Kurzer Umriß der Lebensgeschichte des Herrn Joseph von Fraunhofer, Joseph von Urtzschneider, München, 1826. Reproduced by Google Books

4 Ibid., Fraunhofer

5 *A Popular History of Astronomy During the Nineteenth Century*, Agnes M. Clerke, Adam and Charles Black, London, 1908. Republished by Scholarly Press, St. Claire Shores, Michigan, 1977, p. 133

6 Ibid., Fraunhofer, pp. 28–29

7 Ibid., Fraunhofer, p. 30

8 Ibid., Kirchhoff, 2010, pp. 117 ff

9 Ueber die Winkel der optischen Axen des Aragonits für die verschiedenen Fraunhofer'schen Linien, G. Kirchhoff, *Annalen der Physik* 08, 567–75, 1859

10 Ibid., Hübner, 2010, pp. 117

11 Ibid., Hübner, 2010, pp. 118–19

12 Kirchhoff's letter of September 17, 1860 to the physicist Christian Ludwig Gerling, cited by Hübner, ibid., Hübner, 2010, p. 119

13 Chemische Analyse durch Spektralbeobachtungen: Erste Abhandlung, G. Kirchhoff & R. Bunsen, *Annalen der Physik und Chemie* 110, 161–89, 1860. Translated into English in the *Philosophical Magazine*, series 4, 20: 131, pp. 88–109, 1861

14 Chemische Analyse durch Spectralbeobachtungen: Zweite Abhandung, G. Kirchhoff & R. Bunsen, *Annalen der Physik und Chemie* 113, 337–81, 1861. Translated into English in two issues of the *Philosophical Magazine*, series 4, 22: 148. pp 329–49, and 4, 22: 150, 498–510, both published in 1862

15 Ibid., Hübner, 2010, p. 135

16 Ibid., Hübner, 2020, p. 125

17 Ibid., Kirchhoff & Bunsen, 1861, Zweite Abhandung

18 Ibid., Hübner, 2010, p. 128

19 Bemerkungen über die Kräfte der unbelebten Natur (Remarks on the Forces of Inanimate Nature), J. R. Mayer, *Liebigs Annalen der Chemie*, pp. 233 ff., 1842

20 On the Relation between the Radiating and Absorbing Powers of Different Bodies for Light and Heat, G. Kirchhoff, *Philosophical Magazine*, Series 4 20: 131, 1–21, July 1860, see p. 17. This publication was a translation of the German article Kirchhoff had submitted to the *Annalen der Physik* in January 1860: Ueber das Verhältnis zwischen dem Emissionsvermögen und dem Absoptionsvermögen der Körper für Wärme und Licht, *Annalen der Physik* 109, 275–301, 1860

21 Ibid., Kirchhoff, 1860, p. 15

22 Ibid., Kirchhoff & Bunsen, 1861, Erste Abhandlung

23 Ibid., Kirchhoff, 1860, p. 17

24 Ibid., Kirchhoff, 1860, pp. 6 and 12

25 Ueber das Gesetz der Energieverteilung im Normalspectrum, M. Planck, *Annalen der Physik*, series iv 4, 553–63, 1901

26 Über das Farbige Licht der Doppelsterne und einiger anderer Gestirne des Himmels, C. Doppler, *Abhandlungen der königlich böhmischen Gesellschaft der Wissenschaften, zu Prag*, Folge V, 2, 3–18, 1842

27 Untersuchungen über das spektroskopische Doppelsternsystem β Aurigae, H. C. Vogel, *Astronomische Nachrichten* 165, Nr. 3944, 113–22, 1904

28 Hermann Carl Vogel & E. B. Frost, *Astrophysical Journal* 27, 1–11, 1908

29 Julius Scheiner, E. B. Frost, *Astrophysical Journal* 41, 1–9, 1915

30 Electrical Disturbances Apparently of Extraterrestrial Origin, K. G. Jansky, *Proceedings of the Institute of Radio Engineers* 21, 1387–98, 1933

31 I have described these developments in greater detail in my book, *In Search of the True Universe: The Tools, Shaping, and Cost of Cosmological, Thought*, M. Harwit, Cambridge University Press, 2013, Chapter 7, pp. 121–56

32 Calibration of Low-Frequency Radio Telescopes Using the Galactic Background Radiation, G. A. Dulk, et al., *Astronomy & Astrophysics* 365, 294–300, 2001

33 Observations of Giant Pulses from Pulsar B0950+08 Using LWA1, Jr-Wei Tsai, et al., *The Astronomical Journal* 149:65 (10 pp.), February 2015

34 A Single Fast Radio Burst Localized to a Massive Galaxy at Cosmological Distance, K. W. Bannister, *Science* 365, 565–70, 2019

35 Über Gravitationswellen, A. Einstein, *Sitzungsberichte der Preussischen Akademie der Wissenschaften*, 154–67, 1918

36 Observation of Gravitational Waves from a Binary Black Hole Merger, B. P. Abbott, et al. (LIGO Scientific Collaboration and Virgo Collaboration), *Physical Review Letters* 116, 061102, February 12, 2016

37 GW170104: Observation of a 50-Solar-Mass Binary Black Hole Coalescence at Redshift 0.2, B. P. Abbott, et al. (LIGO Scientific and Virgo Collaboration), *Physical Review Letters* June 1, 2017

38 Multi-messenger Observations of a Binary Neutron Star Merger, co-authored by teams of authors all over the globe, *Astrophysical Journal Letters* 848, L12, 59 pp., October 20, 2017

39 Theory of Communication, D. Gabor, *Journal of the Institute of Electrical Engineers* 93, 429–41, 1946

40 Radio Emission Signatures in the Crab Pulsar, T. H. Hankins & J. A. Eilek, *Astrophysical Journal* 670, 693–701, 2007

41 The Cosmic Energy Inventory, M. Fugukita & P. J. E. Peebles, *Astrophysical Journal* 616, 643–68, 2004

42 Cosmological Constraints from the SDSS Luminous Red Galaxies, M. Tegmark, et al., *Physical Review D* 74, 123507, 2005

43 Ibid., Fukugita & Peebles, 2004

44 The Baryon Census in a Multiphase Intergalactic Medium: 30% of the Baryons May Still Be Missing, J. M. Shull, B. D. Smith, & C. W. Danforth, *Astrophysical Journal* 759(23), November 1, 2012

45 A Cosmology Calculator for the World Wide Web, E. L. Wright, *Publications of the Astronomical Society of the Pacific* 118, 1711, 2006, revised 2014

46 Nine-Year Wilkinson Microwave Anisotropy Probe (WMAP) Observations: Cosmological Parameter Results, G. Hinshaw, et al., *Astrophysical Journal Supplement Series* 208, 19, October 2013

47 Planck 2015 Results. XIII. Cosmological Parameters, Planck Collaboration, *Astronomy & Astrophysics* 594 A, 13, 2016

48 An account of these observations is given in "Über Sternschnuppen," F. W. Bessel, *Astronomische Nachrichten* 16, 321,1839

49 Observations of Extremely Cool Stars, G. Neugebauer, D. E. Martz, & R. B. Leighton, *Astrophysical Journal* 142, L399–401, 1965

50 Observations of Infrared Galaxies, D. E. Kleinmann & F. J. Low, *Astrophysical Journal* 159, L165–72, 1970

51 On a New Kind of Rays, W. C. Röntgen, *Nature* 53, 274–76, 1896

52 Evidence for X Rays from Sources Outside the Solar System, R. Giacconi, H. Gursky, F. R. Paolini, & B. Rossi, *Physical Review Letters* 9, 439–43, 1962

53 X-Ray Sources in the Galaxy, S. Bowyer, E. T. Byram, T. A. Chubb, & H. Friedman, *Nature* 201, 1307–8, 1964

54 On the Radioactivity of K^{40}, H. E. Suess, *Physical Review* 73, 1209, 1948

55 Ueber Eletricitaätszerstreuung in der Luft, J. Elster & H. Geitel, *Annalen der Physik*, 4th series vol. 2, 421–46, 1900

56 The Azimuthal Asymmetry of the Cosmic Radiation, T. H. Johnson, *Physical Review* 43, 834–35, 1933

57 Evidence That Protons Are the Primary Particles of the Hard Component, T. H. Johnson, *Reviews of Modern Physics* 11, 208–10, 1939

58 The Atomic Weight of "Thorium" Lead, Frederick Soddy, *Nature* 98, 469, February 15, 1917

59 Quantised Singularities in the Electromagnetic Field, P. A. M. Dirac, *Proceedings of the Royal Society A* 133, 60–72, 1931

60 The Apparent Existence of Easily Deflectable Positives, C. D. Anderson, *Science* 76, 238–39, 1932

61 Letter dated December 4, 1930 addressed to attendees of a conference held in Tübingen, Germany, a few days later. A copy of the letter is kept in the Pauli archives at CERN

62 Detection of the Free Antineutrino, F. Reines, et al., *Physical Review* 117, 159–74, 1960

63 Observation of a Neutrino Burst from the Supernova SN 1987A, K. Hirata, et al., *Physical Review Letters* 58, 1490–93, 1987

64 Observation of a Neutrino Burst Coincident with Supernova 1987A in the Large Magellanic Cloud, R. M. Bionta, et al., *Physical Review Letters* 58, 1494–96, 1987

65 Die Rotverschiebung von extragalaktischen Nebeln, F. Zwicky, *Helvetica Physica Acta* 6, 110–27, 1933, see p. 126

66 Extended Rotation Curves of High-Luminosity Spiral Galaxies V. NGC 1961, the Most Massive Spiral Known, V. C. Rubin, W. K. Ford, Jr., & M. S. Roberts, *Astronomical Journal* 230, 35–39, 1979

67 Observational Evidence from Supernovae for an Accelerating Universe and a Cosmological Constant, A. G. Riess, et al., *Astronomical Journal* 116, 1009–38, 1998

68 Discovery of a Supernova Explosion at Half the Age of the Universe, S. Perlmutter, et al., *Nature* 391, 51–54, 1998

69 On Gravitational Waves, A. Einstein & N. Rosen, *Journal of the Franklin Institute* 223, 43–54, January 1937

70 Discovery of a Pulsar in a Binary System, R. A. Hulse & J. H. Taylor, *Astrophysical Journal* 195, L51–53, 1975

71 A New Test of General Relativity: Gravitational Radiation and the Binary Pulsar SR 1913+16, J. H. Taylor & J. M. Weisberg, *Astrophysical Journal* 253, 908–20, 1982

72 Observations of Post-Newtonian Timing Effects in the Binary Pulsar PSR 1913+16, J. M. Weisberg & J. H. Taylor, *Physical Review Letters* 52, 1348–52, 1984

73 Astrophysical Implications of the Binary Black Hole Merger GW150914, B. P. Abbott, et al. and the LIGO/Virgo consortium, *Astrophysical Journal* 818, L22 (15 pp.), 2016

2

Primordial Messengers and Their Interpretation

2.1 Primordial Messengers

The earliest moments in the history of the Universe are largely unknown. Most of the messengers that could have reached us from those founding instants were systematically erased along the way. We are left with a few surviving markers, wayposts on a lengthy trail whose traces may still be discerned thanks to a few particularly robust carriers of information. Some have already been reliably detected. Others are being pursued and may someday be located.

By the early 1940s, most astronomers were persuaded that the Universe had long been expanding. Spectra of remote galaxies consistently showed the most distant among them exhibiting highly redshifted spectra. These could readily be explained if the galaxies had been mutually receding, expanding away from each other since the dawn of time. A number of studies attempted to discern whether such notions could be verified by messengers still reaching us from those early epochs. None were found to date back to a defining instant of creation, but a sufficient number arriving from succeeding epochs provided a measure of continuity.

Historically, the advances came in two phases. The first was motivated by a drive from 1948 to the mid-1950s to account for the origins of the heavy chemical elements, all of whose nuclei are rich in neutrons. These elements were thought to have emerged in an early universe largely populated by free neutrons. Though this notion was later found to have been wrong, it correctly led to the conclusion that the temperature of the Universe must have been extremely high at age $t \sim 1$ second, and provided not only a credible temperature estimate for those times but also a well-defined mix of particles and radiation permeating the Cosmos at that epoch.

Agreement on this was sealed by the discovery in 1965 of a pervasive cosmic microwave background radiation, shown three decades later to emit a ubiquitous blackbody spectrum at a temperature $T = 2.725$ K, reasonably close to a value extrapolated from conditions believed to have existed at cosmic age 100 seconds.

The knowledge gained through this early drive raised a series of new questions seeking to provide a seamless history ranging from the first few seconds in the existence of the Cosmos, down to the present. Answers to these had to be found before it became possible to determine conditions that had existed even earlier, quite possibly a mere 10^{-43} seconds after the birth of time and ranging up to age $t \sim 1$ second.

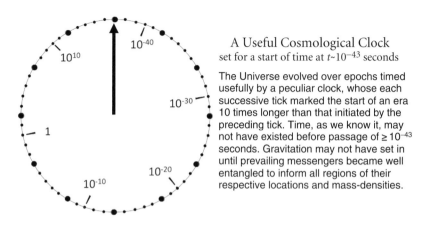

A Useful Cosmological Clock
set for a start of time at $t \sim 10^{-43}$ seconds

The Universe evolved over epochs timed usefully by a peculiar clock, whose each successive tick marked the start of an era 10 times longer than that initiated by the preceding tick. Time, as we know it, may not have existed before passage of $\geq 10^{-43}$ seconds. Gravitation may not have set in until prevailing messengers became well entangled to inform all regions of their respective locations and mass-densities.

This earlier part of cosmic history, whose assembly started around 1968, has recently received remarkable support from observations made with NASA's Wilkinson Microwave Anisotropy Probe, WMAP, and the European Space Agency's *Planck* mission, as we will see later in this chapter.

For now, however, let us start at the beginning of time, and continue from there.

2.2   A Useful Cosmological Clock

Because successive cosmic events occurred on drastically escalating time scales, as we will see, it may be useful to trace them with a peculiar *cosmic clock*. Set in motion at the launch of time, this clock has ticked 60 times by now. Each successive tick has taken 10 times longer for the pointer to move than the preceding tick had required! The first tick took place at cosmic age 10^{-43} seconds; the last at age 10^{17} seconds or 3 billion years.

By now, the clock's single pointer has completed a full circle around the dial. Its most recent tick marks the start of the present era, following the culmination of a *cosmic dawn*, during which the first stars, galaxies, and massive black holes had already begun to form. Our most powerful telescopes today enable us to look out to the far reaches of the Universe to still reveal this earliest generation of stars just beginning to form.

This chapter will follow two parallel tracks. Its main text provides the historical trends preoccupying the astrophysicists of the times. Meanwhile, the cosmic clock will track significant cosmological junctures as we now see them in retrospect.

2.3   The Very Earliest Times

Some 14 billion years ago, at the dawn of time, the Universe began transmitting messages.

The evidence gathered to date indicates a rapid sequence of initial events succeeding each other at breakneck speed before relaxing to a slower pace as the Cosmos expanded.

Today's consensus is that the Universe originally was highly compact, extremely hot, and has been continually expanding since earliest times. The moment of creation nevertheless remains undefined. Einstein's general relativity teaches us that gravity and geometry are inseparable aspects of space. Neither property can be instantaneously determined. Awareness of the distribution of ambient mass at some distance d from any location requires a minimal time interval $\Delta t \geq d/c$ for this information to arrive, even at the speed of light, c. In the absence of some form of prehistory, it appears doubtful that gravity could have already existed at the very start of time, and unclear how long an epoch may have been required before a final level of gravitational interaction was in place.

A common assumption is that no coherent dynamics could have set in before the Universe had attained an age of at least $\sim 10^{-43}$ seconds. This is the shortest time required for information traveling at the speed of light to traverse the span, or diameter D, of the tiniest geometric structure relativity envisages – a Planck-mass black hole.[1,2]

Although Einstein's general theory of relativity for many decades suggested that no form of matter or energy can ever escape a black hole, Stephen Hawking was able to show in 1974 that this is incorrect. In a groundbreaking letter to *Nature*, he wrote:[3] [a]

---

[a] The Appendix provides numerical values of c, h, G, and other fundamental constants of nature, as well as the Solar mass $M_\odot$ and the Kelvin temperature scale K.

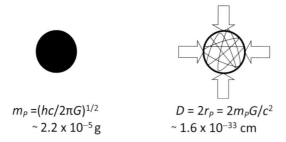

$m_p = (hc/2\pi G)^{1/2}$ $D = 2r_p = 2m_p G/c^2$
$\sim 2.2 \times 10^{-5}\,g$ $\sim 1.6 \times 10^{-33}$ cm

Fig. 2.1 The Earliest Conceivable Coherent Structures and the Smallest Black Holes. The smallest conceivable black holes have a mass called the Planck mass, m_p, and a radius, r_p, known as the Planck length, $\ell_p \equiv r_p$.

It seems that any black hole will create and emit particles such as neutrinos and photons at just the rate that one would expect if the black hole was a body with a temperature of ... $\sim 10^{-6} M_\odot / M$ K. As a black hole emits this thermal radiation one would expect it to lose mass. This in turn would increase the surface gravity and so increase the rate of emission. The black hole would therefore have a finite life of the order of $10^{71}(M_\odot/M)^{-3}$ s.... Near the end of its life the rate of emission would be very high.

If a particle, even one with no mass at all when at rest, is squeezed into a tiny box, as in Figure 2.1, the quantum mechanical *uncertainty principle*, first advocated by Werner Heisenberg in 1927, says it must acquire a compensating increased momentum and thus also an increased mass–energy density. The smallest volume capable of enclosing any finite amount of mass–energy is that of a black hole; and among black holes, the smallest exhibit the highest mass–energy densities.

Hawking concluded that these miniature black holes emit blackbody radiation at a temperature that dramatically increases with diminishing black hole mass. The tiniest conceivable black holes thus promptly explode, suicidally ejecting their mass–energy in the form of radiation.[4]

The earliest conceivable instant during which such a black hole could act coherently is roughly the time required to traverse a distance equivalent to the diameter of the Planck mass, at the speed of light, $t_P \sim 2r_P/c \sim 10^{-43}$ seconds, as Figure 2.2 indicates. That way, one part of the black hole can remain informed of physical processes occurring in any other of its parts, and the entire black hole can act coherently. The Planck time thus is the earliest time at which our current laws of physics can make some causal sense.

The lowest conceivable mass of a black hole is the *Planck mass*, $m_P \sim 2.2 \times 10^{-5}$ gram. Its radius is the *Planck length*, $\ell_P \sim 1.6 \times 10^{-33}$ centimeters, and it exhibits

$$D/c = 2t_P = 2r_P/c$$

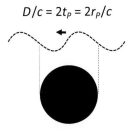

Fig. 2.2 The time required to traverse the diameter D of a Planck black hole at the speed of light is $D/c = 2t_P = 2\ell_P/c \sim 0.5 \times 10^{-43}$ seconds, where t_P is the Planck time, and the Planck length $\ell_P \equiv r_P$ is this black hole's radius.

the highest-conceivable temperature, $T_P \sim 10^{32}$ K, at which it evaporates in at least a *Planck time* $t_P \sim 0.5 \times 10^{-43}$ seconds, the time required for information to transit the radius of this smallest realizable physical span.

This, roughly speaking, is why the earliest epoch in cosmic evolution guided by physical laws is judged to have set in no earlier than expiration of twice the Planck time, say, $2t_P \sim 10^{-43}$ seconds.

2.4   Entanglement and Gravity

The least massive non-rotating, electrically neutral black holes vanish by ejecting oppositely directed quanta of radiation – pairs comprising a particle and its antiparticle, each exhibiting an identical mass–energy but oppositely directed linear and angular momentum. Among these are photons which are their own antiparticles, as well as neutrinos, ν, and their dedicated antiparticles labeled $\bar{\nu}$.

Such paired radiations and particles are said to be mutually *entangled*. If you determine the mass–energy, linear momentum, and angular momentum of one of these quanta, you automatically also know its paired antiparticle's oppositely directed properties. Arriving in the mutually remote regions they penetrate, each member of the pair would provide information on events at their common point of origin.

Emitted in sufficient abundance at earliest times, these paired radiations conceivably could have spread the information required to infuse dynamic and thermal equilibrium and homogeneity across sizeable domains. This is important because modern theories worry, quite generally, that without pervasive entanglement, neither geometry nor gravity could exist, nor could general relativity be expected to be the primary theory governing early cosmic evolution.

Entanglement is a quantum mechanical means of creating pairs of messengers capable of conveying mutually complementing messages to recipients physically far apart: If you capture one of the two messages, and a remote colleague captures the other, each of you precisely knows the message received by the other.

The very first cosmological task of photons, neutrinos, gravitons, or any other messengers potentially existing at earliest epochs may thus have been to establish a level of coherence among mutually adjacent regions that could permit gravity to extend its range of influence. Once gravity was established, the prevailing high temperatures and energy densities might naturally have been expected to cause the Cosmos to expand.

However, such speculations did not make much sense until a number of basic questions could be answered, and these were sufficiently enigmatic that no clear way of asking them seemed at hand.

2.5   A First Enigma

The earliest epochs as just sketched suggest that the primordial universe would have contained identical numbers of particles and their paired antiparticles emitted by primordial black holes. Yet today's Universe consists largely of matter and a near-absence of antimatter. A potential resolution of this enigma had long been buried in data, some of which we earlier encountered in Table 1.2. The entry for *Cosmic Microwave Background Radiation* reads 4.64×10^{-34} grams per cubic centimeter and represents the energy density of a cosmic blackbody radiation bath at a temperature $T = 2.725$ K pervading the Universe today. Each cubic centimeter of this radiation bath contains $\sim$400 of these photons. Later in this chapter we will encounter their origin. Although photons have zero rest-mass, the mean mass–energy of each of these 400 photons is $\sim 1.16 \times 10^{-36}$ gram.

The mass of a hydrogen atom, the most abundant atomic constituent, is 1.67×10^{-24} gram, and Table 1.2 lists the total baryonic rest-mass-density in the Universe as 4.4×10^{-31} grams per cubic centimeter, corresponding to a number density of atomic nucleons around 2.6×10^{-7} per cubic centimeter, a factor of 1.5×10^9 lower than the number density of microwave photons.

In 1972, Yakov Borisovich Zel'dovich (Figure 2.3), at the Institute of Applied Mathematics in Moscow, addressed the enigma of why the Universe today is filled with atomic matter but almost no antimatter. He showed that if the early Universe contained a small fractional excess of matter over antimatter, most of the matter and antimatter would have mutually annihilated, giving rise to photons.[5] The number density of photons observed today is about a billion times more abundant than the surviving initial excess of matter over antimatter.

Fig. 2.3 Yakov Borisovich Zel'dovich, the most influential Soviet astrophysicist and cosmologist of the era following World War II. He worked on a wide range of theoretical problems including some of the most secret military projects, such as the design and control of Soviet atomic and hydrogen bombs. His insatiable curiosity, amazing versatility, and joy in solving difficult problems inspired a whole generation of young Soviet astrophysicists.

Zel'dovich interpreted the relative number density of cosmic microwave background photons and the corresponding relative number density of observed nucleons as conveying the message that nucleons and antinucleons present in the Universe at its birth had mutually annihilated to give rise to the observed number density of the background radiation. Their currently available ratios suggest a primordial particle–antiparticle asymmetry amounting to roughly one part in a billion of the total matter plus antimatter density of the Universe at the start of time.

Just when this widespread annihilation had taken place, Zel'dovich did not say, and we still do not know; but his suggestion hinted at a resolution of the dilemma.

Today, the matter–antimatter asymmetry still leaves many astronomers and physicists yearning for a better explanation. Why, for example, was the initial discrepancy between matter and antimatter not as small as that between the number of positively and negatively charged particles in the Universe, which by all indications is zero?[6; 7]

2.6   The Second Enigma

A second troubling question about the early Universe concerned the density fluctuations one might anticipate to have existed at the dawn of time. In

1902, the English astronomer James H. Jeans had shown that a gaseous assembly of particles in mutual gravitational attraction should become unstable when the temperature of the gas sufficiently cooled. Based on Newtonian mechanics and known as the Jeans criterion, it appeared to explain why today's Universe is far from bland. High-density patches abound as clusters of galaxies separated by intervening voids. The clusters further break down into individual galaxies often containing billions of individual stars.[8]

Given these inhomogeneities, an article the theoretical physicist Evgeny Mikhailovich Lifshitz in Moscow wrote in 1946 caused considerable concern. It showed that Einstein's general relativity would prohibit any fully homogeneous model of an expanding universe from ever collapsing gravitationally to form the abundance of galaxies observed, nor any other density inhomogeneities.[9] How could the very existence of galaxies then be explained? The article's general conclusions appeared sound.

Yet Zel'dovich remained untroubled. He found no reason to believe that the Universe had lacked any and all density inhomogeneities at birth. He did not know the distribution of density fluctuations at the onset of time, but thought a good guess might be that they were *scale invariant*, meaning that, when examined under ever higher magnification their size distribution would always appear precisely identical, independent of the degree of magnification.

He concluded that the magnitude of primordial mass–energy density fluctuations at levels of order 10^{-4} times the mean density at that epoch would ultimately result in a patchy distribution of galaxies comparable to those observed in the Universe today.[b]

If one agrees with the conclusions Zel'dovich reached, we can assert that a number of known messengers reaching us already are conveying information about the very early Universe. The first message discussed above is conveyed by the relative numbers of atoms and photons detected in today's Universe. The second is conveyed by the patchiness of today's Universe. The significance of these messages, however, is based on their interpretation. The data reaching us by virtue of the messengers gathered are uncontested! Whether we also interpret their significance correctly is not as clear.

---

[b] Zel'dovich did not mentioned it, but the Einstein–Fowler equation of the kinetic theory of gases in thermal equilibrium predicts a level of inhomogeneity of the same order as Zel'dovich now cited.[10]

The point of emphasizing this difference is that the ability of messengers to convey information reliably is no guarantee that we are always able to correctly interpret the message – though astronomers by and large have found Zel'dovich's interpretations increasingly convincing as the spectrum of fluctuations still observable in deep maps of the cosmic microwave background radiation have revealed that the spectrum indeed is scale-free.

2.7   The Third Enigma

Two decades after the audacious pronouncements of Zel'dovich, the Princeton University physicist Robert Dicke began giving talks about two other vexing conundrums: If the Universe had indeed been cooling through adiabatic expansion since earliest epochs, as had been assumed ever since 1927 when the Abbé Georges Lemaître had proposed a relativistic model of the Universe that had expanded adiabatically since birth, then it was difficult to avoid a problem that Dicke now explained in a paper co-authored with his Princeton colleague Jim Peebles.[11] [c]

Dicke pointed out that, at the enormously high temperatures and accompanying thermal mass–energies existing at the earliest conceivable cosmic epochs, general relativity predicted that the Universe would have undergone a rapid expansion under *adiabatic conditions* – conditions neither infusing energy into the expanding region nor draining energy from it. This expansion would then have continued, though at systematically declining speeds.

Under these conditions, how could two mutually remote regions of the Universe we view today along antipodal directions in the sky appear to have largely identical properties? Surely, they must all along have been well beyond each other's relativistic horizons as they mutually receded at velocities far exceeding the speed of light, c, throughout early expansion. Many of these regions are only now entering our range of view as the cosmic expansion rate has steadily dropped, leading to ever-expanding cosmic horizons.

Dicke pressed for a clarification of this apparent impasse.

2.8   The Fourth Enigma

Dicke then also pointed out that, looking out to regions in the Universe at the greatest distances, we conclude that many stars shining today have been shining for billions of years. General relativity indicates that for the Universe

---

[c] The term *adiabatic* means "without the addition or reduction of energy."

to have existed that long, it should have initially formed with essentially no inherent curvature. Any initially positive curvature would have led the Cosmos to first expand, within a tiny fraction of a second, and then promptly collapse. Similarly, any significant initial negative curvature would have rapidly and irreversibly expanded the Universe to far lower densities than those anywhere observed. To avoid either calamity, Dicke showed that the cosmic curvature would early on have had to be precisely zero to an accuracy within roughly one part in 10^{55}. This would have been difficult to either establish or subsequently maintain.

Marc Holman, a post-doctoral associate at Western University in London, Ontario, has for some years been investigating the foundations of physical theories and their inherent implications. In a recent analysis he challenges Dicke's surprise at finding that the Universe must have had a cosmic curvature indistinguishable from zero at the dawn of time. Rather, Holman maintains, any cosmological model based on Einstein's formulation of general relativity would inherently exhibit zero curvature at earliest times – not as some arbitrary feature, but as an inescapable consequence of the theory's foundations.[12]

While neither Dicke's arguments, nor those of Zel'dovich thus had sound foundations, they filled a philosophical vacuum in their time and inspired astrophysicists to look for new approaches.

2.9   The Inflationary Universe

The two questions Dicke had raised remained unanswered until around 1979–80. Then, two independently proposed but remarkably converging approaches began to resolve them.

The first was proposed by Aleksei A. Starobinsky at the Landau Institute of Theoretical Physics of the Soviet Union's Academy of Sciences.[13; 14] The second was due to Alan Guth, at the time a post-doc at the Stanford Linear Accelerator (SLAC).[15]

Starobinsky recognized that the Universe could not have expanded adiabatically throughout its entire existence because it would then have excessively cooled, despite the enormous temperatures postulated to have existed at earliest times. Instead, he suggested that, shortly after its creation and some stabilizing adiabatic expansion, the expansion of the Universe could have switched to be driven by a cosmological constant Λ, which Einstein had initially introduced to keep the Universe forever static, as he and his contemporaries assumed the Universe to be. But rather than adopting Einstein's value for this constant, Starobinsky opted for a value more akin to a mass–energy density the Dutch astronomer Willem de Sitter had proposed decades earlier, in 1917. It fit general relativity just as readily as Einstein's cosmological constant had but gave rise to an expanding universe instead.

The relative size of the Universe at any given epoch may be denoted by a *scale factor a*, an arbitrary measure of the spatial separation between any given two points in the Universe. Once this arbitrary value is adopted for any chosen epoch, the distance between any two firmly embedded regions increases in direct proportion to a as the Universe expands.

For an expansion driven entirely by a cosmological constant Λ, with no noticeable contribution by the mass-density of matter or radiation, general relativity dictates that the expansion rate, the Hubble constant, $H = [da/dt]/a$, be given by $H = (\Lambda c^2/3)^{1/2}$. Under these conditions, the cosmic expansion increases exponentially with time, growing from an initial scale factor a_0 to $a(t)$ at any later time t given by

$$a(t) = a_0 e^{Ht}.$$

Starobinsky's exponential expansion nowadays is referred to as *inflation*. He designed it to eventually terminate and blend into a slower adiabatic expansion starting at a cosmic temperature comparable to that at the start of inflation but thereafter gradually cooling as the Universe continued to expand.

Alan Guth, who apparently was unaware of Starobinsky's work, came to the same conclusion, but for two quite different reasons: Theoretical physicists had been conjecturing about *Grand Unified Theories* which involved two theoretical predictions of particle physics.

The first of these was that *magnetic monopoles* should exist in the Universe in sufficiently high abundances that they could not have evaded astronomical detection by now. Yet none had ever been found! Guth now proposed that the absence of magnetic monopoles could be explained if the Universe had at early times undergone a sufficiently high inflationary expansion to dilute the number density of primordial magnetic monopoles to a level at which only a tiny fraction, or none at all, might currently be located within today's speed-of-light horizon.

The second was that, at temperatures of order $T \sim 10^{27}$ K, implying particle energies of order $\sim 10^{23}$ eV, the strengths of gravitational, electromagnetic, and the weak as well as the strong nuclear interactions among particles would begin to approach each other. This would lead to thermal equilibrium between photons, neutrinos, gravitons, and energetic cosmic-ray particles. Guth suggested that if the temperature at the end of inflation was indeed as high as $\sim 10^{27}$ K, a transition to an adiabatic phase of the Universe would emerge, initially filled throughout with high-energy particles and radiation.

Both Starobinsky's and Guth's sets of reasoning agreed that a de Sitter expansion in the early Universe could later segue into an adiabatic expansion of the Universe to provide a coherent history that appeared consistent with all we currently know.

2.10   A History Conveyed by Surviving Messengers

Between them, the four enigmas just discussed had reached us in four mutually complementing ways.

The first had been relayed to us by the prevalence of photons over nucleons, the respective sets of messengers relaying the message.

The second had been raised by the inhomogeneity exhibited by the distribution of clusters of galaxies across the sky, as well as by the ranges of galaxy masses. Here, the messengers directly reaching us were the photons the galaxies emitted; the message we received emanated from the galaxies' mass and spatial distributions.

The third was brought to our attention by photons arriving from all over the celestial sphere to report the homogeneity and isotropy with which galaxies appeared to be distributed across the sky. The pervasive homogeneity had been troubling Bob Dicke, but went away by postulating two early cosmic epochs of the Universe. The first comprised an initial period lasting about 10^{-35} s during which a tiny patch of the early Universe could come into thermal equilibrium. This meant that even the regions we see in remote antipodal parts of the Universe today could have been in close mutual contact at those primeval times.

Dicke's second worry, about the extremely low cosmic curvature required for our Universe to have survived for the billions of years we observe stars to have been existing, was settled not by actual observation of such a low curvature but by postulating a sequence of expansionary phases so vast as to straighten out any systematic curvature the Universe might have exhibited at the dawn of time.

A region stretching across $\sim 10^{-32}$ cm at birth, and adiabatically expanded to some 10^{-25} cm at the start of inflation, would then have grown by another factor of $\sim 10^{26}$ to span roughly 10 cm. If, following inflation, the dimensions of the Universe grew by yet another factor of $\sim 10^{24}$ in the adiabatic expansion following inflation, for a total expansion over these three phases by a factor of $\sim 10^{57}$, this cumulative expansion readily exceeded Dicke's required total factor of $\geq 10^{55}$ to smooth out any initial curvature.

None of these dimensions, nor the time spans involved, are as yet well defined. They are mutually consistent and fit currently available observations, but will need to be adjusted as we learn more. I cite them here mainly to provide a rough sense of scale and sequence.

Readers should also be aware that the inflationary theory has undergone myriad changes over the past four decades, and not all the proposed variants can be clearly distinguished from one another through available observations. Individual variants have arisen as new observational features have become established, so that the inflationary theory does not strike one as emanating from a single set of principles, but rather as a way of describing a universe that has been tailored and fitted step by successive step in response to observational data emerging over time.

Nothing is wrong with such an approach; but it has been driven by successive messages the Universe has been found to convey – Dicke's realization that the isotropic appearance of the Universe, as well as the observed age of some of the oldest stars, required explanation; Zel'dovich's recognition that the degree of patchiness of the Universe and the observed dominance of matter over antimatter needed to be understood as well.

No doubt, the theories put forward to explain these observations have a satisfying ring of relevance and possibly truth. Particle physicists, however, continue to search for a more fundamental rationale for the dominance of matter over antimatter, and for the scale-free pattern of density fluctuations that Zel'dovich proposed more than half a century ago.

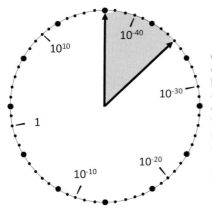

A Planck Era Expansion
from $\geq 10^{-43}$ to 10^{-35} seconds

Once a well-defined energy density and existence of any inherent curvature had been established, an expansion begun any time after $t \geq 10^{-43}$ s could have continued for 10^{-35} s, dropping the temperature from $\sim 10^{32}$ to $\sim 10^{28}$ K, and permitting expansion at nearly the speed of light to bring regions $\sim 10^{-25}$ cm across into partial equilibrium, while leaving a scale-invariant fluctuation spectrum intact.

A de Sitter Inflation
10^{-35} to 10^{-33} seconds

By 10^{-35} s a local energy density defined by a cosmological constant Λ began to drive an exponential expansion marked by scale-invariant fluctuations named inflatons, the primary messengers outliving that epoch. Ending at ~10^{-33} s after a ~10^{26}-fold expansion to ~10 cm, inflation gave way to myriad radiations and particles, yielding its energy density to heat them to temperatures of 10^{26}–10^{27} K.

2.11   The Earliest Post-Inflationary Messengers

For want of a better approach, we now need to further delve into unobserved realms to portray the Universe largely on the basis of hints provided by high-energy particle theory not yet experimentally tested over many of the energy ranges of prime cosmological interest.

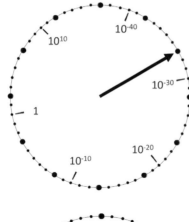

The New Cosmic Constituents
$t = 10^{-33}$ seconds

Inflation expanded the cosmic horizon receding at the speed of light, from an initial 10^{-25} cm at 10^{-35} s to a final baseball-sized ~10 cm at t ~10^{-33} s, before giving rise to an *adiabatically* expanding phase of diminishing temperatures. At the start of this era, antiparticles and particles may have formed in equal numbers; but the nature of these original messengers remains unknown.

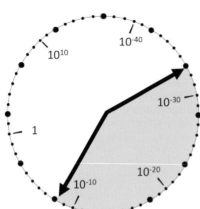

Extinction of Particle Species
From 10^{-33} to 10^{-8} seconds

Following inflation, the cosmic expansion continued. Temperatures fell from $\leq 10^{27}$ to $\leq 3 \times 10^{14}$ K, lowering particle energies to $\leq 2.5 \times 10^{10}$ eV. These temperatures and energies are uncertain because unknown generations of messengers may have formed and then annihilated faster than the cooling gas could replace them. During this interval – we do not yet know just when or how – all but one antiparticle out of every 10^{10} annihilated.

The number of particle species or types of radiation – in short, the number of distinct species of messengers – among which the released inflationary energy density was divided at onset of adiabatic expansion, remains unknown. Their temperatures of order 10^{26} to 10^{27} K, corresponding to particle energies of the order of 10^{22} to 10^{23} eV apiece, would have been roughly 10^2 to 10^3 times higher than even the highest cosmic-ray particle energies observed anywhere in today's Universe.

The most energetic cosmic-ray particles incident on Earth today appear to top out at $\sim 10^{20}$ eV, a factor of a hundred or a thousand below those at the end of inflation. The most powerful particle accelerator built to date, the *Large Hadron Collider* at the European Council for Nuclear Research, CERN, in Switzerland, probes the properties of particles and their collision products, at energies no higher than $\sim 1.5 \times 10^{13}$ eV, or temperatures of order 10^{17} K. Because CERN data, to date, provide no hint of how antimatter may have selectively been eradicated, any systematic destruction of cosmic antimatter would have had to occur at temperatures in the range $10^{17} \leq T \leq 10^{27}$ K. Cosmic particle energies dropped to the lower of these temperatures only after the age of the Universe had reached $\geq 10^{-12}$ seconds.

2.12   Early Subatomic Messengers

Atomic nuclei consist of *quarks* and *gluons*, collectively referred to as *partons* because they appear solely as parts of the protons and neutrons constituting atomic nuclei. In today's Universe they never appear isolated as individual quarks or individual gluons. The quarks have rest-masses and electric charges listed in Table 2.1. None of the gluons do.

The six types or *flavors* characterizing different quarks exhibit a wide range of mass–energies difficult to establish accurately because quarks never appear totally isolated. Two of the quarks, the *up (u)* and the *down (d)* quarks, are the

TABLE 2.1

Quarks and Leptons

Quark Name	Mass–Energy (eV)	Electric Charge (electron charges)	Lepton Name	Mass–Energy (eV)	Electric Charge (electron charges)
up (u)	$\sim 2.3 \times 10^6$	+2/3	electron (e^-)	5.1×10^5	−1
down (d)	$\sim 4.8 \times 10^6$	−1/3	e-neutrino (ν_e)	< 2	0
strange (s)	$\sim 9.5 \times 10^7$	−1/3	muon (μ^-)	1.06×10^8	−1
charm (c)	$\sim 1.3 \times 10^9$	+2/3	μ-neutrino (ν_μ)	< 2	0
bottom (b)	$\sim 4.7 \times 10^9$	−1/3	tau (τ^-)	1.78×10^9	−1
top (t)	$\sim 1.73 \times 10^{11}$	+2/3	τ-neutrino (τ_ν)	< 2	0

primary constituents of protons and neutrons, and the atomic nuclei they form. Their respective mass–energies are relatively low: $\sim 2.3 \times 10^6$ and $\sim 4.8 \times 10^6$ eV. As Table 2.1 shows, the other four quarks have far higher masses.[16]

Protons consist of two *up* and one *down* quarks; neutrons of two *down* and one *up* quarks. Because the respective electric charges of the *up* and *down* quarks are $+2/3$ and $-1/3$, the electric charges of the resulting protons and neutrons are $+1$ and 0.

Protons, neutrons, as well as all the particles listed in Table 2.1 have corresponding antiparticles of identical mass and opposite charge. Like neutrons, antineutrons carry no charge. Photons are their own antiparticles; we do not currently know whether neutrinos also are their own antiparticles.

At cosmic age around 10^{-10} s when the temperature had dropped to about 2×10^{15} K, the top quarks and antiquarks should have begun to annihilate more rapidly than collisions among ambient particles by then could replenish.

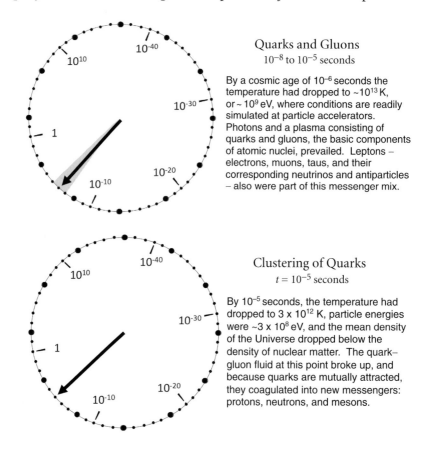

Quarks and Gluons
10^{-8} to 10^{-5} seconds

By a cosmic age of 10^{-6} seconds the temperature had dropped to $\sim 10^{13}$ K, or $\sim 10^9$ eV, where conditions are readily simulated at particle accelerators. Photons and a plasma consisting of quarks and gluons, the basic components of atomic nuclei, prevailed. Leptons – electrons, muons, taus, and their corresponding neutrinos and antiparticles – also were part of this messenger mix.

Clustering of Quarks
$t = 10^{-5}$ seconds

By 10^{-5} seconds, the temperature had dropped to 3×10^{12} K, particle energies were $\sim 3 \times 10^8$ eV, and the mean density of the Universe dropped below the density of nuclear matter. The quark–gluon fluid at this point broke up, and because quarks are mutually attracted, they coagulated into new messengers: protons, neutrons, and mesons.

At epochs around 10^{-5} s, temperatures above $\sim 3 \times 10^{12}$ K, and particle energies $\sim 3 \times 10^8$ eV, the quarks and gluons constituted a seamless *quark–gluon plasma*. As the cosmic mass–energy density dropped further, a *freeze-out density* was reached, at which the quarks and gluons lacked the energy to move apart. They clustered into *hadrons*, distinct types of particles consisting of specific combinations of quarks and gluons. As the temperature of the Universe further dropped, the most frequently encountered hadrons were the *baryons*, particles that like protons and neutrons contain three quarks, or *mesons* containing only a pair of quarks. At this stage, the Universe was filled with particles naturally detected only in *cosmic ray showers*, though recreated also in lower-energy ranges at high-energy particle accelerators.

2.13   The Buildup of Light Elements in the Early Universe

By the early 1930s, hydrogen was known to be the most abundant element in the atmospheres of stars, with helium the only other significant component. All the other elements jointly account for no more than 2% of cosmic atomic mass.

In 1938 Hans Bethe, at the time a 32-year-old theorist at Cornell University, and Charles L. Critchfield, a graduate student at George Washington University in the District of Columbia, showed that if this mix permeated the entire body of a star like the Sun, fusion of hydrogen to form helium in the star's high-temperature central regions could account for the vast amounts of energy the Sun radiates.[17] They found a number of parallel pathways, each opened by fusion of two hydrogen nuclei – protons, P – to form a deuterium nucleus D, a neutrino v, and a positron e^+, the antiparticle to an electron:

$$P + P \rightarrow D + v + e^+.$$

Deuterium then combined with an ambient proton to form the nucleus of the helium *isotope* ^{3}He, whose various interactions with other ambient nuclei would eventually lead to the formation of the stable helium isotope ^{4}He through successions of more than half a dozen potential reactions involving protons, neutrons, deuterium, the helium isotope ^{3}He of atomic mass 3, and the hydrogen isotope tritium ^{3}H. Some of these are

$$D + P \rightleftharpoons {}^3He + \gamma, \qquad D + N \rightleftharpoons {}^3H + \gamma,$$
$$^3He + {}^3He \rightarrow {}^4He + 2P, \quad {}^3H + P \rightarrow {}^4He + \gamma.$$

Here γ represents the emission of an energetic photon.[d]

[d] The chemical properties of an atom are specified by the number of protons in its nucleus. Where atoms with an identical number of protons differ in mass because the number of neutrons in their respective nuclei differs, one speaks of the atom's distinct *isotopes*, each designated by the number of *nucleons*, the total number of protons plus

Once ^{4}He is formed through the indicated reactions, it either radiates away excess energy through the emission of a photon γ, or is accompanied by the ejection of two protons. Neither reaction is readily reversed. The directions of the arrows indicate the potential outcomes of each reaction; the oppositely directed arrows remind us that in equilibrium the reactions can proceed in either direction.

Bethe and Critchfield's work thus solved a prime problem of astrophysics. We now knew at least one source of energy that could account for the steady outpouring of energy from stars. This process, however, also raised two new questions: Was the conversion of hydrogen into helium at the centers of stars the sole source of all helium in the Universe? And, if so, were vastly heavier elements, such as uranium, similarly produced in stars through the further fusion of a mix of hydrogen and helium to form all the more massive elements? No such reactions appeared in sight. If none existed, could at least some of the heavy elements have been primordially produced in an extremely hot and dense early phase of the Universe before the Cosmos expanded to its current dimensions and far lower temperatures?

Just before onset of World War II, the young German theorist Carl Friedrich von Weizsäcker identified these two possibilities but came to no firm conclusions.[18; 19] A potential difference between the two mechanisms was that, in most stars, all conceivable nuclear reactions ultimately had to take the predominant abundance of protons and electrons into account. In contrast, as far as one could tell in the late 1930s, the most abundant constituent in the early Universe could well have been neutrons. This was significant because the distinguishing feature of the nuclei of heavy elements was their high ratio of neutrons to protons. The nuclei of the naturally occurring uranium isotope, ^{238}U, for example, comprise 146 neutrons and only 92 protons.

2.14   Audacious Predictions

Not until the end of World War II did concerted attempts to resolve these alternatives resume. The new charge was led by George Gamow – a man of enormous spontaneity, enthusiasm, and deep physical intuition. He charted a new, broadly encompassing world view breathtaking in its intellectual brilliance and sheer audacity that still guides cosmology today.[20]

---

neutrons the isotope's nucleus contains. ^{3}He is the helium isotope of mass 3, whose nucleus contains two protons and one neutron. The hydrogen isotope *tritium*, ^{3}H, has the same number of nucleons, but only one proton and two neutrons. Its single proton endows it with the chemical properties of both ordinary hydrogen H whose nucleus consists of one proton, and deuterium D whose nucleus contains one proton and one neutron.

Gamow's vision could frequently also run afoul of important considerations, and two much younger astrophysicists, Ralph Asher Alpher and Robert C. Herman in the United States, working on cosmological problems in their spare time at night, and the 29-year-old Chushiro Hayashi working in isolation half a globe away at Naniwa University in Japan, contributed most of the important theoretical advances that ultimately led to clarity and conviction.

Gamow began with a stumble. It still confounds his otherwise brilliant efforts. For a first paper, jointly written with his extraordinarily talented graduate student, Ralph Alpher, Gamow also invited Hans Bethe to participate.[21] Always playful and ready for an April Fools' joke, Gamow wanted to promote an article with the byline Alpher, Bethe, Gamow – reminiscent of the letters α, β, γ initiating the Greek alphabet, and by inference perhaps also suggesting a brand new way of looking at cosmic history and the formation of heavy elements.

To have it ready in time for the April 1 issue of the 1948 *Physical Review*, the paper, barely the size of one journal page and thus light on detail, reached the journal on February 18, 1948, probably just in time to still be included for its designated issue.

The pizzazz of this paper quickly gained notoriety among physicists, who celebrated the jocular byline but rarely found time to read further – which might have been just as well. Gamow's rush to publish introduced several errors at odds with the genuinely important cosmological papers which he, and separately Alpher and Herman, published in the subsequent weeks and months of their most fruitful year, 1948–49.[e]

Bethe had not been materially involved in writing the paper, but had gone along for the joke. The badly shaken young Alpher, and his more senior and probably less easily ruffled thesis supervisor Gamow, picked up the pieces and started all over again.

Based on some of the most obvious corrections, Gamow continued his cosmological sketches, in one of which he predicted the formation of galaxies as the Universe steadily cooled.[22] By now, he had also come to the important realization that element production in the early Universe could proceed only through the formation of helium, presumably along the lines Bethe and Critchfield had already indicated in 1939. However, all along, Gamow was assuming that neutrons, whose half life for decay into protons was about 10 minutes, had been

---

[e] The α, β, γ paper had several failings: Most important, it had assumed that the density of the Universe and thus also its expansion rate would have been primarily determined by the density of matter during the first minutes of existence. The authors failed to notice that, at this early epoch, the mass–energy density of photons – let alone the comparable mass–energies of the accompanying neutrinos, antineutrinos, and highly energetic electrons and positrons – had been millions of times higher than the mass–energy density of cosmic protons and neutrons. These oversights produced a cascade of misconceptions and urged an entirely new start.

the sole form of matter initially occupying the Universe. A high abundance of neutrons would thus be available after an initial formation of deuterium and subsequently ^{4}He through the process of Bethe and Critchfield, and continue from there to form the neutron-rich heavy elements through fusion of less massive nuclei and residual neutrons.

Using the expansion rate general relativity provided, Gamow correctly estimated that the interval during which deuterium or heavier elements might have formed could not have lasted more than about 100 seconds. Gamow was largely correct also in his predictions of helium formation. But, as the University of Chicago physicists Enrico Fermi and Anthony Turkevich were soon to show, the path to formation of heavier elements was blocked under conditions prevailing at these early times. Their oft-quoted, but never-published calculation had shown that the most likely reactions between two nuclei somewhat more massive than helium would inevitably lead to unstable nuclei, which would quickly decay back to helium or less massive nuclei.

This finding of Fermi and Turkevich was later quoted in detail by Alpher and his co-author Robert C. Herman in a review published in 1950.[23] It rang a death knell on the idea of formation of heavy elements at early epochs. Beyond small quantities of a few light isotopes of hydrogen, beryllium, helium, or lithium, none of the heavier chemical elements could have been produced at primordial epochs.

2.15   Alpher and Herman's Prediction of the Microwave Background

Late in 1948, Gamow wrote one more, somewhat longer, cosmological article in which he summarized much of his and Alpher's recent work. It appeared in the October 30, 1948 issue of *Nature*.[24]

The response to this was swift, and dramatic. Two weeks after the appearance of Gamow's article, *Nature* published a short rebuttal by Ralph Alpher and Robert Herman, in which they pointed out three rather elementary errors in Gamow's submission.[25] All were rather minor, but one phrase in their short letter came as a jolt! It reads, "the temperature in the universe at the present time is found to be about 5 K."

The implication of Alpher and Herman's statement was clear: Evidence of an originally hot Universe must still be around today! If found it would verify some of the crucial assumptions that Gamow had sketched.

The prediction of Alpher and Herman's letter to *Nature* presaged a more formal discussion, a few months later, in which they showed in greater detail how a universe that had been sufficiently hot to initiate nuclear reactions at early times should have left its imprint on a thermal radiation bath still permeating the Universe today. Their paper also corrected for the mistaken pressures in the α, β, γ paper, written a few months earlier. However, they too were still assuming that neutrons were the dominant constituent of the early Universe and that protons would be under-abundant until the neutrons gradually decayed with a half life of order 10 minutes.[26]

Importantly, their work also provided a sketch of the epoch at which radiation had decoupled from matter. Full decoupling meant that the Universe would become transparent to photons, enabling them to traverse the Universe to great distances.

2.16   Hayashi's Insights

Responding to the work of Gamow, and Alpher and Herman, Chushiro Hayashi (Figure 2.4) published a seminal paper in 1950, in which he noted that, at the high temperatures T prevailing at early times, particle energies kT would have greatly exceeded the electrons' rest-mass energies, $m_e c^2$.[f] This meant that electrons e^-, positrons e^+, neutrinos ν, and their antiparticles, the antineutrinos $\bar{\nu}$, would all be moving at relativistic speeds and remain about as abundant as photons.[27] At these temperatures, neutrons would not dominate the atomic mass-density. Rather, neutrons N and protons P would rapidly reach thermal equilibrium with their ambient mix of particles and radiation through reactions such as

$$N + e^+ \rightleftharpoons P + \bar{\nu}; \quad N + \nu \rightleftharpoons P + e^-; \quad N \rightleftharpoons P + e^- + \bar{\nu}.$$

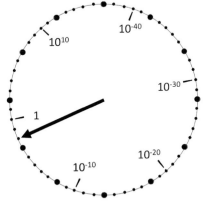

10⁴⁰ → 10^{-40}

10¹⁰ → 10^{10}

10^{-30}

1

10^{-20}

10^{-10}

Nuclear Particle Equilibrium
$t = 10^{-2}$ seconds = 10 milliseconds

At 10 milliseconds, the temperature dropped below 10^{11} degrees. The number of neutrons declined below the number of protons because neutrons are more massive than protons plus electrons, and readily convert into these lighter particles. Throughout, electrons, positrons, antineutrinos, neutrinos, and photons remained in thermal equilibrium.

Hayashi found that, at temperatures higher than $\sim 2 \times 10^{10}$ K, the interactions of protons and neutrons with electrons and positrons proceeded far more

---

[f] Here k is the *Boltzmann constant*.

Fig. 2.4 Chushiro Hayashi, 1920–2010, photographed in early 1961, roughly 10 years after completing his cosmological work. He was now developing a theory of star formation.

rapidly than the drop in temperature due to cosmic expansion. At cosmic age around 10 milliseconds, as temperatures dropped below $\sim 10^{11}$ K, the proton-to-neutron ratio P:N began to rise irreversibly as electron and antineutrino energies dropped, reducing the reverse conversion of protons into neutrons.

By $t \sim 0.1$ seconds, at a temperature $T \sim 3 \times 10^{10}$ K, equilibrium between electrons, neutrinos, and their respective antiparticles could no longer be maintained either. The neutrino and antineutrino energies had sunk too low to form electrons and positrons in the reaction

$$e^+ + e^- \rightleftharpoons \nu + \bar{\nu}.$$

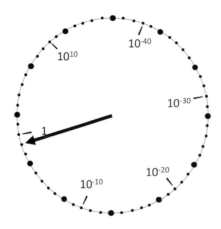

Neutrino Decoupling
$t = 10^{-1}$ seconds

Cosmic expansion now rapidly cooled neutrinos and antineutrinos, whose dropping energies and weak interactions decoupled them from the electrons and positrons remaining in thermal equilibrium with photons. The temperatures of the neutrinos and the photons, two critical messengers from this epoch, would henceforth differ.

As the cosmic expansion proceeded, and the distances between particles increased, their mutual collisions dwindled, and temperatures also dropped. By cosmic age 1 second, and a temperature $T \sim 10^{10}$ K, particle energies had dropped to $\sim 10^6$ eV, where neutrons and protons no longer could maintain thermal equilibrium, and the ratio of protons to neutrons froze.

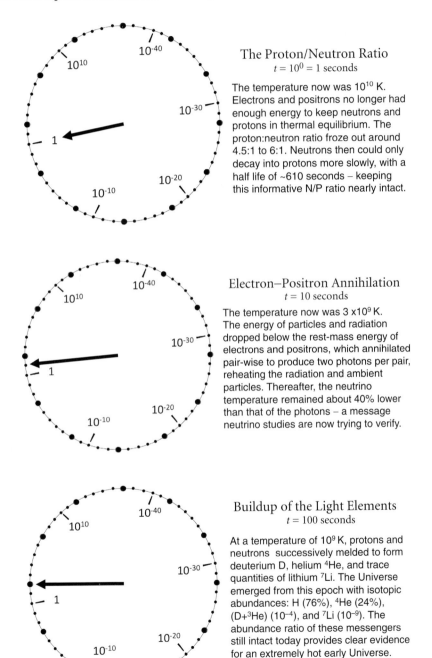

The Proton/Neutron Ratio
$t = 10^0 = 1$ seconds

The temperature now was 10^{10} K. Electrons and positrons no longer had enough energy to keep neutrons and protons in thermal equilibrium. The proton:neutron ratio froze out around 4.5:1 to 6:1. Neutrons then could only decay into protons more slowly, with a half life of ~610 seconds – keeping this informative N/P ratio nearly intact.

Electron–Positron Annihilation
$t = 10$ seconds

The temperature now was 3×10^9 K. The energy of particles and radiation dropped below the rest-mass energy of electrons and positrons, which annihilated pair-wise to produce two photons per pair, reheating the radiation and ambient particles. Thereafter, the neutrino temperature remained about 40% lower than that of the photons – a message neutrino studies are now trying to verify.

Buildup of the Light Elements
$t = 100$ seconds

At a temperature of 10^9 K, protons and neutrons successively melded to form deuterium D, helium ^{4}He, and trace quantities of lithium ^{7}Li. The Universe emerged from this epoch with isotopic abundances: H (76%), ^{4}He (24%), (D+^{3}He) (10^{-4}), and ^{7}Li (10^{-9}). The abundance ratio of these messengers still intact today provides clear evidence for an extremely hot early Universe.

2.17   Primordial Synthesis of Light Chemical Elements

Hayashi's most important conclusion was that the proton-to-neutron ratio should have inevitably frozen out at a proportion of about 4:1 if the initial temperature of the Universe had ever been sufficiently high to establish thermal equilibrium. Thereafter, the neutron fraction would gradually have declined through nuclear decay of neutrons into protons dictated by the neutron's relatively long half life of $\sim$10 minutes.

Assuming this P:N freeze-out ratio, and recognizing that nuclei other than helium ^{4}He were unlikely to form in significant numbers during primordial epochs, Hayashi calculated the ratio of hydrogen to helium nuclei at these early times. Helium production required the fusion of two neutrons and two protons. For every eight initial protons, and the corresponding two initial neutrons, there would be six residual protons for each ^{4}He nucleus formed. Hayashi noted that this ratio of number densities of 6:1 was close to the H:^{4}He abundance ratios in stellar atmospheres and meteorites, at the time estimated, respectively, as 5:1 and 10:1.[g]

Hayashi's detailed calculations and the conclusions he drew from them were a first quantitative indication that at least much of the helium observed in the Universe today appeared to have formed at primordial cosmological times, rather than in the hot central regions of stars. This provided confirming evidence that the temperature in the Universe must at one time have been of the order of $\geq 10^{10}$K.[29]

Hayashi had assumed that the primary contributor to the mass–energy density of the Universe was radiation. Although this was correct for the epoch of helium formation, the assumption fails at temperatures above 10^{10} K, where collisions between photons produce electron–positron pairs whose further collisions would have given rise to neutrino–antineutrino pairs. Collectively, the mass-densities of these four types of particles then exceed the mass-density of radiation, and dominated the rate at which the Universe was expanding.

2.18   More Detailed Calculations of Alpher, Follin, and Herman

In 1953, three years after the publication of Hayashi's paper, the problem of including this interconversion of radiation and matter was further addressed by Alpher, Herman, and their colleague James W. Follin, Jr. at

---

[g] Today's best estimate of the freeze-out ratio is $\sim$6:1 and the ratio of hydrogen atoms to helium atoms in the Solar System is $\sim$10:1 – still in rough agreement with Hayashi's conclusions.[28]

the Johns Hopkins Applied Physics Laboratories.[30] Their calculations, which expanded on Hayashi's work mainly in applying relativistic quantum statistics more suitable to calculating processes at early epochs and their adoption of a somewhat different neutron lifetime that had recently become available, led to small but significant changes in the proton-to-neutron ratio at freeze-out, placing it roughly in the range of 4.5:1 to 6:1. This was in somewhat better agreement with observed abundance ratios of hydrogen to helium, adding confidence that the theory at least was consistent with this most firmly established ratio of observed chemical abundances.

The amounts of helium and other light elements, and their stable isotopes that may have been formed when the age of the Universe was about 100 seconds, depends sensitively on the rapidly declining cosmic mass-density at that epoch. This density determined the instantaneous expansion rate and thus the diminishing temperature. The mass-density, in turn, depended on the number of particle species abundant at the time.

Today, we know of the existence of a thermal gamma-radiation bath and three species of neutrinos, v_e, v_μ, and v_τ, all of which appear inevitably to have been primordial ingredients of the Cosmos around the epoch of primordial nucleosynthesis. These four primary sources of mass–energy driving the expansion would have determined the duration of nuclear synthesis and thus the ratio of the isotopes produced.

Given these densities and those of the prevailing protons, the observed light element abundances in the Universe today are in good agreement with primordial light-element production. The number of ^{4}He atoms in the local Universe is almost precisely one-tenth the number of ^{1}H atoms. Helium atoms thus make up $\sim 28\%$ of the baryonic mass of the Universe. This is far more helium than could have been produced in stars and ejected into their environs over the age of the Universe. A primordial production of these atoms alone, therefore, was quite likely in any case, but that production also had to be consistent with an anticipated production of deuterium D, the helium isotope ^{3}He, and the lithium isotope ^{7}Li by the same primordial processes.

Figure 2.5 shows that these light-element isotopes indeed appear consistent with hydrogen densities extrapolated backward in time to that epoch. The main uncertainty in the figure appears to be the value η, the ratio of baryons to photons in the Universe today.

The relative abundances of primordial deuterium D, helium isotope ^{3}He, and lithium isotope ^{7}Li, all of which can be destroyed at temperatures encountered in the outer layers of stars, or potentially also produced in stars, are somewhat more difficult to establish, but have been painstakingly modeled and calculated with such processes in mind. Figure 2.5 shows the best estimates we have today

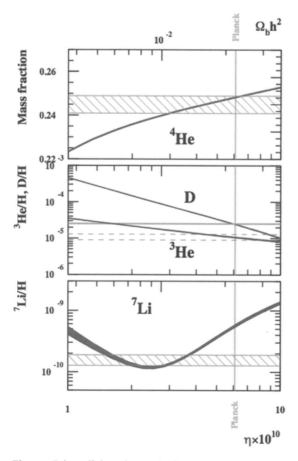

Fig. 2.5 Primordial Nucleosynthesis. The abundances of the light isotopes of helium ^{4}He, ^{3}He, deuterium ^{2}D, and lithium ^{7}Li are thought to have been primordially produced. The labeled curves show the theoretically predicted isotopic abundances as a function of today's cosmic baryon-to-photon ratio $\eta \sim 6 \times 10^{-10}$ indicated on the abscissa where it has been multiplied by 10^{10}. Superposed horizontal lines or cross-hatched areas indicate uncertainties in the measured abundances attributed to primordial synthesis. The vertical line labeled "Planck" indicates the best estimates of today's photon-to-baryon ratio derived from background observations obtained by the space mission *Planck*. $\Omega_b \sim 0.05$ is the cosmic baryon mass–energy, as a fraction of the total cosmic mass–energy, and $h \sim 0.7$ is the best current estimate of the Hubble constant measured in units of 100 km s^{-1} Mpc^{-1}. (From "Primordial Nucleosythesis," Alain Coc and Elisabeth Vangioni, *International Journal of Modern Physics E* 26, No.8, 20 pp., Fig. 5, 2017, with permission of the authors and the journal's publishers.)

for all these abundances for a range of assumed ratios of number densities of baryons n_b and photons n_γ and the *baryon fractions* $\eta \equiv n_b/n_\gamma$, which should have remained constant since primordial times. Although the value of η is still difficult to accurately determine because the cosmic ratio of baryons to photons in intergalactic space is still poorly understood, Figure 2.5 shows that for a value of $\eta \sim 6 \times 10^{-10}$ the mass fractions of ^{4}He, D, and ^{3}He all fall into line with theoretical expectations. Solely, the observed abundances of ^{7}Li do not fit well. They are too high by a factor of ~3 to 4. Though this discrepancy remains unexplained, it is worth noting that the observed cosmic mass fraction of ^{7}Li is ~20,000 times lower than that of ^{3}He and ~50,000 times lower than that of D, so that minor sources of lithium production in stars could account for the observed excess.[31] The close mutual agreement of the mass fractions of D, ^{3}He, and ^{4}He evident today thus also argues for a baryonic fraction $\eta \sim 6 \times 10^{-10}$, providing further evidence for primordial production of these three isotopes. Note also that η is the same ratio of nucleons to photons in the Universe today, which Zel'dovich had emphasized in his explanation of why the Universe today is composed primarily of matter, rather than equal amounts of matter and antimatter. It indicates that the two sets of explanations based on entirely distinct observations, at the very least, are mutually consistent.

2.19   Cross-Over to a Matter-Dominated Expansion

In one of his most inspired 1948 papers, Gamow had made one more important prediction.[32] He noted that the cosmic expansion would lead to a drop in matter density proportional to the third power of cosmic expansion, whereas the mass-density of radiation would decline as the fourth power. The photons' more rapid decline with cosmic expansion would be due to their correspondingly expanding wavelengths, the accompanying drop in the radiation's temperature, and thus also in the mass–energy per photon.

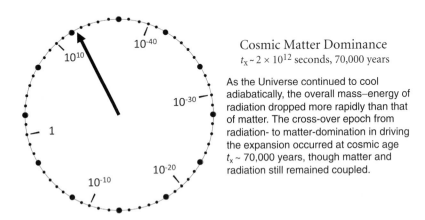

Cosmic Matter Dominance
$t_x \sim 2 \times 10^{12}$ seconds, 70,000 years

As the Universe continued to cool adiabatically, the overall mass–energy of radiation dropped more rapidly than that of matter. The cross-over epoch from radiation- to matter-domination in driving the expansion occurred at cosmic age $t_x \sim 70,000$ years, though matter and radiation still remained coupled.

For a drop in the radiation's temperature inversely proportional to the expansion of the Universe, this would mean that the coupled matter and radiation densities would eventually equal each other at a level of 10^{-24} g cm^{-3}, at a temperature of order 10^3 K. Gamow estimated that this *cross-over* would occur when the Universe was $\sim 10^7$ years old. Today's estimates for the age $t_\times$, temperature $T_\times$, and redshift $z_\times$ at cross-over of mass-densities are $t_\times \sim 70,000$ years, $T_\times \sim 8700$ K, and $z_\times \sim 3200$, reasonably close to Gamow's conjectures, considering that he was envisioning wholly unexplored cosmic territories.

As the cosmic expansion continued and temperatures dropped even further, radiation would decouple from matter and cool more rapidly. Today's best estimates for the temperature at decoupling is $\sim$3000 K, corresponding to a redshift $z \sim 1100$, and cosmic age $\sim$375 million years.

2.20   The Cosmic Background Radiation of Penzias and Wilson (1965)

In 1965, Arno A. Penzias and Robert W. Wilson at the Bell Telephone Observatories discovered a ubiquitous radio flux pervading all of extragalactic space. A horn-reflector antenna at the Bell Telephone Laboratories' Crawford Hill facility in Holmdel, New Jersey, which they were attempting to calibrate, kept producing a pervasive signal that the two young scientists originally thought might be spurious. It simply would not go away, no matter where the telescope was pointed in the sky. Its strength was compatible with a ubiquitous radiation at a temperature of $\sim$3.5 K and thus remarkably close to Alpher and Herman's coarse prediction of 5 K of which Penzias and Wilson had been unaware.[33] Once its authenticity and origin had been established, however, it convinced astrophysicists in a way that no other observation had previously matched that the Universe indeed had once been extremely hot!

Even firmer confirmation came in late 1989 when NASA launched the Cosmic Background Explorer mission, COBE. It established that the mean temperature of the radiation bath permeating the Cosmos today is $\sim$2.73 $\pm$ 0.01 K. Along any given line of sight the spectrum was that of a blackbody of fixed temperature smoothly changing with direction across the celestial sphere. This provided a measure of both the direction and the speed at which our planet Earth, our Solar System, and our Milky Way galaxy are traversing a homogeneous and isotropic radiation bath exhibiting a single ubiquitous blackbody temperature.

The conclusion drawn from Figure 2.6 was clear: At the epoch of decoupling of radiation from matter, the Universe had been in true thermal equilibrium![34]

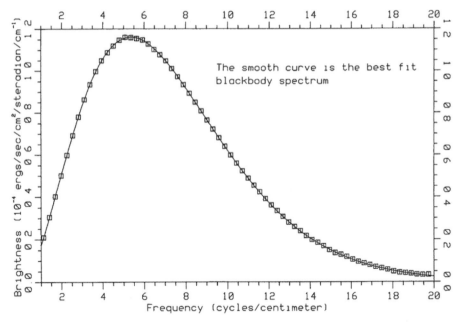

Fig. 2.6 The spectrum of the cosmic microwave background radiation, obtained in 1990 with the Cosmic Background Explorer, COBE. The data are exhibited with a best-fit curve of a blackbody spectrum at temperature ~ 2.73 K. These observations were undeniable proof that the radiation was thermal, deviating from that of a blackbody by no more than 1% of peak emission, anywhere within the spectral range covered. A more robust confirmation that the Universe had once been in thermal equilibrium at high temperatures, as well as opaque, could hardly have been imagined! (From "A Preliminary Measurement of the Cosmic Microwave Spectrum," J. C. Mather, et al., *Astrophysical Journal Letters* 354, L37–40, May 10, 1990, with permission of the NASA/COBE Science team and the American Astronomical Society.)

2.21   The Anticipated Neutrino Background Radiation

Like the 1950 paper of Hayashi, the Alpher–Follin–Herman paper had confined itself to the era during which most of the light elements were forming. It no longer speculated about the formation of heavy elements because the prospects for heavy-element production in the early Universe had begun to look dim. Its major contribution was its anticipation of a thermal neutrino cosmic background, analogous to that of the microwave background.[35]

If observed as predicted, these neutrinos would reflect conditions in the Universe ~0.1 seconds after the birth of time. The cosmic temperature at that epoch was $\sim 3 \times 10^{10}$ K, where the weak interaction rates were becoming too slow to compete with the cosmic expansion rate, and the neutrinos decoupled from matter.

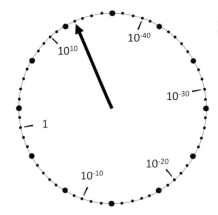

Matter and Radiation Decouple
$t = 1.2 \times 10^{13}$ seconds = 375,000 years

From 10^2 until 10^{13} seconds, the Universe continued to expand and cool. By 10^{13} seconds the temperature had fallen to 4000 K. Electrons and ions formed atoms decoupling from photons. The Cosmos became transparent. Radiation emitted at that time is now revealed as a microwave background, whose spatial fluctuations mimic the mass distribution during the inflationary 10^{-35} to 10^{-33} seconds.

By cosmic age 10 seconds, and temperatures $T \sim 3 \times 10^9$ K, radiation energies are expected to have dropped below the electron and positron rest-masses. Annihilation of these particles would then heat the ambient radiation field irreversibly through the reaction

$$e^+ + e^- \rightleftharpoons \gamma + \gamma,$$

raising the temperature of the radiation bathing the Universe to a higher temperature than that of the prevailing neutrino bath that had decoupled from matter and radiation at cosmic age 0.1 seconds.

Alpher, Follin, and Herman predicted that the ratio of photon temperature T_γ to neutrino temperature T_ν should thereafter remain constant, leading to an observable ratio $T_\gamma/T_\nu \sim 1.40$ today. For the microwave background photon temperature measured by the COBE mission, ~ 2.73 K, their predicted thermal neutrino spectrum should exhibit a temperature ~ 1.9 K, corresponding to neutrino energies $\sim 2 \times 10^{-4}$ eV.

Laboratory efforts aimed at ultimately finding this additional background component are currently underway at the KATRIN project in Karlsruhe, Germany, and at the PTOLEMY experiment at the Princeton Plasma Physics Laboratory in the United States.[36]

This ambitious neutrino observation can be carried out in ground-based laboratories because the neutrino background is ubiquitous. The expected signal for a detection would come from the derived spectrum of electron neutrinos absorbed in the transformation of the hydrogen isotope tritium, ^{3}H, into the helium isotope ^{3}He, through the reaction

$$v_e +^3 H \rightarrow^3 He + e^-.$$

Verifying whether or not we correctly understand conditions at a time when the Universe was only ~0.1 seconds old would provide access to the history of the Cosmos at an epoch considerably earlier than that probed by the abundance of primordial light elements at cosmic age ~100 seconds.

When the neutrino background is eventually measured, the detected neutrinos are expected to convey a message confirming their decoupling from the pervasive photon bath, well before a cosmic age of 10 seconds!

2.22   A Stored Imprint of the Inflationary Phase

In the wake of the COBE mission, NASA's *Wilkinson Microwave Anisotropy Probe, WMAP*, launched in June 2001, and the *Planck* mission, initiated by the European Space Agency in May 2009, mapped a wealth of faint microwave background fluctuations that the inflationary phase of the Cosmos had imprinted on the sky.

These tiny deviations were found to have amplitudes of the order of one part in ten thousand of the mean surface brightness of the radiation and to be scale invariant – both traits amazingly confirming the predictions of Zel'dovich three decades earlier in 1972![37]

Figure 2.7 shows these minute fluctuations in the radiation as first revealed by WMAP. Though clearly visible on these two antipodal hemispheres on the sky, the fluctuations in the background radiation range from a deviation of only one part in 10^4 above the mean microwave surface brightness to one part in 10^4 below.

As we saw earlier in this chapter, the inflationary expansion of the Universe from 10^{-35} to 10^{-33} seconds was exponential, with an invariant Hubble constant throughout, yielding a colossal dilation from an initial scale factor a_i to a final value a_f corresponding to a ratio

$$\frac{a_f}{a_i} = 10^{26}.$$

This expansion was so rapid and so pervasive that the irregular pattern of the primordial mass distribution which Zel'dovich had predicted should be scale

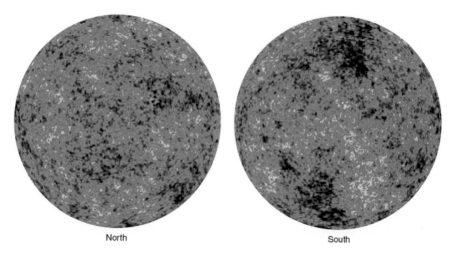

North   South

Fig. 2.7 Fluctuations in the Microwave Background Radiation recorded by the Wilkinson Microwave Anisotropy Probe Mission, WMAP. The peak to peak temperature deviations, from a mean value of 2.725 K, are confined to a range of -2×10^{-4} K below the mean in dark shaded regions to $+2 \times 10^{-4}$ K above the mean in the lightest shaded domains. Measuring these tiny deviations, and assuring that they can reliably be attributed to the microwave background, rather than some arbitrary noise, required a heroic decade-long effort. (From "Three-Year Wilkinson Microwave Anisotropy Probe (WMAP) Observations: Temperature Analysis," G. Hinshaw, et al., *Astrophysical Journal Supplement Series* 170, 288–334, June 2007, Fig. 21, p. 323. with permission of the NASA/WMAP Science Team and the American Astronomical Society. I thank Chuck Bennett, lead of the WMAP team, for transforming this figure into the black and white format shown.)

invariant became frozen in place. Not that gravitational attraction between neighboring masses could not bring them slightly closer to one another once inflation had stopped, but those gravitationally induced mass flows were negligibly minuscule on the scale of the preceding expansion.

This structural stability persisted throughout the subsequent radiation-dominated adiabatic phase, during which the rapid expansion continued for some time, though progressively slowing down as the dominating radiation density continued to diminish in response to a drop from an initial temperature of $\sim 10^{27}$ K to a final 3000 K at decoupling.

As the temperature dropped, the scale factor a – the linear dimension of the Universe – expanded, but the product aT remained constant:

$$\frac{a_{final}}{a_{initial}} = \frac{T_{initial}}{T_{final}} \sim \frac{10^{27}}{(3000)} \sim 3 \times 10^{23}.$$

Meanwhile, the expansion rate given by the Hubble constant dropped in inverse proportion to the square of the scale factor, and inversely with time,

$$H \propto a^{-2} \propto t^{-1}.$$

Throughout these two phases, inflation followed by adiabatic expansion, the Universe displayed a horizon receding at the speed of light, c, beyond which an observer could not see. This *Hubble radius* determined by the ratio $R_H \equiv c/H$ defines the extent of the observable part of the Universe, the *Hubble sphere*, at any given epoch.

Two consequences arise:

(i) During inflation, the Hubble constant is independent of the systematically increasing scale factor $a(t)$, and so is the distance to the horizon R_H, even though each phase in the expansion reveals successively smaller fluctuations initially present in the pre-inflationary universe. So, during inflation, the radius of the observable fraction of the Universe keeps shrinking as the Universe expands,

$$R_H/a(t) \propto a(t)^{-1},$$

leading to a decline of this ratio by up to a factor of 10^{26} by inflation's end.

(ii) During adiabatic expansion, $R_H(t)$ is proportional to $a(t)^2$, and the ratio

$$R_H(t)/a(t) \propto a(t).$$

Accordingly, during adiabatic expansion up to decoupling, $R_H(t)/a(t)$ grows by a factor of $\sim 3 \times 10^{23}$.

If, as in Figure 2.8, we now look at the inflationary phase of the Universe at a time when the Universe had already expanded, say, by a factor of 300 since the onset of inflation, only a further expansion by a factor of 3×10^{23} would remain to achieve a total inflationary expansion by a factor of 10^{26}. Amplification during the succeeding adiabatic phase leading up to decoupling can exactly compensate for this earlier reduction in the Hubble radius. This means that at decoupling between matter and radiation the appearance of the frozen-in mass distribution visible to us at that epoch can appear identical to that in force after inflation had already expanded the Universe by a factor of ~ 300.

If, as Zel'dovich had proposed, the fluctuation pattern was scale-free but random at the start of inflation, it will have continued to be scale-free and random throughout inflation. No matter which epoch during inflation precisely coincides with the pattern at decoupling later on, the fluctuations we observe in the microwave background radiation today will correspond to the fluctuations

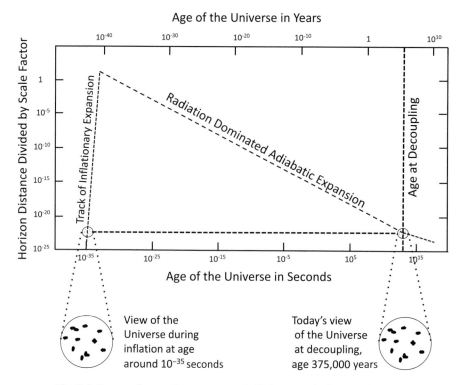

Fig. 2.8 Because fluctuations present at inflation remain frozen within the rapidly expanding Universe throughout the inflationary phase and the subsequent radiation-dominated expansion, the fluctuations observed in the microwave background radiation faithfully image the primordial fluctuations at an epoch when the ratio of the *Hubble radius*, $R_H \equiv c/H$, to the the scale factor a, c/Ha, was identical for the two epochs. The evolution illustrated here shows the rapid ascent of the ratio during inflation, and the subsequent more gradual decline during adiabatic expansion remaining in force for a much longer time interval. This growth and decline in R_H/a, which just cancel each other, make the comparison of the two views identical. This is indicated by the magnified, hypothetical maps of fluctuations shown within the larger two circles at bottom but attributed to the two smaller circles, respectively shown at early and late epochs, at lower left and right in the framed diagram.

dominating some unspecified epoch during inflation that just happens to be re-emerging into full view right now.

If all this seems magical, one needs only to note the crucial factor enabling this mapping of an early cosmic epoch onto a later epoch. It is that any identifiable single interval lasting only $\sim 10^{-35}$ seconds during inflation, today maps onto a far longer period lasting millions of years exhibiting hardly any change. This is because after the end of inflation the cosmic scale factor $a(t)$ expanded by another factor of 3×10^{23}. As stated above, time intervals t expanded as the square of this factor. Consequently, at its re-emergence today, any time interval of the inflationary period would appear to be changing $\sim 10^{47}$ times more slowly. An inflationary interval of 10^{-35} seconds during inflation would by now persist for $\sim 10^{12}$ seconds, or thirty thousand years.

2.23   The Dark Ages and the Formation of Stars, Galaxies, and Quasars

The decoupling of matter and radiation occurred at a temperature of $\sim$3000 K. Thereafter matter and radiation cooled at separate rates as the Universe further expanded. This epoch is generally referred to as *the Dark Ages* because to date it has not been directly observed.

We do not yet know just how low the temperature of atomic matter had to drop before the distribution of gases no longer could resist mutually attracting gravitational forces and matter began to clump and form stars. This may only have happened after atomic temperatures had dropped to around 20 K, where atomic hydrogen H could form molecules H_2. These would readily radiate away heat produced in sustained contraction so that galaxy and star formation could begin. Such a low temperature may only have been reached a few hundred million years after decoupling, around redshift $z \sim 10$.

The Dark Ages
$t = 10^{14}$ to 10^{16} seconds, 3 to 300 Myr

The mass-density of particles in the Cosmos by now exceeded the radiation density, and began to dominate the cosmic expansion rate. Though density fluctuations contracted into ever-denser clouds, the Cosmos remained dark. A faint atomic-hydrogen radio glow may soon serve as a messenger telling us more about this still-mysterious epoch preceding the formation of stars.

The term *Dark Ages* may not much longer remain accurate because hydrogen atoms are expected to have emitted energy, albeit faintly, at rest frame wavelengths of 21 cm at temperatures prevailing during those Dark Ages. Photons emitted at this wavelength should now be reaching us redshifted by a factor of $z \geq 10$ into a broad wavelength band at wavelengths $\lambda_z \sim 21(z + 1)$ cm of the radio continuum. Attempts to detect this radiation are currently being pursued internationally by several groups.

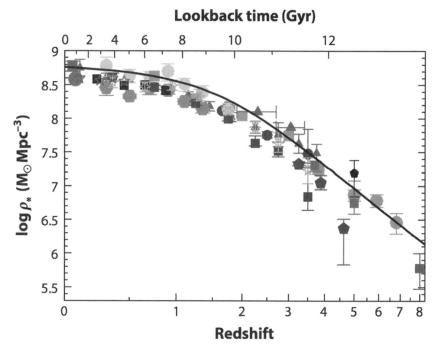

Fig. 2.9 Plot of the rising mass-density accumulating in stars as the Universe evolved. The density shown refers to the total mass of matter condensed in stars, within a volume that is expanding at the same rate as the entire Universe. The solid line shows a best fit to the data. (From "Cosmic Star-Formation History," P. Madau & M. Dickinson, *Annual Review of Astronomy and Astrophysics* 52, 415–86, Fig. 11, p. 464, 2014.)

Molecular hydrogen is able to cool more rapidly than hydrogen atoms. The ability of molecules to convert collisional energy into induced molecular vibrations and rotations, whose energy is then radiated away, appears to have been critical to the ultimate formation of stars, galaxies, and quasars. Though the historical order in which these different bodies may have formed is still quite uncertain, direct observations of the emitted molecular hydrogen spectrum should, sooner or later, clarify the order of these sequences.

The epoch at which stars began to form in the Universe is not yet precisely known. Observations now can reach back to an era at which the redshift of galaxies is $z \geq 8$. The gradual increase of atomic matter accumulating into stars is still being determined, but follows, as shown in Figure 2.9, an evolution indicated by Pierro Madau at the University of California at Santa Cruz and Mark Dickinson at the National Optical Astronomy Observatory in Tucson, Arizona.[38]

2.24   The Evolution of Cosmic Elemental Abundances (1996)

In the mid-1990s, Antoinette Songaila and Lennox L. Cowie of the University of Hawaii discovered that faint traces of triply ionized carbon and silicon atoms could be detected in extragalactic space at redshifts as high as $z \sim 3$, when the Universe was roughly 2 billion years old.[39] Their sensitive spectroscopic techniques permitted them to detect these ions through their absorption of radiation emitted by more distant quasars.

These data showed that, by a cosmic age of 2 Gyr, the abundance of carbon relative to hydrogen was already about 1% of the density currently observed in the Sun, while the ratio of silicon to carbon was three times higher than today's. These abundance ratios, Songaila and Cowie pointed out, were reminiscent of the chemical abundances in stars formed many billions of years ago and now populating the *Galactic halo*, a tenuous spherical region enclosing the Milky Way disk. The halo accounts for most of the Galaxy's mass, partly in the form of these old stars, but predominantly in the yet-to-be-defined *dark matter*. Presumably the halo stars originated in gas clouds containing the same silicon/carbon mix Songaila and Cowie had observed and date back to these same early epochs. A history of heavy-element production was about to begin.

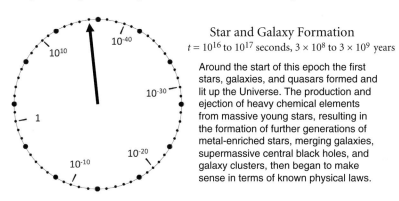

Star and Galaxy Formation
$t = 10^{16}$ to 10^{17} seconds, 3×10^8 to 3×10^9 years

Around the start of this epoch the first stars, galaxies, and quasars formed and lit up the Universe. The production and ejection of heavy chemical elements from massive young stars, resulting in the formation of further generations of metal-enriched stars, merging galaxies, supermassive central black holes, and galaxy clusters, then began to make sense in terms of known physical laws.

2.25   A Closing Note on Primordial Element Production

Many of the findings cited in the present chapter resulted from a hypothesis that heavy elements might have formed at primordial epochs. This notion eventually proved to have been misleading. In the meantime, the laboratory work of William A. Fowler at Caltech, and the theoretical work of Edwin E. Salpeter at Cornell University and Fred Hoyle in Cambridge, UK, showed that massive stars could find a way to produce carbon by overcoming the barrier at 5 atomic mass units, which Fermi and Turkevich had shown to be impenetrable through successive binary collisions of stable isotopes. This could be achieved in the dense interior of massive stars where two colliding ^{4}He nuclei could briefly form a metastable state of beryllium ^{8}Be, which an additional collision with a third ^{4}He nucleus could convert into a stable carbon atom ^{12}C.[40]

From there a succession of further interactions could build many of the other heavy elements.[41]

By now, some of the more massive nuclei appear to be generated just before onset of supernova explosions; but the most massive relatively stable nuclei, ranging up to uranium ^{238}U, are currently believed to form in the explosively ejected remnants of merging pairs of neutron stars collapsing to form a black hole.[42]

2.26   Do Dark Energy and Dark Matter Messengers Exist?

Chapter 1 briefly mentioned dark energy and dark matter, and our ignorance of whether or not they produce messengers of their own to interact with other forms of cosmic energy in ways that go beyond the mutual gravitational attraction of all massive bodies. To date, all attempts to detect dark matter messengers directly have failed.[43] Dark energy messengers have fared no better.

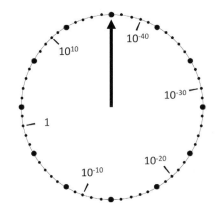

Our Local Environs Emerge

$t = 10^{17}$ seconds = 3 billion years

At 10^{17} seconds, a novel "dark energy" began to drive cosmic expansion. Stars, galaxies, and giant black holes now formed in dark matter haloes, exploding heavy elements into their surroundings at record rates. The Sun and Solar System formed at cosmic age ~ 9×10^9 years. Life on Earth emerged a billion years later. A suite of messengers, now including cosmic rays, revealed an ever-richer panorama.

2.27   Interpreting the Information Messengers Convey

The strongest evidence we have for the existence of a hot early universe is provided by a list of messengers still reaching us today. These include:

The Matter–Antimatter Asymmetry

The ratio of microwave background photons to nucleons in the Universe suggests the mutual annihilation of matter and antimatter at early times.

Thermal Information

Photons exhibiting the thermal spectrum and temperature of the cosmic microwave background radiation indicate that the Universe once was opaque to all electromagnetic wavelengths before matter decoupled from radiation. The messengers are microwave photons.

Chemical Information

The ratio of hydrogen atoms H to helium ^{4}He atoms observed in the Universe today indicates that the ratio of protons to neutrons, which at highest temperatures must have been close to unity, had risen to 4:1. This establishes the temperature, density, and time at which most of the helium in the Universe first formed. The messengers are hydrogen and helium atoms.

The relative abundances of the hydrogen isotope deuterium D, the helium isotope ^{3}He, and the lithium isotope ^{7}Li displayed in the spectra of the earliest stars ever formed alert us to the ratio of photons to nucleons at the epoch of primordial nucleosynthesis; Figure 2.5 shows this relation graphically. The messengers here are the photons emitted by the three atomic isotopes.

The near absence of spectral signatures of heavy elements in intergalactic gases observed at high redshift, and the gradually increasing strength of these spectral features with diminishing redshift, tells us that the early Universe was devoid of these elements. They were only produced once the first stars and quasars had formed. The messengers are photons emitted by the atomic species whose spectra exhibit these elemental trends with redshift.

Neutrinos

The discovery that neutrinos come in three different flavors, ν_e, ν_μ, and ν_τ, also is consistent with the significant mass-densities they collectively contributed at early epochs to account for the overall mass-densities required to produce the observed ratios of hydrogen-, helium-, and light-isotope abundances, as well as the temperature of the microwave radiation observed today. The messengers are the isotopes of various atoms.

Mass and Velocity Distributions

Spatial fluctuations in the surface brightness of the microwave background radiation reflect the mass-distribution spectrum of matter during the epoch of inflation. The messengers conveying this information today are photons streaming out of slightly enhanced density domains originally formed around *inflatons*, scale-invariant mass fluctuations prevailing throughout inflation. Although not directly observed, the inflatons now are inferred from this clustering of microwave photons.

Systematic spectral frequency shifts across arrival directions of microwave background photons indicate the isotropy of radiation in a derived cosmic rest frame relative to which the Earth's motion is being observed. The messengers are microwave photons.

Currently Sought Messengers Likely to Be Detected Within the Next Decades

These are:

The neutrino cosmic background radiation. The messengers here will be the decay products of tritium ^{3}H.

Redshifted 21-cm neutral hydrogen emission radiated during the Dark Ages, providing evidence of the age of the Universe when the temperature had dropped sufficiently for electrons and protons to form atomic hydrogen. The messengers here will be radio-wave photons.

Redshifted spectra of molecular hydrogen indicating further cooling during the Dark Ages to a temperature at which hydrogen molecules H_2 could form, a step likely to have been involved in the formation of the earliest stars. The messengers will be far-infrared photons.

First indications of a gravitational-wave cosmic background spectrum. The messengers here are not yet known but could be photons, dispersed by the gravitational waves.

Looking over the examples just listed it seems clear that verifying such predictions should lead to establishment of a convincing history, rather than a mere set of findings dependent on how the individual messages are interpreted. Are our current interpretations definitely correct? Could they merely be revealing a set of misleading coincidences? At some point the combined evidence available reaches a tipping point at which a mere coincidence can be discounted and we become increasingly confident that our interpretation is on the right track.

Most astrophysicists today appear to agree that this tipping point has already been passed, though our ignorance of the first moments of creation or the nature of dark matter and dark energy warn us to be prepared for major surprises revealing a far richer setting.

Notes

1 The Large N Limit of Superconformal Field Theories and Supergravity, J. Maldacena, *Advances in Theoretical and Mathematical Physics* 2, 231–52, 1998

2 A more formal discussion of emergence may be found in: Equilibrium to Einstein: Entanglement, Thermodynamics, and Gravity, A. Svesko, *arXiv: 1818.12236v1 [hep-th]* October 29, 2018

3 Black Hole Explosions? S. W. Hawking, *Nature* 248, 30–31, 1974

4 Ibid., Hawking, 1974

5 A Hypothesis, Unifying the Structure and Entropy of the Universe, Ya. B. Zel'dovich, *Monthly Notices of the Royal Astronomical Society* 160 (Short Communication) 1P – 3P, 1972

6 The Asymmetry between Matter and Antimatter, H. R. Quinn, *Physics Today* 56/2, 30, 2003

7 Violation of CP Invariance, C Asymmetry, and Baryon Asymmetry of the Universe, A. D. Sakharov, *Soviet Journal of Experimental and Theoretical Physics Letters* 5, 24, 1967

8 The Stability of a Spherical Nebula, J. H. Jeans, *Philosophical Transactions of the Royal Society, Series A* 199, 1–53, 1902

9 On the Gravitational Stability of an Expanding Universe, E. M. Lifshitz, *Journal of Physics of the USSR* 10, 116, 1946

10 See e.g., *Astrophysical Concepts* 4th edition, M. Harwit, Springer Verlag 2006, pp. 556 ff

11 The Big Bang Cosmology – Enigmas and Nostrums, in R. H. Dicke & P. J. E. Peebles, *General Relativity: An Einstein Centenary Survey*, Cambridge University Press, pp. 504–17, 1979

12 How Problematic Is the Near-Euclidean Spatial Geometry of the Large-Scale Universe?, M. Holman, *Foundations of Physics* 48, 1617–47, 2018

13 Spectrum of Relict Gravitational Radiation and the Energy State of the Universe, A. A. Starobinsky, *Soviet Physics, JETP Letters* 30, 682–85, 1979

14 A New Type of Isotropic Cosmological Models without Singularity, A. A. Starobinsky, *Physics Letters B* 91, 99–102, 1980

15 Inflationary Universe: A Possible Solution to the Horizon and Flatness Problems, A. H. Guth, *Physical Review D* 23, 347–56, 1981

16 Review of Particle Physics, K. A. Olive, et al. (Particle Data Group), *Chinese Physics C* 38(9): 090001, 726, 732–36, and 757, 2014

17 The Formation of Deuterons by Proton Combination, H. A. Bethe & C. L. Critchfield, *Physical Review* 54, 248–54, 1938

18 Über Elementumwandlungen im Innern der Sterne I, C. F. v. Weizsäcker, *Physikalische Zeitschrift* 38, 176, 1937

19 Über Elementumwandlungen im Innern der Sterne II, C. F. v. Weizsäcker, *Physikalische Zeitschrift* 39, 633–46, 1938

20 Expanding Universe and the Origin of the Elements, G. Gamow, *Physical Review* 70, 572–73, 1946; erratum: 71, 273, 1947

21 The Origin of the Chemical Elements, R. A. Alpher, H. Bethe, & G. Gamow, *Physical Review* 73, 803–4, April 1, 1948

22 The Origin of the Elements and the Separation of Galaxies, G. Gamow, *Physical Review* 74, 505–6, 1948

23 Theory of the Origin and Relative Abundance Distribution of the Elements, R. A. Alpher & R. C. Herman, *Reviews of Modern Physics* 22, 153–212, April 1950; see p. 194

24 The Evolution of the Universe, G. Gamow, *Nature*, 162, 680–82, 1948

25 Evolution of the Universe, R. A. Alpher & R. Herman, *Nature* 162, 774–75, 1948

26 Remarks on the Evolution of the Expanding Universe, R. A. Alpher & R. C. Herman, *Physical Review* 75, 1089–95, 1949

27 Proton-Neutron Concentration Ratio in the Expanding Universe at the Stages Preceding the Formation of the Elements, C. Hayashi, *Progress of Theoretical Physics* 5, 224–35, March–April, 1950

28 Abundances of the Elements: Meteoritic and Solar, E. Anders & N. Grevesse, *Geochemica et Cosmochimica Acta* 53, 197, 1989

29 Ibid., Hayashi, 1950

30 Physical Conditions in the Initial Stages of the Expanding Universe, R. A. Alpher, J. W. Follin, Jr., & R. C. Herman, *Physical Review*, 92, 1347–61, 1953

31 Primordial Nucleosythesis, A. Coc & E. Vangioni, *International Journal of Modern Physics E* 26, No. 8, 20 pp. 2017

32 Ibid., Gamow, 1948

33 A Measurement of Excess Antenna Temperature at 4080 Mc/s, A. A. Penzias & R. W. Wilson, *Astrophysical Journal* 142, 419–21, 1965

34 A Preliminary Measurement of the Cosmic Microwave Spectrum by the *Cosmic Background Explorer (COBE)*, J. C. Mather, et al., *Astrophysical Journal* 354, L37–40, May 10, 1990

35 Ibid., Alpher, Follin, & Herman, 1953

36 Can One Measure the Cosmic Neutrino Background? A. Faessler, et al., *arXiv:1602.03347v1 [nucl-th]* February 10, 2016

37 Ibid., Zel'dovich, 1972

38 Cosmic Star-Formation History, P. Madau & M. Dickinson, *Annual Review of Astronomy and Astrophysics* 52, 415–86, see p. 464, Fig. 11, 2014

39 Metal Enrichment and Ionization Balance in the Lyman α Forest at $z = 3$, A. Songaila & L. L. Cowie, *Astronomical Journal* 112, 335–51, 1996

40 Nuclear Reactions in Stars without Hydrogen, E. E. Salpeter, *Astrophysical Journal* 115, 326–28, 1952

41 Synthesis of the Elements in Stars, E. M. Burbidge, G. R. Burbidge, W. A. Fowler, & F. Hoyle, *Reviews of Modern Physics* 29, 547–650, plus 4 plates, 1957

42 Multi-messenger Observations of a Binary Neutron Star Merger, jointly published by workers with 951 distinct affiliations, *Astrophysical Journal Letters* 848, L12, 59 pp., October 20, 2017

43 Viewpoint: The Relentless Hunt for Dark Matter, D. Hooper, *Physics* 10, 119, October 30, 2017

Part II

The Bounded Energies of Nature's Messengers

3

Cosmic-Ray Particles, Photons, and Leptons

3.1 Enrico Fermi and the Acceleration of Cosmic Rays

In 1949, Enrico Fermi began an inquiry into the nature and origins of cosmic rays. Like Kirchhoff, a century earlier, Fermi was that rare combination of physicist equally gifted in experimental and theoretical work. In the mid-1920s his theoretical studies had laid many of the foundations of quantum statistics.[1]

Early in 1934, Fermi published a theory describing how nuclear reactions leading to the emission of energetic electrons e^-, generally referred to as *beta particles*, could proceed.[2] Beta decay may most simply be illustrated by the spontaneous emission of free neutrons, particles normally found to be stable only in the nuclei of atoms. Removed from the nucleus, a neutron denoted by the letter N typically survives only 15 minutes before decaying into a proton, P, an electron, e^-, and an antineutrino, $\bar{\nu}$:

$$N \rightarrow P + e^- + \bar{\nu}.$$

Neither the neutrons nor the antineutrinos carry an electric charge, and the positive charge of the proton exactly cancels the negative charge of the electron. The name *beta decay* for this process dates back to earlier times, before physicists had realized that the negatively charged *beta particles* they had detected in their nuclear investigations were simply high-energy electrons.

That same year, 1934, Fermi and his colleagues at the Physical Laboratory of the University of Rome were also experimenting with the transmutation of a wide variety of chemical elements. Bombarded with low-energy neutrons many of these atoms could readily be converted into neighboring elements in the *periodic table*.[3]

Deliberately cautious about identifying the new radioactive isotopes they had created, Fermi's group assumed that the total number of nuclear protons, and also the total number of neutrons, had to be conserved in any reaction. Jointly referred to as *nucleons*, this conservation of protons and neutrons implied the conservation of nucleons as well. The total number of nuclear protons involved in a reaction and emanating from it, respectively denoted by the subscripts on the right and the left side of the arrow, needed to equal each other. Similarly, the sum of the superscripts on each side of the arrow denoting the total number of nucleons had to be conserved.

$$^1N_0 + {}^{27}Al_{13} \rightarrow {}^{24}Na_{11} + {}^4He_2$$

and

$$^1N_0 + {}^{24}Mg_{12} \rightarrow {}^{24}Na_{11} + {}^1H_1.$$

Note that the subscripts, the number of protons in an element, redundantly also identified the chemical element involved, determined solely by its nuclear charge: Subscript 0 identified the bombarding neutrons; subscripts 13 and 12, respectively, corresponded to the elements aluminum Al and magnesium Mg. Among the reactants formed, 2 stood for helium He, and 1 for hydrogen H. Because the subscripts on each side of the reaction had to equal each other, the radioactive element they had produced through each of these two reactions had to have subscript 11, standing for sodium Na.

Reassuringly, in both reactions, they found the decay period of this radioactive isotope they had produced to be "about 15 hours." Today, the half life of the sodium isotope ^{24}Na is known to be 14.96 hours.

Impressive!

3.2   A Spate of New Particles

In the 1930s the neutrons with which Fermi and his coworkers were bombarding atoms were still a novelty. Neutrons had only been discovered two years earlier, in 1932, when James Chadwick at Cambridge University had first noted and identified them. The potential existence of neutrinos also was entirely new, having been suggested only four years earlier, in 1930, by the 30-year-old Wolfgang Pauli trying to make headway on a theory of beta decays.

Although no neutrinos had ever been directly observed, physicists of the late 1930s had little doubt that they must exist. Laboratory studies had indicated that energy and momentum conservation in nuclear reactions would otherwise be violated, and without these conservation laws physics would simply fall apart. Just as the conservation of energy had been the one guide Kirchhoff had felt he could trust when nothing else was known about the Sun, back in 1860, it was easier for Pauli in 1930 to believe in the principle that energy is always conserved rather than doubting the existence of a hypothetical new particle which he felt would ultimately be found to exist.

At the seventh Solvay Conference on Physics held in Brussels from October 22 to 29, 1933, Pauli recalled:[4]

> In June 1931, at the occasion of a conference in Pasadena, I proposed the following interpretation.... the expulsion of β-particles is accompanied by a very penetrating radiation of neutral particles which has not yet been observed. The sum of the energies of the β-particle and the neutral particle ... emitted by the nucleus in a single process would be equal to the energy which corresponds to the upper limit of the β-spectrum... [This would] admit not only the conservation of energy but also of momentum, angular momentum... The hypothetical particle had to have spin 1/2.

Although in 1934 neither the neutrino, ν, nor its antiparticle, the antineutrino, were even known to exist, Fermi's insight and daring led him to adopt the existence both of neutrinos and of freely floating neutrons – while at the same time stimulating the process of beta decay in his laboratory by bombarding atoms with neutrons in order to artificially produce new atomic species.

Even the concept of antiparticles was new at the time. Their existence had been presaged in 1930 and clarified in 1931 by the 29-year-old Cambridge University theorist Paul A. M. Dirac. Dirac's results had been of interest in themselves. But what had excited many physicists most was a surprising finding in 1932 by Carl D. Anderson, a 27-year-old post-doc at the California Institute of Technology.

Anderson had been conducting cosmic-ray studies with a cloud chamber. A number of early photographs he obtained showed puzzling trails, apparently produced by "a positively charged particle having a mass comparable to that of an electron."[5] Six months later, with 15 such tracks in hand, one due to a particle with an energy as high as $\sim$200 million electron-volts (MeV), Anderson was increasingly confident that his findings were correct.[6] He named these new particles *positrons*. Soon these were interpreted as dramatic evidence for the

Fig. 3.1 Portrait of Enrico Fermi (1901–1954). (With permission from the AIP Emilio Segrè Visual Archives, Segrè Collection Photograph A33.)

existence of antiparticles that Dirac's interpretation of quantum physics had predicted, as "a new kind of particle, unknown to experimental physics, having the same mass and opposite charge to an electron."[7]

With the continuing rise of Fascism in Italy in 1938, Enrico Fermi (Figure 3.1), his wife Laura, and their two children embarked for the United States, where Fermi first joined the physics faculty of Columbia University in New York before moving on to the University of Chicago. In 1942, shortly after Japan's raid on Pearl Harbor and the entry of the United States into World War II, Fermi set up the first controlled self-sustaining nuclear reactor in Chicago. He also joined the Manhattan Project, where he served as one of the leading scientists devising the first atomic bombs.

Fermi's earlier experiments on the transmutation of elements may not have led to the long-cherished alchemists' dream of turning lead into gold; but by 1945, a variant of the process he and his colleagues in Italy had pursued in the mid-1930s had led to the production of the plutonium isotope ^{239}Pu from uranium ^{238}U, and to the construction of atomic bombs.

3.3   The Origins of Cosmic Rays

By the late 1930s, some of the most energetic particles and radiations found in nature had been clearly identified as arriving from above the atmosphere. Although the particles' identities were largely unknown, physicists studying their properties expected that these *cosmic rays* might eventually provide information on highly energetic processes originating well beyond Earth's confines. A number of telling facts had already been established.

The flux of cosmic rays at sea level varied with latitude, suggesting that the declining strength of Earth's magnetic field at low latitudes played a role.[8] Because neutral particles pass through a magnetic field unscathed, this indicated that a significant fraction of the cosmic rays consisted of electrically charged particles. An east–west asymmetry in the flux at low latitudes further suggested that the particles carried a positive charge.[9; 10; 11] By 1941, balloon observations carried out at altitudes ranging above 20 km had further singled out protons as a dominant fraction of these energetic cosmic rays.

In the meantime, a gamma-ray component of the cosmic rays was also receiving attention. Gamma rays are high-energy photons – electromagnetic radiation – which, in contrast to protons, carry no charge and remain unaffected by Earth's magnetic field. Agreement was emerging that, at balloon altitudes or below, these were mainly secondary components produced by primary protons interacting with atmospheric atoms and molecules at even higher altitudes.

Gradually, cosmic-ray researchers were also gaining confidence that new types of charged particles must exist. They would need to be far more massive than electrons, but less massive than protons because they exhibited charge-to-mass ratios significantly higher than those of protons. Two classes of these particles emerged, pions and mesons, respectively designated π and μ.[12; 13]

By 1949, four years after the end of World War II, Fermi had turned his attention to the puzzling question of how and where the highest-energy cosmic rays might originate.[14] At the time, the high-energy extremes these particles might reach had not yet emerged but were known to be far higher than any produced in the most powerful accelerators physicists had by then constructed.

Fermi surmised that ionized nuclei could be accelerated to high energies through collisions with randomly moving interstellar clouds which, as the Swedish astrophysicist Hannes Alfvén had proposed, should be permeated by magnetic fields.[15] Most astrophysicists of those days dismissed Alfvén's work as enigmatic and eccentric. Referees at leading journals often rejected his theories out of hand to prevent publication of his work. Fermi, however, found Alfvén persuasive. He took the potential existence of large-scale magnetic fields seriously and recognized that they could efficiently accelerate incident charged particles through persistent collisions between the particles and high-velocity interstellar magnetized clouds.

Fermi's proposed acceleration of cosmic rays was somewhat analogous to the accelerations a ping-pong ball undergoes when repeatedly bounced between a ping-pong paddle and a table top as the paddle is rapidly lowered toward the table. The ball's velocity increases on each successive bounce off the descending paddle, so that its final speed, when the paddle can come no closer to the table, far exceeds the speed at which the paddle was lowered.

The only problem Fermi foresaw was that this mechanism for producing high cosmic-ray particle energies would work only if the nascent cosmic rays' injection energies into interstellar space exceeded a high initial threshold. This would have to be of order 2×10^8 eV for protons, rising to $\sim 2 \times 10^{10}$ eV for the nuclei of atomic oxygen and $\sim 3 \times 10^{11}$ eV for the nuclei of iron.

Initial energies of this magnitude were required if a cosmic ray's energy gain through successive collisions with encountered interstellar clouds was to exceed its energy losses through myriad minor collisions with individual atoms in the tenuous inter-cloud medium which the particle would need to traverse between collisions with magnetized clouds. Although such high injection energies appeared available for protons, Fermi worried that no mechanism seemed to exist to endow iron nuclei with the much higher injection energies they would require. Yet, prevailing measurements indicated that iron nuclei were significant constituents of the cosmic rays incident on Earth's upper atmosphere.

Fermi's worries may have arisen from a paper published a year earlier by H. L. Bradt and B. Peters of the University of Rochester.[16] Their investigation of the primary cosmic-ray particles incident on the upper atmosphere showed that these were not only protons but frequently also carbon, nitrogen, or oxygen nuclei with energies ranging up to ~ 10 GeV, even though, as the authors noted, "absorption of heavy primaries due to collisions in the residual air above [the photographic plates Bradt and Peters were using] is not negligible." Under these circumstances, how could Fermi account for the extreme injection energies his process for generating the more massive cosmic-ray particles appeared to demand? As we now know, *supernova explosions* can accelerate nuclei to high injection energies. And the distances the nuclei cover between successive collisions with a rapidly moving magnetized shock front can be relatively short in the supernova's immediate surroundings.

Somewhat surprisingly, Fermi appears to have been unaware of the very notion of supernovae, although their potential role in the generation of cosmic rays had already been proposed 15 years earlier. In 1934, Walter Baade and Fritz Zwicky of the Mount Wilson Observatory noted that an explosion observed in the Andromeda Nebula in 1885 had exhibited a luminosity ten thousand times higher than that of this galaxy's periodically exploding ordinary novae.[17]

Within our own galaxy, the Milky Way, supernova explosions have been rare. None have been visible to the unaided eye since Tycho Brahe and Johannes Kepler, respectively, wrote about the two most recent of these explosions in 1572 and 1604. In a companion article immediately following this first paper Baade and Zwicky quantitatively pointed out that cosmic rays might well originate in the extremely violent explosions of these rare stars.[18] Their paper, however, was not widely cited, and Fermi may have been unaware of their proposal.

Fermi's paper correctly pointed out that, because magnetized interstellar clouds were so many orders of magnitude more massive than individual cosmic-ray particles, his picture of continual acceleration of those particles until their energies reached equilibrium with those of the clouds would lead to unimaginably high cosmic-ray energies. It remained unclear, for another 17 years, whether the rise to such fantastic energies could somehow be thwarted once the process Fermi was predicting had commenced. If so, what were the highest energies cosmic rays would be prevented from exceeding?

3.4   The Microwave Background Radiation

A realization that a maximum cosmic-ray energy might actually exist arose in 1965 through the far-reaching discovery of the cosmic microwave background radiation by Penzias and Wilson. Once news of the background radiation was accepted, it took only a few weeks for astrophysicists everywhere to realize that this radiation placed upper limits on the energies that gamma rays and cosmic-ray protons could maintain for any length of time, and thereby also placed limits on the energies Fermi's cosmic-ray acceleration could establish.

By 1965–66, Fermi no longer was alive, but he probably would have been pleased to see how these various processes could conspire to produce a self-regulating universe in which the energies of available messengers were kept within predictable bounds that, as we will see, gradually relaxed as the Universe expanded – one of the most fascinating features the expanding Cosmos exhibits.

3.5   The Highest-Energy Cosmic and Gamma Rays

Penetration of Gamma Rays through Extragalactic Space

Within months of the discovery of the cosmic microwave background radiation, Robert J. Gould and Gerald Schréder, two young astrophysicists at the University of California at La Jolla, realized and predicted that high-energy cosmic gamma-ray photons would not be able to penetrate appreciable distances across the Universe.[a]

Mutually destructive collisions among photons had never been directly observed at the time. But, citing the well-understood mutual annihilation of electrons e^- and positrons e^+ giving rise to a pair of identical-energy photons γ in the electron–positron rest frame,

$$e^- + e^+ \rightarrow 2\gamma \, ,$$

Gould and Schréder invoked the potential existence of the reverse process to predict that, on colliding with a microwave background photon γ_{MW}, a high-energy gamma ray γ_{HE} at 10^{14} eV or higher would form an electron–positron pair,

$$\gamma_{MW} + \gamma_{HE} \rightarrow e^- + e^+.$$

The two photons would thus disappear; the newly formed electron and positron would drift apart, unlikely to recombine. The original high-energy photon would thus be destroyed. The electron and positron most likely would subsequently encounter other particles and further dissipate the original gamma ray's energy.[19]

Some years later, Gould and the Israeli astrophysicist Yoel Rephaeli extended this early work to include even higher energy cosmic-ray photons, noting that the collision of ever-higher-energy gamma rays with the microwave background radiation tended to favor one of the collisionally created secondary particles, either the newly created electron or its companion positron, to further propagate the original gamma-ray's momentum along its initial direction. Subsequent collisions of this charged particle with another background photon might then regenerate a photon that would continue conveying a further fraction of this energy along the same direction.

Gould and Rephaeli concluded that at extremely high gamma energies, the effective loss of energy per unit path length over a long succession of such collisions tended simply to a constant energy loss of 10^{-7} eV per centimeter of distance traveled, regardless of the energy of the initial gamma ray. This meant that the descendants of a gamma ray having, say, an initial energy of 10^{19} eV, would lose half that energy in propagation across a distance of order 50 Mpc. Correspondingly, the descendants of a gamma ray of energy 10^{16} eV would lose half its initial energy across a distance of only 50 kpc. Both these distances are

---

[a] On two other joint papers with Gould, Schréder's name is given as *Gérard P. Schréder*.

relatively small compared to the overall dimensions of the Universe, though at the very highest energies they are comparable to distances separating major clusters of galaxies.[20]

Extragalactic space is crowded with roughly 400 microwave background photons per cubic centimeter. The energy of any highly energetic gamma rays that might be generated through especially violent processes could thus cross distances of at most ~100 Mpc through this dense fog of low-energy photons before succumbing to total degradation.

This prediction of Gould and Rephaeli has shown itself consistent with observations indicating that the most distant galaxies with highly active nuclei from which photons at energies $\geq 10^{13}$ eV do appear to reach us are no more than 100 Mpc away.

We have no indication that distant active galactic nuclei (AGN) fail to generate photons at energies exceeding this 10^{13} eV limit. Within our own rather quiet Milky Way galaxy we detect a number of sources, among them the relatively nearby Crab Nebula supernova remnant, which routinely radiate photons at energies exceeding the 10^{14} eV limit.[21] Within the Galaxy, distances from which these photons arrive are only of order 10 kpc, roughly ten thousand times less distant than ~100 Mpc. This leads to an extremely low probability for any of the high-energy photons emitted by them colliding with a microwave background photon before reaching Earth.

We can only conclude that remote galaxies similar to ours do generate gamma rays at energies above 10^{13} eV but that these simply do not reach us across the vast distances they would have to traverse.[22] Across a span of ~100 Mpc, essentially all higher-energy photons are collisionally destroyed.

A remarkable aspect of the papers by Gould and Schréder, and Gould and Rephaeli, was that no cosmic gamma rays with energies anywhere near the original 10^{13} eV limit they considered had ever been observed at the time. Nor were such highly energetic gamma rays observed until decades later. Not until the launch of the Orbiting Space Observatory OSO-3, by William L. Kraushaar and his MIT colleagues in March 1967, were the first-ever cosmic gamma rays at energies exceeding 5×10^7 eV detected – and this was at an energy level still a factor a hundred thousand times lower than the Gould/Schréder limit.[23]

Their's was a leap of pure faith in the fundamental laws of physics. And it worked!

Penetration of Energetic Protons through Extragalactic Space

Gould and Schréder were not the only astrophysicists who, early on, realized that the microwave background radiation would set up barriers to the passage of high-energy particles and radiation. Almost as quickly, Kenneth

Greisen at Cornell University in the United States and, independently, Georgiy T. Zatsepin and Vadim A. Kuz'min of the Lebedev Physics Institute of the Soviet Academy of Sciences predicted a similar destruction of high-energy protons colliding with these pervasive microwaves – although the destruction of the protons would occur at significantly higher proton energies around and above 10^{20} eV, and lead to the production of neutral pions π^0 and ultimately also gamma rays through a process described in Section 3.6, immediately below.[24; 25]

The distances the highest-energy protons and heavier-element nuclei can traverse before assured destruction is only of the order $\sim$30 Mpc. The $\sim 10^{20}$ eV energy limit at which the particles are destroyed is now referred to as the Greisen–Zatsepin–Kuz'min or GZK cut-off.

Figure 3.2 depicts the relative penetration depths of gamma rays and protons over a range of energies.[26]

The article by Penzias and Wilson had appeared in the July 1, 1965 issue of the *Astrophysical Journal*. The responding article by Gould and Schréder arrived at the *Physical Review Letters* office by December 29, that same year; Greisen's article was received at the same journal just three months later, on April 1, 1966; and the article by Zatsepin and Kuz'min was received by its Soviet publishers on May 26 that year. All this, in less than 11 months! The speed with which the succession of these papers responded attests to the tremendous interest the discovery of the microwave background radiation generated and the almost instant realization that this background might seriously limit our long-term quest to fully understand the Universe!

Penetration of Electrons through Extragalactic Space

For the highest electron energies in the 10^{14} eV range, collisions with a single microwave background photon can reduce the electron's total energy by a large fractional amount. Such collisions, known as Compton scattering, can restrict the electron's total penetration of extragalactic space to distances of the order of a few kiloparsecs before it loses most of its kinetic energy. The apparent photon energy in the electron's rest frame is so high that each photon can appear to have an energy far higher than the electron's rest-mass.[27] [b] At lower electron energies, in the 5×10^{10} eV range, the energy reduction is more gradual and the penetration through extragalactic space can be of order 100 Mpc, though not necessarily along a constant direction if extragalactic magnetic fields randomly deflect the electrons' trajectories over shorter distances.

[b] An electron's *rest frame* is a coordinate system in which the electron is considered to be at rest and defines the motions of all other particles and radiations relative to this reference frame.

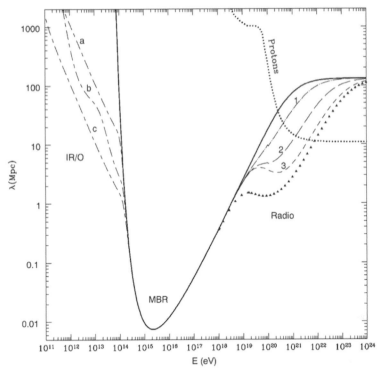

Fig. 3.2 Mean penetration depth λ of high-energy photons and particles traversing the extragalactic medium. Collisions of highly energetic cosmic-ray particles and gamma rays with cosmic microwave photons limit their traversal to distances measured in megaparsecs, Mpc, shown on the scale at left. The dark curve at center represents the limits the cosmic microwave background pervading the extragalactic medium today erects to damp the transmission of gamma rays, at energies above ~ 10^{14} eV, at left, and to protons above ~10^{20} eV, at right. Dot-dashed lines at left indicate how the transmission of energetic gamma rays is further curtailed by different levels of visible light from stars and infrared radiation by dust, both within the galaxy generating the gamma rays and by radiation encountered along an extragalactic trajectory. The dashed lines at right indicate how highly energetic protons suffer a similar fate through collisions with radio waves at different energy densities distributed across regions a proton's trajectory may traverse. The curve labeled with small triangles is a worst case positing extreme radio flux densities. The finely dotted curve at top right represents an extragalactic penetration depth assuming a weighted radio flux at a mean of levels labeled 1, 2, and 3. (From "Constraints on the Very High Energy Emission of the Universe from the Diffuse GeV Gamma-Ray Background," Paolo S. Coppi & Felix A. Aharonian, *Astrophysical Journal* 487, L9–12, 1997, with permission of the authors and the American Astronomical Society.)

3.6   Neutrinos and Gamma Rays

Although the astronomical community quickly accepted the cut-off that Greisen, Kuz'min, and Zatsepin proposed, a complementing and more encompassing argument reaching similar conclusions only surfaced in two papers published three decades later. One was by Eli Waxman, at the time an Israeli post-doc at the Princeton Institute for Advanced Study in New Jersey; the other, appearing in print two years later, sprang from the joint efforts of Waxman and John Bahcall who headed astrophysics at the Institute at the time.[28; 29]

In the first of these articles, Waxman noted that three cosmic-ray particles with energies appearing to be higher than 10^{20} eV had been detected in the previous four years. Each had been registered at a distinct observatory, making their fidelity more reassuring. Given the GZK cut-off, the distances from which these particles had arrived could not be greater than $\sim$100 Mpc. He proposed that the rarely occurring gamma-ray bursts, GRBs, whose origins had been a mystery for two decades but had recently been postulated to originate at intergalactic distances across the Universe, might be likely sources.

One clue was the extremely fast rise time of these bursts, developing on time scales of just milliseconds, meaning that the regions emitting them could not have expanded to dimension spanning more than $\sim$100 kilometers, even at velocities approaching the speed of light. Their high luminosities, if they indeed were exploding at distances of order 100 Mpc, would have to be of order $\sim$10^{51} erg per second, indicating that the outburst had to entail a dense plasma in which photons, electrons, positrons, and protons all were coupled in thermal equilibrium. The high energy of the gamma rays emitted by this heated plasma further confirmed that the expansion of the emitting regions had to be relativistic.

Although the dense GRB plasma would convert a significant fraction of the explosive energy into heat, another fraction was bound to also end up in shocks, large-scale motions within the burst's expanding fireball. Repeated recoil of protons off these shocks and their rapid acceleration through the Fermi process could then raise the energies of at least a fraction of the protons to the observed 10^{20} eV levels. However, recoil sequences at these shock fronts would have to be rapid to drastically raise the protons' energies, and this could only happen if strong magnetic fields were permeating the shocks to shorten the proton trajectories between successive recoils, accelerating them within a matter of minutes to the highest observed energies.

The acceleration of the protons would, however, be hampered through energy losses from collisions with ambient gamma rays to produced pions. In this

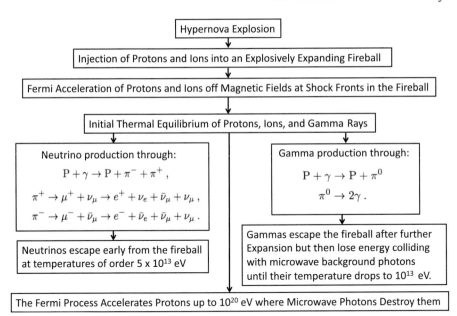

Fig. 3.3 The production of energetic particles and radiation in the fireball generated by a hypernova producing a gamma-ray burst. The maximum energies protons can attain through Fermi acceleration are ultimately bounded by collisions with microwave background photons. But those maxima may not be universally reached if high-density radio waves expected to be generated by energetic electrons spiralling in the magnetic fields at the fireball's shock fronts prematurely intercept the Fermi-accelerated protons.

process half of a proton's energy would go into neutral pions π^0, which would quickly decay into two gamma rays, as shown on the right of Figure 3.3.

The rate at which the protons' energies were dissipated through this process thus could not be faster than the rate at which the protons were being accelerated, and this placed further constraints on the minimum required magnetic field strengths.

Considering all these constraints, Waxman concluded that the three 10^{20} eV protons observed in previous years were likely compatible with their having been generated in GRBs. Most of the energy released in the explosion would have been dissipated by the time the outburst had expanded to a radius of about 10^8 kilometers – comparable to the radius of Earth's orbit about the Sun – a distance the outburst would have spanned within a few minutes after the explosion's initiation.

Two years later, in the article Waxman published jointly with John Bahcall, the two astrophysicists further noted that the GRBs now appeared likely to originate at far greater distances even than supernovae, making them orders of magnitude more powerful than any explosions previously observed. The explosions generating the GRBs thus came to be called *hypernovae*. An important contribution of Waxman and Bahcall's re-examination of these explosions was their prediction that GRBs should be accompanied also by the copious emission of neutrinos with energies on the order of 10^{14} eV.

Once the proton energies in the expanding fireball reached roughly 10^{16} eV, their strong interactions with gamma-ray photons would give rise to two distinct sets of interactions. As already noted, half the proton energy would go into generating gamma rays. The other half would give rise to neutrinos through successive reactions listed in the panel on the left in Figure 3.3.

The bar above a symbol, such as in $\bar{\nu}_\mu$, indicates an antiparticle, in this case a muonic antineutrino; ν_e and $\bar{\nu}_e$ denote electron neutrinos and antineutrinos.

Waxman and Bahcall proposed that, through such reactions, protons with energies even as low as 10^{15} eV, colliding with 10^{12} eV photons, characteristic of the fireball temperature before it cooled through expansion, were likely to produce neutrinos that could well have energies of order $\sim 5 \times 10^{13}$ eV, roughly one-twentieth the energy of the protons setting the reactions in motion.

Because the expanding fireball tends to attain its highest expansion velocity at highest density, the gamma rays generated during initial expansion are readily reabsorbed. In contrast, neutrinos once formed can easily escape, so that the energy loss through neutrinos is likely to significantly exceed the gamma-ray flux of a GRB. Individual neutrinos might then have energies of up to $\sim 10^{15}$ eV in part because their thermal energies within the fireball could be further boosted by the high expansion velocity of the emitting plasma. Gamma rays able to escape, once the plasma was further diluted through subsequent expansion, could well have energies ranging to comparable levels.

Whereas the emitted energetic neutrinos would readily propagate across extragalactic space to eventually reach Earth, gamma rays would be able to cross spans of order 100 Mpc only if their energies fell below the Gould and Schréder threshold of $\sim 10^{13}$ eV, potentially by first shedding energy by the means proposed by Gould and Rephaeli.[30; 31]

Just as Waxman's paper had predicted two years earlier, the Waxman and Bahcall article also proposed that protons with energies as high as 10^{20} eV or more could be generated in these GRBs, though their joint paper made clear that, along the way to reaching these extreme energies, the protons would have to give rise to 10^{13} to 10^{14} eV gamma rays as well as neutrinos.

Daring as they were, these theoretical predictions of the 1960s and 1990s needed observational verification. Much of this had been slow in coming. A number of major new cosmic-ray observatories were required, and it took a further quarter century to come close to completing them.

These new observatories also exhibited the existence of extragalactic neutrinos at higher energies, $\geq 10^{15}$ eV, potentially accelerated by *magnetars*, rapidly rotating, highly magnetized neutron stars – or possibly by *blazars*, supermassive black holes in galactic nuclei ejecting highly energetic outflows in oppositely directed twin jets.[32; 33]

To understand where matters thus stand, we now need to turn to observational efforts that had been pursued in parallel all along.

The Cherenkov Effect

If we wished to intercept and capture any of these most energetic particles individually, before they enter Earth's atmosphere, they would be almost impossibly difficult to register unless we were able to construct detectors with collecting areas spanning hundreds or thousands of square kilometers. We are saved from this prospect by the strong interaction between high-energy particles and Earth's atmosphere through the *Cherenkov effect*. The effect enables the detection of highly energetic, individually arriving particles and photons, as well as the precise registration of their arrival times and indications of the direction from which they arrived.

The effect arose from a puzzling finding first noted in 1934, by two Soviet scientists at the Lebedev Physical Institute in Moscow, Pavel Alekseyevich Cherenkov, at the time about to turn 30, and the 15-years-older head of his laboratory, Sergey Ivanovich Vavilov, an expert on luminescence. Each submitted a short note on this effect published back-to-back in the *Doklady Akademii Nauk, SSSR*, that same year.[34; 35] Three years later, Cherenkov followed up with a brief further note.[36]

As Cherenkov recollected, nearly a quarter of a century later, their experimental investigations, which eventually stretched over a period of close to 25 years, revealed the remarkable properties of radiation arising in a refracting medium traversed by rapidly moving electrically charged particles – motions that could be induced by incident gamma radiation.

In their laboratory studies on liquid solutions of different substances, the two researchers had initially noted that gamma rays of radium, besides causing the liquid solutions to luminesce also provoked their solvents alone to emanate a faint light. This could have been caused by low levels of some impurity and at first appeared unremarkable. Further investigations, however, indicated that this was an entirely new effect whose principle ought to be established, particularly because it also might have practical applications.

Analytic instruments to do this did not exist in the early 1930s. The emitted light was very faint and a researcher could only perceive it by eye after considerable dark adaption. One of the remarkable features of this radiation, not shared by fluorescence, was that its direction of polarization was parallel to, rather than perpendicular to, the direction of the incident gamma radiation's electric vector. By 1936 another new feature, a spatial asymmetry, became apparent. The radiation was emitted solely in the forward direction and at a fixed angle to the direction of the exciting gamma-ray beam.[37] With this information, two theorists, Igor E. Tamm and Il'ya Mikhaylovich Frank, both at the Lebedev Institute as well, soon found a theoretical explanation for the effect, though initially through reasoning that almost led them astray. Writing his recollections in 1958, Tamm first noted:

> [T]he mechanism of this radiation is extremely simple. The phenomenon could have been easily predicted on the basis of classical electrodynamics many decades before its actual discovery. Why then was the discovery so much delayed? I think that we have here an instructive example of a situation not uncommon in science, the progress of which is often hampered by an uncritical application of inherently sound physical principles to phenomena lying outside of the range of validity of these principles.
>
> For many decades all young physicists were taught that light (and electromagnetic waves in general) can be produced only by non-uniform motions of electric charges. When proving this theorem one has – whether explicitly or implicitly – to make use of the fact, that super-light velocities are forbidden by the theory of relativity (according to this theory no material body can ever even attain the velocity of light). Still, for a very long time the theorem was considered to have an unrestricted validity. So much so, that [Igor] Frank and I, even after having worked out a mathematically correct theory of Vavilov-Cherenkov radiation, tried in some at-present-incomprehensible way to reconcile it with the maxim about the indispensability of acceleration of charges. And only on the very next day after our first talk on our theory in the colloquium of our Institute we perceived the simple truth: the limiting velocity for material bodies is the velocity of light in vacuo (denoted by c) whereas a charge moving in a medium with a constant velocity v will radiate under the condition $v > c'$, ... without violating the theory of relativity ... [as long as the velocity of light in the medium c' obeys the condition] ... ($c' < v < c$).[38]

The significance of Cherenkov radiation to astrophysics was that a highly energetic but potentially rare gamma ray, proton, or other energetic particle incident on Earth's atmosphere, or some other refractive medium, could induce a cascade of less energetic particles each of which nevertheless exceeded the speed of light within the refractive medium. The lateral spread of the cascade propagating through the medium could then excite signals in a whole array of individual detectors, sampling the cascade so that not only the initial energy of the primary particle might be deduced, but also its direction of arrival reconstructed – all without ever having to directly capture the rare primary. With an array of detectors suitably deployed over a large area, information about some of the rarest impacting energetic gamma rays or highly energetic nuclei could in this way be reliably recorded.

In investigations of cosmic rays in the late 1930s, the young French physicist Pierre Olivier Auger accidentally stumbled on this effect under conditions that, at first sight, could hardly have differed more from the arduous laboratory experiments Cherenkov and Vavilov had conducted.

At a symposium held half a century later, on March 30–31, 1989, at the University Pierre and Marie Curie in Paris, in honor of his 90th birthday, Auger spoke of some of the first indications he had found of coincident arrival of cosmic-ray photons at different Geiger counters he had arrayed for his investigation of incident cosmic rays. Speaking in English he recalled:[39]

> I wanted to evaluate the extent of these … [coincidences] … in two or three [Geiger] counters as a function of their separation. It was a surprise to observe coincidences when the counters were separated by more than one meter. Suspecting a new phenomenon, I decided to go whole hog, if I may so express myself, and thanks to the technical help of Roland Maze, we placed one of the counters in another building, more than one hundred and fifty meters away on rue Pierre Curie where my laboratory was. And there were still coincidences! It was the discovery of "cosmic ray showers." By pursuing this work at high altitudes, in order to increase the cosmic ray intensity, I showed that the showers covered more than a hectare in extent, and hence the number of particles making up the showers was such that the energy of the primary particle which originated the "giant shower" must have been more than a million billion electron-volts. That was nearly a billion times greater than the energies of particles accelerated by cyclotrons in these days.

The effect that Auger had discovered was nature's own way of making clear the powerful role the Cherenkov effect plays when energetic cosmic rays interact with Earth's atmosphere. Although large detector arrays capable of efficiently analyzing Cherenkov showers were not built until recent decades, they have by now become indispensable. Many of the theoretical predictions of upper bounds on the energies of the different types of messengers can now be tested and appear to be roughly borne out.

3.7   Cosmic-Ray Observatories and the Particles They Detect

A dozen or more cosmic-ray observatories, each constructed to emphasize sensitivity to distinct cosmic-ray components, began to emerge around the turn of the millennium. Their distribution over the globe further enabled detection of energetic radiation from different portions of the sky and at different times of day. One of these, the HESS array of light collectors, is located on a high plateau in Namibia, in South-West Africa. Another, the MILAGRO Observatory, is located in the Jimenez mountains of New Mexico. The Pierre Auger Observatory has been constructed on a vast plane, the Pampa Amarilla, near the city of Malargüe, in the province of Mendoza, in western Argentina. Observatories of comparable importance are located at several other sites around the globe.

The various observatories operate on different principles and thus collect information that can both complement and confirm each other's observations. Many features are common, but as we will see, some provide unique insights. Two primary features common to all are the attempt to trace the directions from which the cosmic-ray primaries generating the observed showers arrive and the need to determine the total energy conveyed by the primary.

Figure 3.4 sketches the ways in which a shower generates different types of nuclear particles and antiparticles, electrons and positrons, gamma rays, and neutrinos, all parts of successive generations of descendants triggered by a single incident primary.

High-Energy Particles

At 10^{20} eV, as seen on the lower right of Figure 3.5, the number densities of the highest-energy nucleons sharply decline. In agreement with theoretical predictions, cosmic-ray particles observed to date exhibit energies rapidly declining around $\sim 3 \times 10^{20}$ eV. Questions do persist on whether the confinement to energies of $\leq 10^{20}$ eV is largely due to the Greisen–Zatsepin–Kuz'min effect, or signifies that the Fermi process cannot accelerate particles to higher energies. But it is not unlikely that both play a role. Ongoing searches to determine the locations from which individual high-energy cosmic-ray particles arrive have, to date, revealed no reliably identifiable galaxies or other sources.

As Figure 3.2 shows, at energies as high as 10^{20} eV cosmic-ray particles can hardly be expected to traverse extragalactic distances of order 100 Mpc, and thus most of the highest-energy primaries actually observed in Figure 3.7 should be

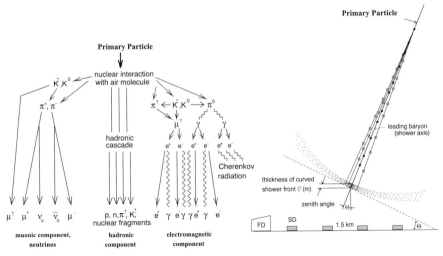

Fig. 3.4 Some of the main processes observed in extensive air showers, and the main hadronic, electromagnetic, muonic, and neutrino components they generate. Typically, a Cherenkov primary particle initiates a cascade, shown on the right, consisting of millions of secondary, tertiary, and subsidiary components generated by the nuclear interaction of the primary with atoms or ions high in the upper atmosphere. As shown on the left, the primary initiates a nuclear reaction generating hadrons, including protons, neutrons, positively as well as negatively charged or neutral pions π, or kaons K, each comprising two quarks. In turn, these particles generate electrons e^-, positrons e^+, positively or negatively charged muons μ^+, μ^-, and their corresponding neutrinos ν_μ and antineutrinos $\bar{\nu}_\mu$, as well as gamma rays and electron neutrinos ν_e and antineutrinos $\bar{\nu}_e$. (Originally published by Bianca Keilhauer in her PhD thesis at the Karlsruhe University, Karlsruhe, Germany, 2003, and the Karlsruher Institut für Technologie, Institut für Kernphysik, it is here reproduced with the author's permission.)

more locally produced, though still believed to originate outside our Galaxy, because magnetic field strengths within the Milky Way and nearby galaxies appear too weak to keep the highest-energy particles from escaping.[40]

By the mid- and late 1990s, considerable insight on naturally occurring highly energetic particles and radiation had already made clear that the Fermi process could only succeed in accelerating a tiny fraction of protons and other ions to the very highest energies observed. More recently assembled information displayed in Figure 3.6 strikingly confirms just how small a fraction of particles ever reaches the very highest energies near the cut-off. Protons with a kinetic energy of order 10^9 eV are roughly 2×10^{14} times more abundant than protons in the 5×10^{14} eV range. The number density as a function of energy, dN/dE, exhibits a terminal slope of about -2.8 at the highest energies where the data closely hug a straight line.

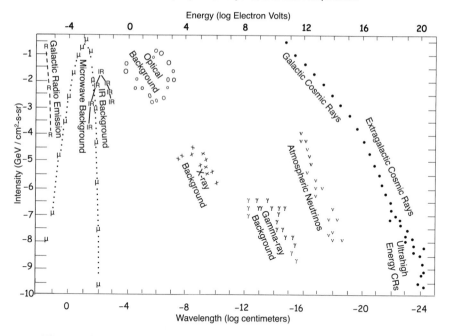

Fig. 3.5 The Intensity of Diffuse Photons, Neutrinos, and Cosmic Ray Particles at Earth. Most of the electromagnetic and cosmic-ray components have an intrinsic Galactic or extragalactic origin. The atmospheric neutrinos, however, are products of Cherenkov radiation initiated by gamma rays or cosmic-ray particles of all kinds incident on the upper atmosphere. (Although it includes additional data, this figure is largely based on a figure originally published by Charles Dermer in "Theory of High-Enegy Messengers," XIV International Conference on Topics in Astroparticle and Underground Physics (TAUP 2015), as part of IOP Publishing's *Journal of Physics Conference Series* 718, 022008, 2016, DOI:10.1088/1742-6596/6717/022008.)

Figure 3.6 shows how far we have progressed by now beyond the early discoveries by Kirchhoff and Bunsen. Today we know not only that the chemical elements composing the Sun's atmosphere and the atmospheres of other stars are similar to those found on Earth, we also find that the Galactic cosmic rays criss-crossing the Milky Way comprise not only these same chemical elements but also some of their rare radioactive isotopes – the very concepts of which would have been foreign to Bunsen and Kirchhoff in 1859–60.

The acronyms at the lower left of Figure 3.6 indicate the impressive number of cosmic-ray observatories at which different portions of the exhibited data were obtained, as well as their mutual overlap. CRN was a cosmic-ray-nuclei detector flown on Space Shuttle Challenger. ATIC, CREAM, and TRACER were instruments flown on long-duration balloon flights reaching altitudes up to 40 km.

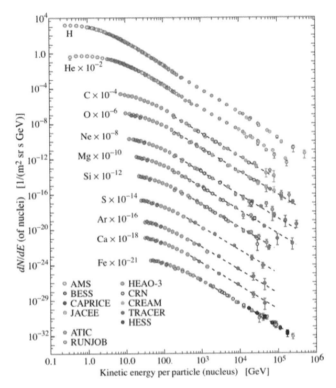

Fig. 3.6 High-energy elemental abundances and energy spectra of Galactic cosmic rays. For clarity in displaying the spectra, the more massive atomic species are successively displaced downward by lowering their arriving number densities per unit energy range, time, collecting area, and solid angle, by such factors as marked by Fe $\times 10^{-21}$ for iron relative to hydrogen. Note the inadvertent error in the original figure, where, for even spacing, the scale entry 10^{-29} at lower left should have read 10^{-28}. (Courtesy of D. Müller & P. Boyle, Particle Data Group, *Chinese Physics C* 40, 100001, 2016, p. 378, Fig. 29.1.)

HESS makes ground-based observations. TRACER was named for the transition radiation it detected. Relativistic charged particles transiting the interface of two media with different dielectric constants emit high-energy photons. Detectors consisting of stacks of such interfaces were used to measure this transition radiation to determine the flux and energies of incident cosmic rays at high atmospheric altitudes.[41; 42]

A glance at Figure 3.6 shows how closely the spectra of all the ions resemble those of protons, persuasively indicating that the mechanism accelerating the protons appears to be responsible also for accelerating the more massive nuclei. But, whereas the turnover point toward a linear descent occurs around 1 GeV for protons, and ~10 GeV for oxygen, this same turnover appears to take place only

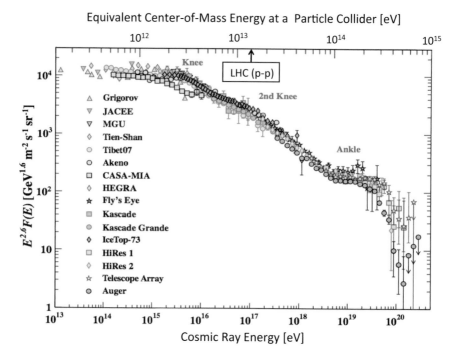

Fig. 3.7 The all-particle spectrum of cosmic rays as a function of energy E per nucleus, obtained from air shower measurements. The vertical range on this plot has been compressed by a factor proportional to $E^{2.6}$, where E is the particle energy indicated on the horizontal scale at bottom. This compression reduces the vertical spread of the plot, making it easier to note amplitude changes in the evolving flux. At low energies, most of the detected particles are thought to have a Galactic origin because the Galaxy's magnetic field prevents the particles from easily escaping. At the highest energies, particles are believed to arrive from extragalactic sources they have escaped. The data shown exhibit the energies of hydrogen, helium, iron, and all other atomic nuclei. As will be seen in Figure 3.10, electron and positron spectra do not extend to the high energies shown on the bottom scale. Open circles, stars, squares, and other symbols present data obtained at different cosmic-ray observatories, generally showing strong mutual agreement. The indicated particle flux depends on theoretical models of the way the cosmic rays interact with the atmosphere; this still is quite uncertain at energies above 10^{18} eV. (I thank Hans Peter Dembinski, of the Max Planck Institute for Nuclear Physics in Heidelberg, for pointing out the significance of identifying the data from specified observatories, both here and at lower left in Figures 3.6 and 3.10, because each observatory uses a different technique in arriving at its conclusions.) The scale at top shows equivalent particle energies in a proton-proton beam collider, such as the Large Hadron Collider denoted LHC(p-p), in Geneva, Switzerland, where two energetic beams of protons collide head on. (Courtesy of J. J. Beatty, et al., Particle Data Group, *Chinese Physics C* 40, 100001, 2016, Fig. 29.8.)

at energies as high as 100 GeV for iron. Fermi's paper of 1949 had predicted that these respective injection energies would lie around 0.2, 20, and 300 GeV. Given how little was generally known about supernovae, the interstellar medium, or its magnetic fields in the late 1940s, his prediction was in remarkably good agreement with the data of Figure 3.6.

One more piece of information emerges from Figure 3.6. If we take the points lying on the curve for iron (Fe) at the very highest energies, and correct for its displacement downward by the indicated factor of 10^{-21} shown to the left of the curve, we find that the abundance of $\sim 10^5$ GeV (10^{14} eV) cosmic-ray iron nuclei approaches the abundance of protons at the same energy – an observation that has more recently been confirmed even more strikingly through the use of the most powerful highest-energy observatory currently available, the Pierre Auger Observatory. The summarizing all-particle spectrum shown in Figure 3.7 shows that higher mass nuclei, mainly iron, closely adhere to the distribution of energies E per nucleus for protons at energies as high as 10^{20} eV.

Figure 3.7 is an all-particle spectrum at energies ranging up to slightly higher than 10^{20} eV. All the particles involved at these high energies are atomic nuclei. Despite the multiplication of the particle flux by a factor of $E^{2.6}$, roughly corresponding to the mean slope at energies below 10^{15} eV, where the spectrum in this presentation indeed appears flat, the particle flux shows a rapid decline at energies above 10^{16} eV per nucleon before precipitously vanishing at and above $\sim 10^{20}$ eV, precisely as Greisen, Zatsepin, and Kuz'min had predicted half a century earlier.

The sharp abundance decline in particle fluxes at these highest energies indicates how rare these cosmic-ray particles are. They strike Earth at rates dropping below one particle per square kilometer per year.

3.8   The Broad Energy Range of Transmitted Photons

Gamma Rays

One of the powerful gamma-ray observatories is the HAWC array located 4100 m above sea level near Mexico's Volcán Sierra Negra. Its data indicate that gamma rays with energies up to at least 100 TeV, i.e., 10^{14} eV, range exist, and can traverse distances of the order of at least several kiloparsecs within the Galaxy.[43; 44] Gamma rays can be distinguished from protons emitted from the same sources at identical energies. Highly energetic arriving cosmic-ray protons yield slightly more blurred images surrounding the sharper images the gamma rays produce – an effect readily explained by the random deflection of the charged energetic particles by their passage through intervening Galactic magnetic fields.[45]

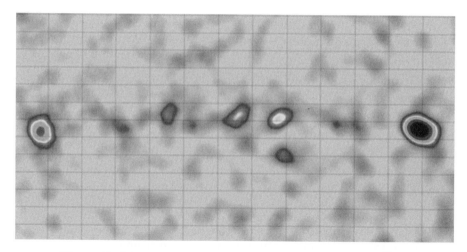

Fig. 3.8 Six Galactic Gamma-Ray Sources with energies above 5×10^{13} eV.
(From "Multiple Galactic Sources with Emission above 56 TeV Detected by HAWC,"
A. U. Abeysekara, et al. (HAWC Collaboration), *Physical Review Letters* 124, 021102,
2020, as reproduced in *Physics–Synopsis*, DOI: 10.1103/PhysRevLett.124.021102.)

Figure 3.8 shows the gamma rays reaching us at these high energies emanating from the Galactic Center. The Crab supernova remnant, much nearer to us, emits such high-energy gammas as well. Undoubtedly, similar sources would generate fluxes at these high energies also in extragalactic sources; but, in apparent agreement with the theoretical work of Gould and Schréder, and Gould and Rephaeli, gamma rays of energy exceeding $\sim 10^{13}$ eV cannot penetrate far through the extragalactic medium.[46; 47]

Most surprisingly, as Figure 3.9 shows, some gamma-ray sources can rapidly flare. Observations of the extragalactic radio source IC 310 obtained with the MAGIC (Major Atmospheric Gamma-ray Imaging Cherenkov) telescopes, at La Palma in the Canary Islands, reveal variability at energies above 3×10^{11} eV, with doubling time scales faster than 4.8 minutes on the night of November 12–13, 2012.[48] The detection of such rapid changes in the emission of a major remote galaxy is not readily explained. The report's authors suggest that the radiation may be induced by the galaxy's central black hole – although, as they point out, "causality constrains the size of the emission region to be smaller than 20% of the gravitational radius of its central black hole."

In apparent confirmation of the existence of rapid extragalactic flares, the VERITAS array has provided similar observations on flaring during an active period, in June 2016, in another extragalactic source catalogued as 1ES 1959+650.

Observatories in space, although gathering information on gamma rays at considerably lower energies, complement such ground-based gamma-ray

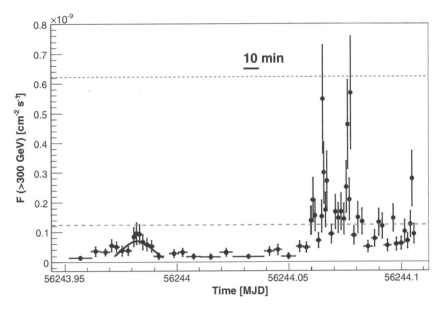

Fig. 3.9 A Gamma-ray Flare in the Galaxy IC310, observed on the night of November 12–13, 2012. The doubling time scale of the flaring was faster than 4.8 minutes. MJD, standing for Modified Julian Date, identifies the precise time of the flare, measured in fractions of a day since noon on January 1 of a selected prehistoric year. (From "Black Hole Lightning Due to Particle Acceleration at Subhorizon Scales," J. Aleksić, et al., *Science* 346(6213), 1080–84, November 28, 2014, see p. 1082; DOI:10.1126/science.1256183.)

observations. Onboard the Fermi Gamma-ray Space Telescope, launched in June 2008, is its Large Area Telescope, which can register photons in the 30 MeV–300 GeV energy range in a field of view covering ~20% of the sky at any given moment. The observatory's other powerful detection system is the Gamma-ray Burst Monitor, whose primary task is to rapidly record gamma rays explosively ejected from distant galaxies.

This monitor detected the first gamma rays from a confirmed binary neutron star merger in 2017, the gravitational waves from which were independently observed by the ground-based Advanced LIGO and Virgo collaboration, whose findings Chapter 4 will present. The gamma-ray data may have been important in providing long-anticipated clues to the formation of some of the most massive elements in the periodic table, whose high observed abundances had previously been puzzling. These heavy elements, however, may also be produced in greater abundance in the central collapse of rapidly rotating massive stars triggering a supernova that explodes much of the star's interior into ambient space.[49]

Extensive air showers can be induced by either gamma rays or energetic cosmic-ray protons and more massive atomic nuclei incident on Earth's atmosphere from space. Such showers produce a complex mix of secondary and tertiary particles, including muons, neutrinos, and nuclear components. Where the sources of the primary gamma rays or cosmic-ray particles are unknown, and individual showers appear to arise in isolation, determining the identity of the primary particle or gamma ray initiating the shower becomes difficult. However, shower primaries with energies higher than $\sim 10^{14}$ eV appear invariably to be triggered not by photons but rather by cosmic-ray particles, whose energies, as seen in Figure 3.7, can range up to the highest energies ever observed, $E \sim 2 \times 10^{20}$ eV, effectively corresponding to the particles' *Compton wavelength* $\lambda = hc/E \sim 6 \times 10^{-25}$ cm. A search carried out with the Auger Observatory indicated that the fraction of photons contributing to the count of cosmic-ray primaries at energies above 10^{19} eV was less than 2%, and potentially far lower.[50]

In contrast, cosmic-ray-particle-induced Cherenkov showers can exhibit energies ranging up to the Greisen–Kuz'min–Zatsepin limit at $\sim 3 \times 10^{22}$ eV where, as we saw earlier in this chapter, collision with the ubiquitous microwave background radiation was predicted to produce just such a cut-off. The most energetic cosmic-ray particles observed thus exhibit energies many million times higher than the gamma-ray cut-off earlier encountered in Figure 1.5.

3.9   Electrons and Positrons

The number counts of protons, iron nuclei, gamma rays, as well as electrons and positrons, all drop rapidly with increasing energy, as Figures 3.6, 3.7, and 3.10 all attest. Although the energies at which the number densities of these different cosmic-ray components rapidly decline are far from identical, the overall patterns resemble each other. Figure 3.10 shows the energy spectra of cosmic-ray electrons and positrons, which, somewhat like the spectra of protons, follow a roughly comparable drop with increasing energy, though their sharp high-energy cut-off occurs at energies E a factor of 10^8 lower – in agreement with the analysis of electron and positron penetration of extragalactic space sketched in Section 3.5.

3.10   The Direct Detection of Neutrinos (1953–60)

Physicists, since the mid-1930s, had been all but convinced of the existence of neutrinos and had carried out quantum mechanical calculations as though there was no doubt about the neutrinos' prevalence. Yet, even two

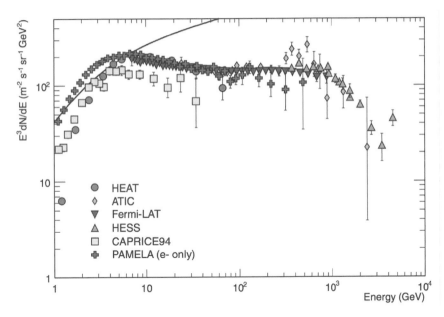

Fig. 3.10 The High-Energy Spectra of Electrons and Positrons. The spectra have been multiplied by a factor E^3 to flatten the display, which otherwise would have shown the number of photons per unit energy range dN/dE rapidly dropping with increasing energy E. Symbols and their designations at lower left indicate the high-energy observatories that generated the respective data. Except for the data obtained by the Pamela station sensitive solely to electrons, all other data points provide the summed spectra of electrons e^- plus positrons e^+. The abundance of positrons is roughly $10 \pm 5\%$ of the total, throughout most of the energy range. The black line at upper left shows the corresponding proton spectrum multiplied by a factor of 0.01. (Courtesy of J. J. Beatty, et al., Particle Data Group, *Chinese Physics C* 40, 100001, 2016, Fig. 29.2.)

decades later, not a single neutrino had ever been directly detected. Somehow, this omission needed to be addressed.

In a succession of increasingly persuasive experiments, conducted between 1953 and 1956, Frederick Reines, Clyde L. Cowan, Jr., and a team of colleagues at Los Alamos finally succeeded in establishing the actual existence of neutrinos. Their experiments involved the reaction

$$\bar{\nu} + P \rightarrow N + e^+$$

in which an antineutrino $\bar{\nu}$ emitted by the Hanford reactor converted a proton P into a neutron N and a positron e^+.[51; 52]

Physicists now could breathe a sigh of relief.

Another two decades later, astronomers had similar reasons for satisfaction.

3.11   Antineutrinos and Neutrinos from a Supernova Explosion (1987)

On February 23, 1987 a burst of neutrinos and antineutrinos was observed at the Kamiokande II neutrino and antineutrino water Cherenkov detector, deep underground in the Kamioka zinc mine in Japan. It consisted of a salvo of 11 neutrinos arriving within an interval of 13 seconds. A precisely coinciding burst of antineutrinos lasting 6 seconds was observed in the United States at the Irvine–Michigan–Brookhaven, IMB detector deep in a salt mine on the shores of Lake Erie, where eight additional antineutrino events were recorded.[53; 54] The detected neutrinos were only a minuscule fraction of all the neutrinos passing through the Kamiokande II or IMB detectors. Most neutrinos, which by then had traversed an entire radius of the exploding star without appreciable absorption, certainly were able to pass through these incomparably smaller detectors without leaving more than a slight, but nevertheless detectable, trace.

Such a short burst of neutrinos and antineutrinos, arriving after a 160,000-year trek from the Large Magellanic Cloud, LMC, showed that the arriving particles had to be moving at, or at least close to the speed of light, implying that their rest-masses could be no higher than $\sim 10^{-32}$ gram, a hundred thousand times lower than the rest-mass of an electron.

The energies of these particles were well over 10^{10} times higher than those expected from primordial cosmic background neutrinos, discussed in Chapter 2. These observations confirmed that astronomical sources indeed emit neutrinos, and that at least some can be detected, albeit with considerable investments of effort. About three hours after the arrival of the neutrino pulse the explosion's visible light was fortuitously recorded photographically as a 6th magnitude star. The delay appeared to reflect the time required for the initial shock of the explosion generated in the star's core to propagate out to the star's surface and for this surface to expand to a radius of $\geq 2 \times 10^{13}$ cm – somewhat exceeding Earth's orbital radius about the Sun – where its emission could have matched that of a star of the observed visual magnitude. To achieve this, the rate of expansion had to be $\sim 40,000$ km/s, more than one-tenth the speed of light and compatible with the supernova's spectral line Doppler shifts observed roughly a day later.[55]

Figure 3.11 identifies the characteristics of the neutrinos observed at Kamiokande. They were detected through their collisions with the ubiquitous water molecules in the detector.

In this reaction the recoil electrons torn out of the molecules by the momentum of a colliding neutrino propagate along the direction of the arriving neutrino.

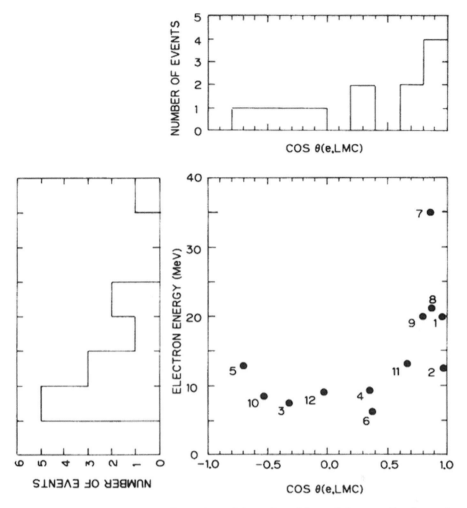

Fig. 3.11 Scatter plot of energies and the cosine of the angle between the observed electron trajectory in the Kamiokande II detector and the direction of the Large Magellanic Cloud. The number to the left of each entry gives its position in the time sequence of observed events. The two projections, above and to the left, provide a complementing display of event statistics. (Reprinted from "Observation of a Neutrino Burst from the Supernova SN 1987," by A. K. Hirata, et al., *Physical Review Letters* 58, 1490–93, 1987, with permission from the American Physical Society.)

The forward-directed *Cherenkov* radiation from the first two arriving neutrinos, respectively labeled 1 and 2 in Figure 3.11, were detected as blue light emitted by these electrons traveling through the water at nearly the speed of light. They indicated the directions from which the neutrinos were arriving, showing that the registered neutrinos came roughly from the direction of the Large Magellanic Cloud.

The nine remaining events appeared to be due to antineutrinos. The characteristic signature of antineutrinos is a roughly isotropic emission of positrons emitted from protons, the hydrogen nuclei in the water, that absorb antineutrinos to form neutrons with the emission of positrons. The antineutrino reaction is

$$\bar{\nu} + P \rightarrow N + e^+.$$

Because most of the recoil momentum is absorbed by the proton, whereas most of the antineutrino energy is carried off by the almost-isotropically emitted positrons e^+, the positrons emit Cherenkov radiation in all directions. This provides a reliable indication that antineutrinos are observed.

Supernovae are not all alike. This one soon showed itself to be a supernova of type SN II, the spectacular death of an isolated, massive star that has used up all its nuclear fuel and must now collapse for lack of sufficient internal pressure to resist gravitational contraction. As the star collapses, its central temperature steeply rises. Mutual collisions among electrons generate vast numbers of neutrinos and antineutrinos which, in complex ways not yet persuasively understood, explosively sweep stellar matter outward even as the stellar core collapses to form a neutron star. The total energy released in such an explosion is of order 10^{53} erg; but only about one percent of this goes into driving matter outward in a flow that will persist for thousands of years. The remaining energy is released mainly in an outpour of neutrinos and antineutrinos lasting no more than about 10–15 seconds.

During those few seconds the rate at which the supernova radiates energy in the form of neutrinos and antineutrinos is more than 10^{18} times the rate at which the Sun radiates visible light, and far exceeds the energy the Sun will radiate in the course of its entire lifetime spanning billions of years.

The Magellanic Clouds are well-studied regions. After the exceedingly luminous initial post-explosion phase had subsided, a blue supergiant star that had previously occupied the site of the explosion was missing – providing strong evidence about the star's pre-explosion characteristics. The earlier observations indicate that its initial mass must have been about $\sim 20 M_\odot$.[56]

3.12   Solar Neutrinos and Their Oscillations (1994)

In 1939, when Bethe and Critchfield resolved the long-standing question of how our Sun was able to steadily shine for billions of years without running out of energy, they also estimated that a small fraction of the energy radiated by the Sun must be transmitted by neutrinos.[57; 58] This was not a trivial amount. If our eyes were able to see these neutrinos, the Sun would appear roughly ten thousand times brighter than the visible light of the full Moon. Neutrinos escape from the center of the Sun with negligible absorption and travel on to reach Earth. The feeble interaction of neutrinos with matter made it difficult to envisage how the solar neutrino flux would ever be detected. Early calculations, by John Bahcall and colleagues in the United States, however, indicated that neutrinos emitted by nuclei of the unstable boron isotope ^{8}B, a minor byproduct of nuclear reactions of hydrogen and helium in the Sun, might produce an observable flux of energetic neutrinos.[59]

Raymond Davis, Jr. at the Brookhaven National Laboratories in the United States, had already developed a technique for detecting these neutrinos, by converting a chlorine isotope ^{37}Cl into an argon isotope ^{37}Ar, both comprising 37 nucleons.[60] For the conversion of solar neutrinos from the decay of boron ^{8}B, the anticipated detection proceeded through the absorption of a neutrino and the emission of an electron,

$$^{37}\text{Cl} + \nu \rightarrow {}^{37}\text{Ar} + e^-. \tag{3.1}$$

The new theoretical predictions by Bahcall and his co-authors encouraged Davis to begin a more comprehensive search for this solar neutrino flux.

Over the years Davis and his colleagues doggedly pursued their search. Their experimental apparatus consisted of 390,000 liters of carbon tetrachloride, at a depth of 4850 ft (1478 m) underground in the Homestake Gold mines in Lead, South Dakota, shielded from cosmic rays by an overhead layer of nearly half a ton of rock per square centimeter. By the early 1990s, their apparatus had reliably been yielding an annual bounty of about 25 argon ^{37}Ar isotope atoms indicating a detection of 25 solar neutrinos a year. This constituted roughly 1/3 the flux predicted by the solar models developed by Bahcall and coworkers.[61]

Why only 1/3?

A few more years had to pass before an answer was at hand:

While Davis had been toiling with chemical techniques, Masatoshi Koshiba, professor of high-energy physics at Tokyo University, had constructed a quite different type of neutrino detector, the Kamiokande detector described above. The Kamiokande observations enabled the first direct observations of the same Solar neutrinos Davis had been collecting, emitted by the boron isotope ^{8}B at energies ranging up to 14 MeV.[62] Observations from 1985 to 1993, however, still indicated a shortfall compared to theoretical predictions. The shortfall was not as severe because, as later realized, not only electron neutrinos but also other neutrino *flavors* could produce the forward scattering observed.[63]

Neutrinos, by then, were known to come in three *flavors*, ν_e, ν_μ, and ν_τ, respectively counterparts to the electron (e), muon (μ), and tau (τ) particles, with which they constitute the family of known *leptons*. All along, these three neutrino flavors had been assumed, like photons, to have zero rest-mass. But if they did possess small rest-masses and all those rest-masses were almost identical, the three flavors of neutrinos would oscillate, i.e., successively cycling from flavor to flavor.

The electron neutrino flux emitted at the Sun might then appear depleted on arrival at Earth, because the ν_e neutrinos would have converted into ν_μ, and/or ν_τ along the way. This MSW effect, named after the Soviet physicists S. P. Mikheyev and A. Smirnov, who had suggested this possibility in 1986, and L. Wolfenstein in the United States, who had made a similar suggestion earlier, in 1979, has by now been fully confirmed quantitatively by many experiments.

While early observations were conducted in the Kamioka mine with ever-larger and more powerful detectors, the most persuasive observations were obtained at the Sudbury Neutrino Observatory set up in a 2-kilometer-deep nickel mine near Sudbury, Ontario, Canada. The heart of the detector consisted of 1000 tonnes of ultra-pure heavy water D_2O in an oval tank surrounded by a tank of ultra-pure H_2O.

D_2O can distinguish different neutrino flavors, and generally reacts differently from H_2O to the arriving neutrinos so that the mix of neutrino flavors arriving from the Sun could, for the first time, be unambiguously documented.[64]

The finding that neutrinos have finite rather than zero rest-mass was arguably the most significant contribution to fundamental particle physics that astronomy provided in the second half of the twentieth century.

3.13   Primordial Neutrinos

In Chapter 2, we encountered a predicted cosmic neutrino background not yet detected but expected to be eventually revealed with equipment progressively refined.

Just as primordial neutrinos originated through equilibrium with other particles and radiations at earliest high-temperature epochs, we saw in the present chapter that comparably high temperatures also generate neutrinos through nuclear syntheses in stars and in supernova explosions. The anticipated spectra of these cosmogenic sources, as well as those of a variety of weaker predicted neutrino sources are shown in Figure 3.12.

IceCube, a South Pole cosmic-ray observatory, comprises a total of 5160 optical sensors respectively mounted at vertical separations of 17 meters on 86 individual strings, each embedded in its own vertical drill hole permeating a cubic-kilometer-sized volume of the crystal-clear Antarctic ice shelf at depths

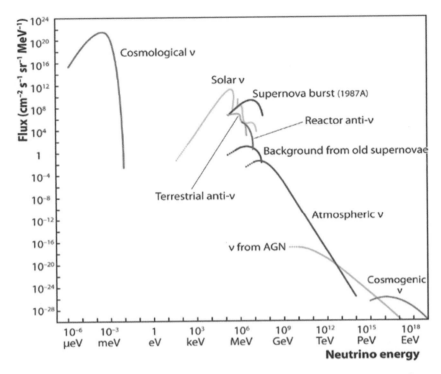

Fig. 3.12 Spectra of Anticipated Neutrinos and Antineutrinos from a Variety of Natural Sources and from Reactors. To date, neutrinos potentially existing at energies above 5×10^{15} eV have not been observed. Note the huge difference between the cosmic neutrino background energies and all other neutrinos. (Image credit: IceCube collaboration/NSF/University of Wisconsin, via https://icecube.wisc.edu/masterclass/neutrinos.)

ranging from 1450 to 2450 meters. Completed in 2011, IceCube has been particularly successful in detecting extremely energetic neutrinos.

A critical feature of the observatory is that the array can identify and register neutrinos that arrive from below the local horizon. This directional capability permits the distinction of primary neutrinos arriving from below the horizon from the vast number of secondary neutrinos in an air shower arriving from above the horizon.

In the early years of its operations, the two highest-energy neutrinos arriving from below the horizon in 2426 days of observations respectively registered energies of $7.7 \pm 2.0 \times 10^{14}$ and $2.6 \pm 0.3 \times 10^{15}$ eV. The latter of these two may have had a total original energy around 4.45×10^{15} eV, only somewhat more than half of which was deposited within the array. Although the effort described included a search for neutrinos with energies ranging from 10^{16} to 3×10^{19} eV, none were found.[65; 66]

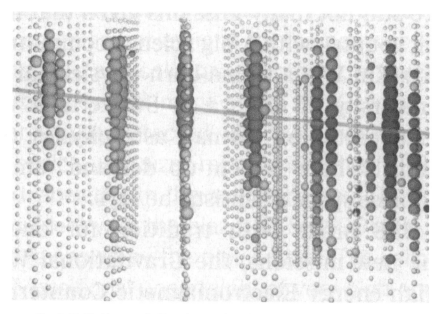

Fig. 3.13 The Energy of a Neutrino Passing through the IceCube Array on September 22, 2017. The energy deposited through secondary radiation is indicated by the shading of the affected optical sensors strung vertically in regularly spaced drill holes in the Antarctic ice. (Image credit: IceCube collaboration/NSF/University of Wisconsin, via https://icecube.wisc.edu/masterclass/neutrinos.)

In Section 3.6, we saw that the interaction of protons with thermal gamma rays in a gamma-ray-burst fireball ultimately leads to the formation of a pair of electron neutrinos and antineutrinos, two pairs of muon neutrinos and antineutrinos, and no tau neutrinos or antineutrinos for a ratio of 1:2:0.

For the so-called muon neutrino events, data for the years from 2009 to 2015 included in this study had often been initiated outside the array, so that a correspondingly smaller fraction of the shower's energy would have been deposited in the array.

Because only neutrinos could successfully traverse the long slant path through Earth, and because the shower-initiating particles were most likely to be neutrino-induced muons able to traverse relatively longer distances through the ambient Antarctic ice before initiating a shower, the majority of these events were likely to have been due to cosmic muon-neutrinos ν_μ or muon-antineutrinos $\bar{\nu}_\mu$ setting off signals in the IceCube array.

Through various tests, the data rule out an atmospheric origin for these energetic neutrinos. An astrophysical origin is strongly favored, in part because neutrinos, unlike protons or gamma rays, avoid collisions with extragalactic radiation or atomic constituents, whereas both high-energy protons and gamma rays interact with microwave background photons preventing their traversing extragalactic distances $\geq$ 100 Mpc.

The simplest way to determine the fraction of the total number of neutrinos these sub-horizon neutrinos represent takes neutrino oscillations into account. Given the long distances the arriving primary neutrinos will have traveled before arriving at Earth, one may safely assume that neutrino oscillations will have redistributed the ratio of electron:muon:tau neutrinos to 1:1:1 – having evolved away from the ratios originally generated in a hypernova fireball, most likely 1:2:0, as proposed in the theory of Waxman and Bahcall. The subterranean neutrinos counted at IceCube may thus be taken to represent one-third of the total number of incident neutrinos arriving at Earth from these directions.

All of these observations, on high-energy gamma rays, neutrinos, protons, and more massive nucleons, when taken together, provide a reasonably coherent picture of the highest-energy messengers that the Universe is able to generate. In particular, observations obtained to date appear to confirm the Waxman–Bahcall theory which predicted that the high-energy cut-off for neutrino energies should be considerably lower than that of protons.

I have presented the theoretical work of Fermi, Gould, Greisen, Kuz'min, Zatsepin, Waxman, and Bahcall, as well as a description of the new cosmic-ray observatories required to test their predictions, in such detail here because, for many years, it appeared that this might be as far as observational tests could ultimately take us. But two or three decades later, the study of the half lives of some of the radioactive material ejected in stellar explosions became further means for reconstructing the astrophysical processes involved, and it is time to turn to these now.

Usually these messengers did not reach us directly but could be identified by their spectra and half lives. Occasionally, however, for nearby supernovae, the actual radioactive material could be swept up by the Solar System and deposited in Earth's oceans.

3.14   Chemical Messengers

In Chapter 1 we noted that the messengers to which Kirchhoff and Bunsen had access were those of visible light whose spectra provided chemical information. Spectroscopic observations, however, may be ill-suited for chemical analyses of low-temperature solids such as meteorites or asteroids. Directly gathered chemical elements may then serve as more suitable messengers to be detected and interpreted in their own right.

The distinction between chemical information gathered by these two means available to astronomers now needs to be emphasized: The first, as originally illustrated by Kirchhoff and Bunsen's approach, is based on the detection and spectral resolution of electromagnetic waves, yielding characteristic spectroscopic displays uniquely identifying radiations emitted by atoms, ions, molecules, or free electrons spiralling in magnetic fields. Spectroscopy is an enormously powerful tool enabling astronomers to gather chemical information often about sources at extreme distances across the Universe.

As a powerful alternative source of chemical information, the direct capture of individual atoms, molecules, or dust grains actually reaching Earth can serve, particularly if their physical or chemical structure can be identified and analyzed. Some of these particles may have been ejected in nearby stellar explosions propelling them through interstellar space and into our Solar System. At other times, material ejected by comets within the Solar System or else mined by space probes landing on the surfaces of asteroids can be directly examined in place by chemical or mass-spectroscopic means.

Asteroids may have remained largely unaltered since the formation of the Solar System, nearly 5 billion years ago. The chemical and mineralogical history imprinted on and preserved by the asteroids and meteorites since primordial times may thus provide useful clues to the origin of the Sun and the planets' early histories.

In 2017 a tumbling, cigar-shaped asteroid or possibly a mildly gas-emitting cometary nucleus unexpectedly arrived from interstellar space, traversed the Solar System at speeds of order 50 kilometers per second, and then resumed its journey through interstellar space. Such bodies ejected from other planetary systems may not be rare. Considerable chemical insight might be gained through dedicated studies of these transients as our Solar System wanders through neighboring regions of interstellar space, at speeds of ~30 km per second, on average transiting a kiloparsec every 40 million years. Given that the Sun lies just 8 kiloparsecs from the Galaxy's center, this is a sizeable span.[67]

3.15   Atoms and Their Isotopes

The chemistry and mineralogy of meteorites – chunks of iron or rock orbiting the Sun – can most readily be determined if a meteorite happens to impact Earth and survives its high velocity transit through the atmosphere without disintegrating. Meteorites have long served as tracers of information on the roles played in the formation of the early Solar System by certain atoms, ions, and their long-lived isotopes.

Microscopic dust grains of interplanetary or interstellar origin can by now also be collected for further analysis. They can be captured through impact on specially prepared low-density materials that decelerate the impinging grains so gently that they come to rest without frictional heating that could alter their chemical composition or mineralogical structure. Though the number of interstellar grains captured in this way has been limited, they show unanticipated diversity and provide at least partial information on their origins and evolution.[68]

3.16   Reading the Deep-Ocean Sediments

A mix of chemical messengers has proved particularly informative on recent supernovae explosions in the Sun's vicinity. Mass spectroscopy of locally captured matter or traces of its radioactivity have been particularly helpful.

Spectroscopic studies across the electromagnetic spectrum have generally yielded mutually complementing information. In 1998, a group of astronomers in Germany, the Netherlands, and the United States led by A. F. Iyudin in Garching, Germany, reported the detection of gamma rays at energies of 1.16 MeV identifying the radioactive decay of the titanium isotope ^{44}Ti from a previously unobserved astronomical source. This had been detected by the Imaging Compton Telescope (COMPTEL) aboard the Compton Gamma Ray Observatory (CGRO) space mission.

^{44}Ti has a half life of 63 years and the emitting source therefore could only be less than a thousand years old. On the basis of known supernova characteristics and complementing X-ray data, the explosion appeared to have occurred roughly 680 years ago at a distance of about 200 pc.[69]

The following year, a group led by K. Knie in Germany pointed out that nearby supernovae of the past few million years should have left traces we could still detect today. The iron isotope ^{60}Fe, whose half life is now known to be 2.6 million years (Myr), appeared to these researchers to be particularly germane for this purpose. As they wrote, "The observation of radioactivities on Earth, which cannot be produced by other processes, would be a clear indication of such an event in the past." [70; 71]

A circumstance making such observations promising is that, for the past few million years, the Sun has been moving through a *Local Bubble* of interstellar gas ionized by hot luminous stars.[72] The most massive of these culminate in core-collapse supernova explosions, so called because the star's central core collapses once the star has used up all the nuclear energy that previously fuelled its radiance. Radiation pressure sufficiently strong to counter the gravitational inward pull of the star's own mass then blows away the star's outer layers.[73]

Supernova ejecta are rich in heavy elements, most notably iron, nickel, and cobalt, each present in mixtures of distinct isotopes. Among these, the iron isotope of atomic mass 60 amu, ^{60}Fe, is of particular interest because, other than core-collapse supernovae, no cosmic sources are known to abundantly produce ^{60}Fe.

Much of the iron in supernova ejecta is expected to ultimately deposit itself on interstellar grains, which the Solar System transiting the Local Bubble may later sweep up. The Sun's radiation pressure and the grains' high surface-to-volume ratio combine to slow down the arriving grains so they gradually spiral inward, some of them reaching Earth's orbit. A majority of grains entering and sinking through Earth's atmosphere finally come to rest on the oceans' floors.

Evidence for the capture of these particles comes from thin discrete layers of deep-ocean deposits laced with the ^{60}Fe isotope, indicating the periodic arrival of supernova-enriched elements conveyed by interstellar dust. Because the half life of ^{60}Fe is as short as 2.62 Myr, and the deep-ocean sediments have accumulated over far longer eons, we can expect detectable traces of this iron isotope to exist only in the sediments' top layers.[74] Further investigations confirm this.[75] Two layers containing ^{60}Fe are evident. A well-documented more recent layer was deposited between 1.7 Myr and 3.2 Myr ago; a less certain earlier influx may date back to between 6.5 Myr and 8.7 Myr. The span of each of these periods of deposition is comparable to the isotope's short half life, making it a clear tracer of supernova explosions in the Solar System's vicinity in the past few million years.

The sensitivity of the ocean-floor sampling is impressive. The total number of ^{60}Fe atoms extracted from the two cores mentioned amounted to a mere 538 atoms. Because the ocean-floor samples also contain considerable admixtures of terrestrial iron, the mass fraction of the ^{60}Fe isotope to total iron summed over all isotopes in the deposits is minuscule, but makes clear that even as astonishingly small a faction as one atom of the isotope ^{60}Fe randomly deposited along with $\sim 6 \times 10^{14}$ atoms of other isotopes of iron can be reliably detected.

Samples of lunar top soil recovered by astronauts on the Apollo 12, 15, and 16 missions of the 1970s provide independent evidence of iron ^{60}Fe deposition on the Moon that can similarly be attributed to the arrival of supernova ejecta from distances of order ~ 100 pc.[76]

3.17   The Isotopic Constituents of Cosmic Rays

Ions are atoms from which one or more outer electrons have been removed. Turbulent magnetized clouds in the ionized Local Bubble tossed around by supernova explosions can accelerate individual ions to energies of the order of several hundred MeV. At these high energies many of these cosmic rays are fully ionized, having lost all or almost all of their electrons.

Nearby cosmic-ray particles from the Local Bubble can diffuse through their turbulent environments to ultimately reach the inner Solar System. As the Sun transits the Local Bubble, many of the cosmic-ray nuclei would be expected to stray from the site of the initial supernova explosion and filter into the Solar System to deeply penetrate the Solar Wind's magnetic field, particularly at times of minimum wind strengths in the Sun's 11-year solar-activity cycle. At these epochs, ions with energies as low as 10^8 eV are detected and can act as indicators of interstellar atomic and isotopic abundances.[77]

Only a relatively small energy range of cosmic rays reaches us directly from the vicinity of recognizable Galactic sources. The deflection of these particles by magnetic fields has had to be sufficiently mild for us to have a fair indication of which regions may have spawned them.

Sensitive detection techniques devised in the past few decades now show that these expectations are borne out. The Cosmic Ray Isotope Spectrometer (CRIS) – a mass spectrometer aboard NASA's Advanced Composition Explorer (ACE) mission launched into near-Earth orbit in 1997 – conducted a census of ambient cosmic-ray particles having energies of several hundred MeV. Mass spectrometers, as their names imply, sort the chemical elements not through recognition of their emitted or absorbed radiation, but rather by singling them out by their electric charge and nuclear mass. This displays not only the relative elemental abundances among the impinging cosmic rays, as shown in Figure 3.15, but also provides a more detailed account of the relative abundances of each element's various isotopes, including those shown for iron in Figure 3.14. The isotope ^{60}Fe is rare: Impressively, the onboard spectrometer CRIS highlighted just 15 – or, more conservatively expressed, 13 ± 4.9 – of these cosmic-ray iron constituents out of a total of 355,000 iron cosmic rays, the overwhelming majority of which had masses in the 54–58 amu range.[78]

Such strikingly sensitive observations highlight how informative the study of individual atomic species can be. The information the number of *nuclides* – distinct nuclei of known chemical elements and their isotopes – can provide as carriers of information is finite. Although slight uncertainties persist, roughly 340 naturally occurring nuclides are found on Earth and presumably in the nearby Universe today. Some 254 of these are believed to be stable; another ~32 have half lives longer than 10^8 yr; and a further ~50 have shorter half lives. The ratios of these nuclides found at different depths in Earth's mantle or oceans, on the surfaces of other planets and their moons, on asteroids and comets, or in meteorites, can provide information of the most varied kind; but the instrumentation required for recording these different types of data remains close to identical for all such observations, and thus the cost of developing even the most sensitive among them has remained affordable.

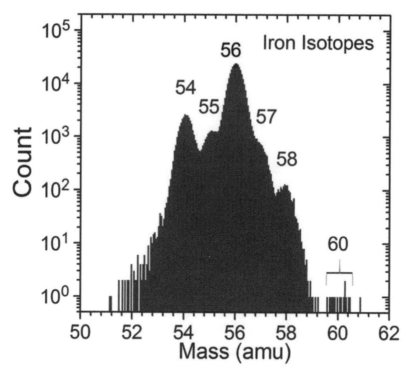

Fig. 3.14 Among isotopes of iron captured by the Cosmic Ray Isotope Spectrometer (CRIS) aboard NASA's Advanced Composition Explorer (ACE) only 15 ^{60}Fe cosmic-ray nuclei were identified among a total of 355,000 iron nuclei ranging over all iron isotopes in the energy range 195–500 MeV. While this number is small, a variety of cross-checks show the data displayed at lower right to be trustworthy. Layered deep-ocean deposits of extraterrestrial origin exhibit comparable isotopic ratios. (From "Observations of the ^{60} Fe Nucleosynthesis-Clock Isotope in Galactic Cosmic Rays," W. R. Binns, et al., *Science* 352(6286), 677–80, May 6, 2016, Fig. 2, DOI:10.1126/science.aad60004.)

Many of these methods yield useful information about recent events in the Solar System's immediate neighborhood. They also provide insight on the chemistry of cosmic-ray particles, but leave us uncertain about where these particles originated. Because of their electric charges, low-energy cosmic rays are readily deflected by Galactic magnetic fields, erasing virtually any trace of where the particles may have originated. The highest-energy cosmic rays are not as readily deflected by magnetic fields, but as a result they readily stream out of the Galaxy and are lost to extragalactic space.

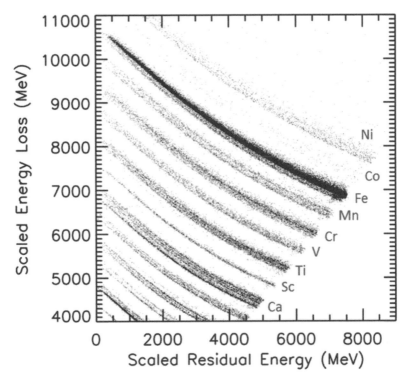

Fig. 3.15 Cosmic-ray elements and their isotopes detected by the Cosmic Ray Isotope Spectrometer (CRIS) aboard NASA's Advanced Composition Explorer (ACE) during 17 years following launch of the spacecraft in 1997. Each of the particles registered in this figure had sustained energy losses on passing through three of the spectrometer's silicon detectors before coming to rest in a fourth detector. The ratio of the summed energy losses in passing through the first three detectors is shown on the ordinate. The energy deposited in the fourth detector is shown on the abscissa. The respective curved traces provide clear identification of the different atomic species, with partial resolution also by isotopic mass. (From "Observations of the ^{60}Fe Nucleosynthesis-Clock Isotope in Galactic Cosmic Rays," W. R. Binns, et al., *Science* 352(6286), 677–80, May 6, 2016, Fig. 1, DOI:10.1126/science.aad60004.)

3.18   Epilogue: The Range of Bounded Messenger Energies

A defining aspect of all these radiations is that the energy of each individually arriving messenger, whether it be a single photon, a neutrino, a gravitational wave, or a cosmic-ray particle, is confined to a bounded energy range. This range differs for each type of messenger, but always remains finite! Figure 3.16 indicates the energy ranges the individual messengers exhibit.

As this chapter has emphasized, the energy bounds of atomic particles, photons constituting electromagnetic waves, and indirectly also cosmic neutrinos in Figure 3.16 are dictated by their interactions with the cosmic microwave background radiation permeating the Universe today, a remnant of a far hotter radiation bath pervading the entire early Cosmos.

At highest energies the bounds are determined by the titanic intergenerational clash pitting energy released by generations of young massive stars, supernovae, and supermassive black holes, against the far larger reservoir of quiescent microwave energy pervading the entire Cosmos today.

As the cosmic expansion eventually lowered the temperature of the Cosmos to ~3200 K this radiation decoupled from atomic matter and continued to

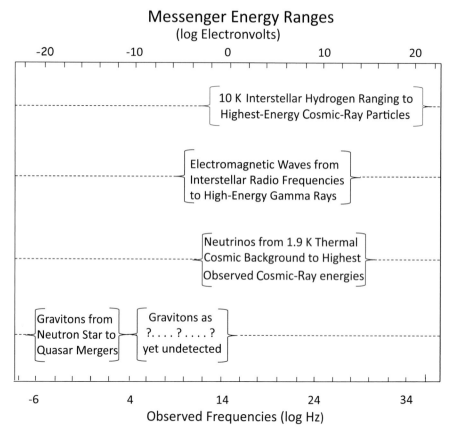

Fig. 3.16 The Mass–Energy Ranges Individual Messengers Exhibit. Although photons, quanta of electromagnetic waves, and gravitons, the analogous quanta of gravitational waves, have no rest-mass, they do exhibit mass–energy E, and corresponding frequencies $\nu \equiv E/h$, where h is Planck's constant. Individual *gravitons*, quanta constituting gravitational waves, are many orders of magnitude less energetic than photons, neutrinos, or cosmic-ray particles, though gravitational wave energy spikes can far outshine instantaneous energy bursts transmitted by any other messengers generated in powerful cosmic explosions.

cool in isolation. By now, 10 billion years later, its temperature has dropped to $T = 2.725$ K.

Meanwhile, the decoupled atomic matter cooled sufficiently quickly on its own to begin forming stars and galaxies. The most massive stars evolved most rapidly, first converting their hydrogen into helium, later converting this helium into some of the more massive atoms the periodic table of elements displays. Both these processes led to the stars' emission of energy.

Then, as even this second source of nuclear energy was depleted, the stars became unable to further resist the strong gravitational forces their own masses were exerting. They rapidly collapsed, their rising central temperatures eventually leading to a sudden outburst of neutrinos and antineutrinos sufficiently powerful to abruptly stop and reverse the collapse in a gigantic supernova explosion blowing the entire star apart.

These powerful stellar explosions generate highly energetic cosmic-ray particles as well as gamma radiation. Both of these forms of energy interact strongly with the ubiquitous microwave radiation bath, amounting to some 400 photons per cubic centimeter throughout the Cosmos today. These confine the energy peaks to which even the most powerful supernovae explosions can accelerate individual cosmic rays – levels of $\leq 10^{20}$ eV, corresponding to roughly 2×10^8 erg per particle.

Highly energetic photons generated in these explosions – gamma rays with individual energies of order 10^{14} eV or higher – similarly clash with microwave background photons to produce electron/positron pairs that ultimately lose energy further through interaction with ambient particles and radiation.

Neutrinos generated in vast cosmic explosions reach energies at least as high as $\sim 10^{15}$ eV. Once generated, these neutrinos can pass through the ubiquitous microwave background radiation virtually unhindered to reach us from even the most distant cosmic realms. They appear to account for some of the highest-energy neutrinos observed today.

This clash between a relatively recent, violent explosion-prone phase of the Universe, and the microwave energy reservoir passed down to us from a much earlier phase of cosmic history, thus restricts the respective energy ranges across which cosmic-ray particles, gamma-ray photons, and neutrinos are observed today.

A clear sign of this clash we may expect and hope to ultimately find is that ejecta from primordially formed massive stars exploding as a first generation of supernovae at epochs possibly as early as redshift $z \sim 10$ would have encountered a far more formidable background radiation energy that would have kept cosmic-ray energies an order of magnitude lower than encountered today.

The theoretical approaches of Gould, Greisen, Zatsepin, and Kuz'min cited earlier in this chapter would all predict this.

Gravitons detected to date have reached maximum energies for rather different reasons, as we will see next in Chapter 4. They are emitted in the collapse of massive bodies, or the merger of mutually orbiting massive bodies, interactions which of necessity occur on time scales given by the radius R of the individual massive bodies, typically *stellar-mass black holes* with radii exceeding 10 kilometers, and by the speed of light, c, which they cannot exceed as they collapse or circle about each other in the last instants before a final merger. These two factors limit the frequency ν of gravitational waves emitted by massive mergers or collapse to a range below $\nu \sim c/R \sim 10^3$ cycles per second, and corresponding graviton energies of order $h\nu \leq 10^{-22}$ erg. Here h is Planck's constant.

Other energetic sources such as fast radio bursts may ultimately be found to emit gravitational waves at higher frequencies. The question marks in Figure 3.16 are meant to indicate this.

With increasing mass, the radius R of merging or collapsing black holes systematically increases, the final frequency before ultimate collapse diminishes, and the energies of generated gravitational waves diminish as well. If supermassive black holes at the centers of merging galaxies are ultimately found to further merge with the emission of colossal bursts of gravitational waves, those bursts will play out on time scales of months, rather than fractions of a second, as Chapter 4 and Figure 4.4 will explain.

Effects of Cosmic Evolution

Today, a limited range of energies confines all known messengers, including cosmic rays, neutrinos, gamma rays, as well as background radiation across the entire electromagnetic and gravitational wave domains. This much can be discerned with a glance at Figures 3.5, 3.7, 3.12, 3.16, 4.4, and Table 3.1. A finite range of energies also appears to confine all other information carriers we know. Our understanding of dark matter or dark energy remains too limited to comment.

The energy bounds on cosmic-ray particles, photons, and neutrinos, predicted by the early work of Gould and Schréder, of Greisen, Zatsepin, and Kuz'min, and of Waxman and Bahcall, not only reflect cosmic conditions today. Their theoretical work places even more stringent bounds on conditions prevailing at earlier cosmic epochs when the microwave radiation temperature and density were far higher than today.

TABLE 3.1

Energy Bounds for Various Messengers

Messenger Name	Upper Bound (eV)	Lower Bound (eV)	Comments
Any and all	$\sim 1.5 \times 10^{28}$	See Figure 3.16	Theoretical bounds
Atomic Matter	$\sim 2 \times 10^{22}$	0	GKZ Theoretical and Observed
Photons	$\leq 10^{14}$	0	Gould & Rephaeli and observed
Gravitons	??	$\sim 10^{-21}$	See Figure 4.4
Neutrinos	5×10^{15}	$kT \sim 1.6 \times 10^{-4}$	Waxman & Bahcall and Observed

The earliest-formed, most distant galaxies we observe today may have formed at redshifts $z \sim 9$, when the Universe was $(z + 1)^3 \sim 1000$ times more compact than it is today, and the number of microwave photons per unit volume was 10^3 times higher than today. This reduced by a factor of a thousand the distances these highest-energy gamma rays and cosmic-ray particles could have traversed without destruction. The energy of the individual microwave background photons at this redshift would also have been $(1 + z) \sim 10$ times higher than they are today and correspondingly able to destroy gamma rays and cosmic-ray particles at energies 10 times lower than those observed today. Limits on the information messengers thus are able to convey are not just a feature of today's Universe. During the epoch of early star, galaxy, and quasar formation, these bounds were even more stringent.

The existence of bounds on the energies messengers can exhibit translate into corresponding bounds on instruments we may need for extracting all the information the Universe is able to convey. This is important in view of the impressive number of novel cosmic phenomena discovered in mid-twentieth century by instrumentalists who looked at the Universe with novel equipment originally designed for military, industrial, or medical use, but now providing fresh perspectives on the Cosmos and cosmic processes.

Today large teams of scientists and engineers continue this work with progressively more powerful, more sensitive instruments pushing into new energy regimes with ever-improving imaging, spectral, and time resolution. These powerful observatories keep informing us about events we never dreamed could exist, and alert us to how much remains to be understood. But, based on the apparently finite number of distinct carriers of information, their finite energy ranges, and the finite ranges of useful resolving powers required to identify distinct messengers, the number of dedicated instruments and observatories we will require to detect the entire cache of information the Universe transmits should also be finite, enabling us to estimate how soon a complete set of observational tools may ultimately be assembled.

An Absolute Upper Energy Bound for All
Conceivable Messengers

We should still clarify whether the Greisen–Zatsepin–Kuzmin bound might ever be exceeded. If so, does an alternative absolute bound exist on the highest energies of messengers that should be considered?

At some level we already answered this question at the start of Chapter 2, where we invoked emission from Planck-mass black holes at the dawn of time. The Planck temperature is $T_P \sim 10^{32}$ K, meaning that emitted radiation, whether it be photons, gravitons or neutrinos, would have energies $kT \sim 10^{16}$ erg or $\sim 10^{28}$ eV, roughly a million times higher than the GZK cut-off.

Hawking considered the likelihood of ever encountering such particles today to be low. Even if a variety of black holes had somehow come into existence at early cosmic epochs and subsequently lost mass by gradually radiating it away, he estimated that their population must be sparse. On average, just 300 such black holes reaching the end of life just now could exist per cubic light year across the entire Universe. Otherwise, he argued, their peculiar gamma-ray spectrum, a recognition sign of their emission, would already have been detected as a dominant gamma-ray background signal exceeding existing gamma-background observations.[79]

We do not yet know whether either dark matter, or dark energy, or some other as yet unknown form of cosmic energy may be conveyed by discrete messengers. None have been detected to date, despite a series of increasingly sensitive searches.[80]

The energies of all the messengers we do currently recognize, however, appear bounded, as indicated in Table 3.1 listing the various messengers' energy bounds. We have not yet enumerated the range of roles that different atoms and their isotopes can play in transmitting information, and should still refer to this, at least briefly. The search for atoms and ions and their isotopes is kept finite by the finite number of atoms and their more stable isotopes. We saw in Section 3.17 that the Universe appears to contains 254 stable isotopes and another 32 that have half lives exceeding 10^8 years, roughly corresponding to a minimum stability required of reliable messengers. These numbers may be large, but they are finite, and modern mass spectrometers can readily identify them.

Once we have insights on all these bounded messengers, have begun to sort out where ambiguities can arise, and have provided the instrumental means to sort them out, we will also have reached a level of confidence that increasingly sophisticated instrumentation will serve little added purpose. We will have arrived at a stage of competence where we can confidently claim that we are able to fruitfully resolve all conceivable ambiguities that could naturally arise in a universe rich in phenomena but limited in the amount of data it can transmit!

This chapter has identified factors that limit the energies of cosmic-ray particles, gamma rays as well as lower-energy photons, and neutrinos. It has also hinted at similar constraints arising for gravitational waves, whose properties we will be examining next. The mass–energies E of individual messengers invariably are bounded as Figure 3.16 shows.

Notes

1 Zur quantelung des idealen einatomigen Gases, E. Fermi, *Zeitschrift für Physik* 36, 902–12, 1926

2 Versuch einer Theorie der β-Strahlen I., E. Fermi, *Zeitschrift für Physik* 88, 161–77, 1934

3 Artificial Radioactivity Produced by Neutron Bombardment, E. Fermi, et al., *Proceedings of the Royal Society of London A* 146, 484–500, 1934

4 *The Solvay Conferences on Physics: Aspects of the Development of Physics since 1911*, Jagdish Mehra, D. Reidel Publishing Company, Dordrecht, Netherlands, 1975, p. 226

5 The Apparent Existence of Easily Deflectable Positives, C. D. Anderson, *Science* 76, 238–39, 1932

6 The Positive Electron, C. D. Anderson, *Physical Review* 43, 491–94, 1933

7 Quantised Singularities in the Electromagnetic Field, P. A. M. Dirac, *Proceedings of the Royal Society of London A* 133, 60–72, 1931

8 Variation of Cosmic Rays with Latitude, A. H. Compton, *Physical Review* 41, 111–13, 1932

9 On the Magnetic Deflection of Cosmic Rays, B. Rossi, *Physical Review* 36, 606, 1930

10 Direct Measurements on the Cosmic Rays Near the Geomagnetic Equator, B. Rossi, *Physical Review* 45, 212–14, 1934

11 Evidence that Protons Are the Primary Particles of the Hard Component, T. H. Johnson, *Reviews of Modern Physics* 11, 208–10, 1939

12 Note on the Nature of Cosmic-Ray Particles, S. H. Neddermeyer & C. D. Anderson, *Physical Review* 51, 884–86, 1937

13 The Nature of the Primary Cosmic Radiation and the Origin of the Mesotron, M. Schein, W. P. Jesse, & E. O. Wollan, *Physical Review* 59, 615, 1941

14 On the Origin of the Cosmic Radiation, E. Fermi, *Physical Review* 75, 1169–74, 1949

15 On the Existence of Electromagnetic-Hydrodynamic Waves, H. Alfvén, *Arkiv för Matematik, Astronomi och Fysik* 29B, 2, 1–7, 1942

16 Investigation of the Primary Cosmic Radiation with Nuclear Photographic Emulsions, H. L. Bradt & B. Peters, *Physical Review* 74, 1828–37, 1948

17 On Super-Novae, W. Baade & F. Zwicky, *Proceedings of the National Academy of Sciences* 20, 254–59, 1934

18 Cosmic Rays from Super-Novae, W. Baade & F. Zwicky, *Proceedings of the National Academy of Sciences* 20, 259–63, 1934

19 Opacity of the Universe to High-Energy Photons, R. J. Gould & G. Schréder, *Physical Review Letters* 16, 252–54, 1966

20 The Effective Penetration Distance of Ultrahigh-Energy Electrons and Photons Traversing a Cosmic Blackbody Photon Gas, R. J. Gould & Y. Rephaeli, *Astrophysical Journal* 225, 318–24, October 1, 1978, see Table 1

21 Multiple Galactic Sources with Emission above 56 TeV Detected by HAWC, A. U. Abeysekara, et al., *Physical Review Letters* 124, 021102, 2020

22 Constraints on the Very High Energy Emission of the Universe from the Diffuse GeV Gamma-Ray Background, P. S. Coppi & F. A. Aharonian, *Astrophysical Journal* 487, L9–12, 1997

23 OSO-III High-Energy Gamma-Ray Experiment, W. L. Kraushaar, G. W. Clark, & G. Garmire, *Solar Physics* 6, 228–34, 1969

24 End to the Cosmic Ray Spectrum?, K. Greisen, *Physical Review Letters* 16, 748–50, 1966

25 Upper Limit of the Spectrum of Cosmic Rays, G. T. Zatsepin & V. A. Kuz'min, *Journal of Experimental and Theoretical Physics Letters* 4, 78–80, 1966. Translated from *ZhETF Pis'ma* 4, No. 3, 114–17, August 1, 1966

26 Ibid., Coppi & Aharonian, 1997

27 Bremsstrahlung, Synchrotron Radiation, and Compton Scattering of High-Energy Electrons Traversing Dilute Gases, G. R. Blumenthal & R. J. Gould, *Reviews of Modern Physics* 42, 237–70, April 1970, see p. 264

28 Cosmological Gamma-Ray Bursts and the Highest Energy Cosmic Rays, E. Waxman, *Physical Review Letters*, 75, 386–89, 1995

29 High Energy Neutrinos from Cosmological Gamma-Ray Burst Fireballs, E. Waxman & J. Bahcall, *Physical Review Letters* 78 (12), 2292, 1997

30 Ibid., Gould & Schréder, 1966

31 Ibid., Gould & Rephaeli, 1978

32 Constraining High-Energy Cosmic Neutrino Sources: Implications and Proposals, K. Murase & E. Waxman, *Physical Review D* 94, 103006, 2016

33 Constraints on Cosmic Ray and PeV Neutrino Production in Blazars, B. Theodore Zhang & Zhuo Li, *Journal of Cosmology and Astroparticle Physics* 3, 24, March 13, 2017

34 P. A. Cherenkov, *Doklady Akademii Nauk, SSSR* 2, 451, 1934

35 S. I. Vavilov, *Doklady Akademii Nauk, SSSR* 2, 457, 1934

36 Visible Radiation Produced by Electrons Moving in a Medium with Velocities Exceeding that of Light, P. A. Cherenkov, *Physical Review* 52, 378–79, 1937

37 Pavel A. Č erenkov, Nobel Lecture of 1958, *Nobel Lectures, Physics 1942–1962*, Elsevier Publishing Company, Amsterdam, 1964

38 Igor Tamm, Nobel Lecture of 1958, *Nobel Lectures, Physics 1942–1962*, Elsevier Publishing Company, Amsterdam, 1964

39 Pierre Auger – A Life in the Service of Science, Lars Persson, *Acta Oncologica* 35(7), 785–87, DOI: 10.3109/02841869609104027

40 The Highest-Energy Cosmic Rays, T. O'Halloran, P. Sokolsky, & Shigeru Yoshida, *Physics Today* 51(3), 31–37, 1998

41 Review of Particle Physics (Particle Data Group), Figure due to D. Müller & P. Boyle in K. A. Olive, et al., *Chinese Physics C* 38, 9, 2014, 090001, p. 378

42 Energy Spectra of Primary and Secondary Cosmic-Ray Nuclei Measured with TRACER, A. Obermeier, et al., *Astrophysical Journal* 742:14 (11 pp.), November 20, 2011

43 Multiple Galactic Sources with Emission above 56 TeV Detected by HAWC, A. U. Abeysekara, et al. (HAWC Collaboration), *Physical Review Letters* 124, 021102, January 17, 2020

44 A Catalog of High-Energy Gamma-Ray Sources, M. Rini, *Physics – Synopsis*, January 15, 2020

45 Ground- and Space-Based Gamma-Ray Astronomy, S. Funk, *Annual Review of Nuclear and Particle Science* 65, 247–49, 2015

46 Ibid., Gould & Schréder, 1966

47 Ibid., Gould & Rephaeli, 1978

48 Black Hole Lightning Due to Particle Acceleration at Subhorizon Scales, J. Aleksić, et al., *Science* 346, 1080–84, 2014

49 Collapsars as a major source of r-process elements, D. M. Siegel, et al., *Nature* 569, 241–44, 2019

50 Upper Limit on the Cosmic-Ray Photon Flux above 10^{19} eV Using the Surface Detector of the Pierre Auger Observatory, J. Abraham and the Pierre Auger Collaboration, *Astroparticle Physics* 29, 243–56, 2008

51 Detection of the Free Neutrino, F. Reines & C. L. Cowan, Jr., *Physical Review* 92, 830–31, 1953

52 Detection of the Free Antineutrino, F. Reines, et al., *Physical Review* 117, 159–74, 1960

53 Observation of a Neutrino Burst from the Supernova SN 1987A, K. Hirata, et al., *Physical Review Letters* 58, 1490–93, 1987

54 Observation of a Neutrino Burst Coincident with Supernova 1987 A in the Large Magellanic Cloud, R. M. Bionta, et al., *Physical Review Letters* 58, 1494–96, 1987

55 *The Supernova Story*, L. A. Marschall, Plenum Press, New York, 1988, pp. 242–43, and 253

56 Supernova 1987A, W. D. Arnett, J. N. Bahcall, R. P. Kirschner, & S. E. Woosley, *Annual Reviews of Astronomy & Astrophysics* 27, 629–700, 1989

57 Energy Production in Stars, H. A. Bethe, *Physical Review* 55, 434, 1939

58 What Do We (Not) Know Theoretically about Solar Neutrino Fluxes, J. N. Bahcall & M. H. Pinneault, *Physical Review Letters* 92, 121301, 2004

59 Solar Neutrino Flux, J. N. Bahcall, W. A., Fowler, I. Iben, Jr., & R. L. Sears, *Astrophysical Journal* 137, 344–46, 1963

60 Attempt to Detect the Antineutrinos from a Nuclear Reactor by the $Cl^{37}(\bar{\nu}, e^-)A^{37}$ Reaction, R. Davis, Jr., *Physical Review* 97, 766–69, 1955

61 A Review of the Homestake Solar Neutrino Experiment, R. Davis, *Progress in Particle and Nuclear Physics* 32, 13–32, 1994

62 Real-Time, Directional Measurements of ^{8}B Solar Neutrinos in the Kamiokande II Detector, K. S. Hirata, et al., *Physical Review D* 44, 2241–60, 1991

63 The Solar Neutrino Problem, W. C. Haxton, *Annual Reviews of Astronomy & Astrophysics* 33, 459–503, 1995

64 The Sudbury Neutrino Observatory, N. Jelley, A. B. McDonald, & R. G. Hamish Robertson, *Annual Review of Nuclear and Particle Science* 59, 431–65, 2009

65 Constraints on Ultra-High-Energy Cosmic Ray Sources from a Search for Neutrinos above 10^{16} eV with IceCube, M. G. Aartsen, et al. (IceCube Collaboration), *arXiv:1607.05886v2 [astro-ph.HE]* November 11, 2016

66 Observation and Characterization of a Cosmic Muon Neutrino Flux from the Northern Hemisphere Using Six Years of IceCube Data, M. G. Aartsen, et al. (IceCube Collaboration), *arXiv:1607.08006v1 [astro-ph.HE]* July 27, 2016

67 A Brief Visit from a Red and Extremely Elongated Interstellar Asteroid, K. J. Meech, et al., *Nature* 552, 378–81, 2017

68 Evidence for Interstellar Origin of Seven Dust Grains Collected by the Stardust Spacecraft, A. J. Westphal, et al., *Science* 345, 786–91, 2014

69 Emission from ^{44}Ti Associated with a Previously Unknown Galactic Supernova, A. F. Iyudin, et al., *Nature* 396, 142–44, 1998

70 Indication for Supernova Produced ^{60}Fe Activity on Earth, K. Knie, et al., *Physical Review Letters* 83, 18–21, 1999

71 ^{60}Fe Anomaly in a Deep-Sea Manganese Crust and Implications for a Nearby Supernova Source, K. Knie, et al., *Physical Review Letters* 93, 171103, 2004

72 The Structure of the Local Interstellar Medium II. Observations of DI , CII, NI, OI, AlII, and SiII Toward Stars within 100 Parsecs, S. Redfield & J. L. Linsky, *Astrophysical Journal* 602, 776802, February 20, 2004

73 The Location of Recent Supernovae near the Sun from Modelling ^{60}Fe Transport, D. Breitschwerdt, et al., *Nature* 532, 73–76, 2016

74 AMS Measurements of Cosmogenic and Supernova-Ejected Radionuclides in Deep-Sea Sediment Cores, J. Feige, et al., *European Physical Journal (EPJ) Web of Conferences* 63, 03003, 2013

75 Recent Near-Earth Supernovae Probed by Global Deposition of Interstellar Radioactive ^{60}Fe, A. Wallner, et al., *Nature* 532, 69–72, April 7, 2016

76 Interstellar ^{60}Fe on the Surface of the Moon, L. Fimiani, et al., *Physical Review Letters* 116, 151104, 2016

77 Cosmic Rays in the Galaxy, P. Meyer, *Annual Review of Astronomy and Astrophysics* 7, 1–38, 1969

78 Observation of the ^{60}Fe Nucleosynthesis-Clock Isotope in Galactic Cosmic Rays, W. R. Binns, et al., *Science* 352, 677–80, May 6, 2016

79 *A Brief History of Time,* S. W. Hawking, Bantam Books, 1988, pp. 115–20

80 Dark-Matter Hunt Comes up Empty, E. Gibney, *Nature* 551, 153–54, November 9, 2017

4

Gravitational Waves

4.1 Early Uncertainties

In late 1915 and early 1916, Einstein finalized his general theory of relativity, which presented gravity in an entirely new way.[1] Two years later, he took up the question of whether the motion of massive bodies would generate gravitational waves analogous to the electromagnetic waves – visible light, radio waves, or X-rays – which the motions of electrically charged particles produce.[2] This indicated that gravitational waves would indeed be produced by the motion of massive bodies orbiting each other and that the propagation velocity of these waves would be the speed of light, c.[a]

Over the next four decades the mathematical foundations of these conclusions were widely critiqued and became progressively murkier. Uncertainties also prevailed on whether sufficiently compact massive objects could even exist in nature to generate observable gravitational waves. These would have to be objects that later came to be known as *black holes*, bodies so massive that not even light could tear itself loose of their gravitational tug.

J. Robert Oppenheimer and one of his graduate students, Hartland Snyder, predicted in 1939 that stellar-mass black holes should exist.[3] If so, these black holes could strip and accrete the outer layers of a nearby *red giant* star, to become even more massive. Conceivably some variants of such processes might also emit gravitational waves.

---

[a] Readers interested in a deeper depiction of gravitational wave emission than outlined in this chapter may find Stuart L. Shapiro and Saul A. Teukolsky's *Black Holes, White Dwarfs and Neutron Stars: The Physics of Compact Objects,* John Wiley & Sons, 1983, especially helpful, particularly their Chapter 16.

133

That same year, 1939, Einstein published a paper in which he argued that black holes could not exist.[4] His paper appeared almost simultaneously with that of Oppenheimer and Snyder's – and nobody quite knew which of the two papers to trust.

Not until 1959 did the combined efforts of theorists Hermann Bondi and Felix Pirani, at Kings College in London, and Ivor Robinson, at the University College of Wales in Aberystwyth, conclusively show that general relativity indeed predicted the existence of gravitational waves.[5]

4.2   Early Attempts at Detecting Gravitational Waves Directly

The direct detection of gravitational waves in nature turned out to be a formidable task. Ultimately, after all lesser attempts and ventures had bogged down, success hinged on massive investments in a number of enormous Michelson interferometers with orthogonally running arms several kilometers long. To check for mutual consistency, two of these interferometers were erected at sites in the United States thousands of kilometers apart; a third was located in Europe; and a fourth was under construction in Japan.

Some background on this is instructive, because it shows how difficult it is to tackle important cosmological problems if the technologies required for their solution do not happen to be useful for either industry or the military, i.e., for anticipated practical use to a nation and its people.

A couple of years before Bondi, Pirani, and Robinson had published their seminal paper, Joseph Weber of the University of Maryland and John Archibald Wheeler of Princeton were simultaneously visitors at Leiden University in the Netherlands, where they began work on a theoretical paper investigating the nature and potential existence of cylindrical gravitational waves.[6] By the outcome of this study, Weber embarked on a search for the actual existence of these waves, a pursuit that dominated the remainder of his professional life. His first steps were to define how gravitational waves would interact with a massive crystalline detector.[7]

After a decade of arduous instrumental work, Weber judged he had made a first detection. He had installed two identical detectors, one at the University of Maryland near Washington, DC, the other at the Argonne National Laboratory 1000 km away on the outskirts of Chicago. Four additional somewhat differently designed detectors were kept in his laboratory at Maryland. By early 1969, he had noted coincident pulses on the identical pair of detectors, as well as an additional pulse on one or two of the other detectors on several days. By his estimates, the random occurrence of such coincidences would have taken place in some cases just once in tens of thousands of years, and in one instance as rarely as once in tens of millions of years. Weber argued that not all of these coincident pulses could be ascribed to independent detector noise. Rather, they should most probably be considered potential detections of gravitational wave pulses. If so, the energy flux of the transiting gravitational wave pulse would have amounted to the passage of 10^4 erg cm^{-2} s^{-1} over the narrow frequency range to which the detectors were tuned.

A few years later, in 1973, J. Anthony Tyson of Bell Laboratories in Murray Hill, New Jersey, who had developed detectors similar to those of Weber, as well as detectors of his own design, wrote a review article on work in the field.[8] There, he listed the gravitational wave work of 15 laboratories world-wide, many of which, including Weber's, were by then operating detectors some orders of magnitude more sensitive than Weber's earlier detectors.

Tyson's paper analyzed a number of subtle sources of noise that could possibly account for whatever signals had by then been detected, and worried that none of the claimed detections might have been registering gravitational waves. In a comment appended to Tyson's paper, Weber gave his reasons for disagreeing, but there was no certain way for either Tyson or Weber to conclusively settle the disagreement. Nobody knew what physical processes might most likely produce gravitational wave pulses in nature, how frequently such pulses should be expected, or how energetic those pulses would be.

An entirely new line of attack appeared to be needed to settle such questions. It gradually revealed itself in the recognition that nature did permit the existence of black holes.

4.3   Disks around Stellar-Mass Black Holes, and Microquasars (1979)

Starting around 1964, *quasars* – highly luminous sources at the centers of galaxies with masses often in the range of 10^8–$10^9 M_\odot$ – were being discovered in significant numbers. Puzzled by their extreme luminosities and masses, Edwin Salpeter at Cornell suggested that a galaxy's interstellar matter accreting on quasars might account not only for the quasars' high masses, but also their astonishingly high luminosities.[9] That same year, 1964, the Soviet astrophysicist Yakov Borisovich Zel'dovich independently proposed a similar accretion-theoretical approach to account for the quasars.[10]

Actual accretion disks, however, were not discovered until almost another two decades later. The mid-1960s had led to the discovery of a series of compact astronomical X-ray sources. One of these was a source in the constellation Scorpius first reported in 1963.[11] Rapid time variations in the X-ray luminosity of this source, as well as of another X-ray source, Cygnus X-2, showed these two sources to probably be compact.[12] Optical observations of Cygnus X-2 further revealed it to be a binary star complex.[13]

Later, as X-ray emitting stars were more frequently found to be members of binaries, the prospects that some of the X-ray emitters could be sufficiently compact and massive to attract and accrete the loosely bound outer layers of a red-giant binary companion gained increasing credence. By then the discovery of pulsars in 1968 had led to the acceptance also that neutron stars exist – as J. Robert Oppenheimer and another of his graduate students, G. M. Volkoff, had similarly predicted in 1939.[14]

One of the first pieces of observational evidence suggesting the existence of stellar-mass black holes came from a sighting of a star registered as entry 433 in a catalog of stars emitting hydrogen-α radiation by ionized gas.[15] C. B. Stephenson and N. Sanduleak of the Case Western Reserve University in Cleveland, Ohio, the authors of this catalog published in 1977, had compiled a list of massive but otherwise quite ordinary, young Milky Way stars whose surface temperatures were sufficiently high to ionize the hydrogen.

A year later, in 1978, David H. Clark in the United Kingdom and Paul Murdin in Australia were engaged in a very different quest. They were studying a point source emitting bursts of radio wavelengths and apparently associated variable X-ray emission near the center of a supernova remnant, labeled W50. In searching for a visible star associated with the radio source, they came across "an unusual emission line star," whose coordinates were just those of SS 433. This source was also strikingly similar to another variable X-ray source, Circinus X-1, a binary star with a 16.6-day period, and also associated with a supernova remnant, indicating more than a chance positional coincidence.[16] Clark and Murdin's paper piqued the curiosity of Bruce Margon at the University of California, Los Angeles, who, with a group of colleagues, undertook a spectroscopic campaign to study SS 433.

They found it "bizarre."

The star's most striking features were two broad spectral-emission lines, one in the red, the other in the near infrared. To Margon's astonishment, these changed wavelength from night to night. As he and his co-authors reported, "[In] spectra spanning eight of nine consecutive nights in the period 1978 October 23–31, the infrared feature moves almost monotonically from $\lambda7400$ to $\lambda7620$, while the red feature changes from $\lambda6120$ to $\lambda5970$, corresponding to +9000 km s^{-1} and −7400 km s^{-1}, respectively, if interpreted as Doppler motion."[17]

As Figure 4.1 showed, observations over more extended periods indicated that the red line "has changed in velocity by at least 12,000 km s^{-1} in 30 days and in

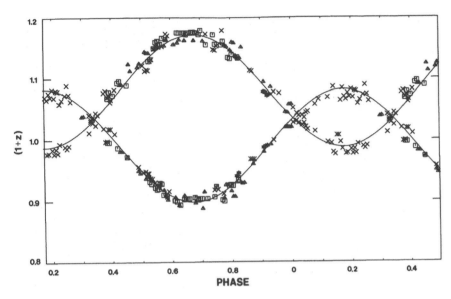

Fig. 4.1 Red- and blueshifts of SS 433 measured by Bruce Margon and coworkers, and a small army of helpful colleagues at various observatories. The data include 155 separate observations over roughly four complete 164-day periods of the precessing jets during the years 1978–79. The data presented here were folded to fit a 164-day period, with somewhat more than one full cycle shown for clarity. (From "The 164 and 13 Day Periods of SS 433: Confirmation of the Kinematic Model," Bruce D. Margon, Steven A. Grandi, & Ronald A. Downes, *Astrophysical Journal* 241, 306–15, Fig. 2, p 309, 1980, with permission of the authors and the American Astronomical Society.)

roughly 60 days has made a complete traverse from λ5850 to λ6120 and return, with evidence of a new cycle to the red beginning immediately thereafter." In the proofs to their paper the authors added, "Further analysis has provided conclusive evidence that the unidentified emission features are Doppler shifted (hydrogen) Balmer lines at a displacement which has varied from 20,000 to 50,000 km s^{-1}."

Later observations revealed SS 433 to be a stellar binary with a total mass approximately 40 $M_\odot$. An evolved star is being tidally stripped of its outer layers by its more compact companion whose mass is ~16$M_\odot$. This star is surrounded by an X-ray emitting disk of gas onto which the accreting material falls. As the star rotates it expels two oppositely directed off-axis jets of matter at a speed ~26% of the speed of light, giving rise to the moving spectral features that Margon and his coworkers had observed. The two beams precess with a 164-day period.[18]

In a book written a few years later, David Clark concluded that SS 433 could be the residual compact remainder of a star that had exploded, somewhat analogous to the Crab Nebula pulsar centered on the Crab Nebula supernova remnant.[19] "Everything so far learned about SS 433 strengthens the conviction that it is a miniature version of the radio galaxies and quasars.... its detailed study should reveal some of the secrets of the radio galaxy and quasar phenomena that we could never hope to unravel through direct observation. The mighty ant stands triumphant among the astronomical elephants."[20]

Another decade would be required to gather further evidence on these mysterious objects. This emanated from a series of papers by Felix L. Mirabel at the Centre d'Études de Saclay in France, Luis F. Rodríguez at Instituto de Astronomia, UNAM, in Mexico, and their coworkers.

In a first paper published in 1992, Mirabel and Rodríguez used the Very Large Array in New Mexico to search for radio emission from the hard X-ray source 1E1740.7–2942, located near the Galactic Center, which had been identified by the X-ray/gamma-ray SIGMA telescope on the Russian GRANAT satellite. Their radio map showed a compact variable core with an emanating, double-sided jet. The two jets were symmetrical about the core, suggesting "that 1E1740.7–2942 is a *microquasar* stellar remnant near the Galactic Centre." This paper appears to be where the name "microquasar" was first coined, with the name's implications correctly identified.[21]

Mirabel and Rodríguez conjectured that the hard X-ray source may be responsible for strong outbursts of 511-keV electron–positron annihilation radiation observed from the Galactic Center region. This suggested that the radio jets they had mapped were emitted by electrons and positrons traveling along the jets more than a parsec before slowing, with the positrons annihilating against electrons in the ambient interstellar gas. The hard X-ray spectrum resembled that of the Galactic X-ray source Cygnus X-1, "one of the best candidates for an accreting black hole of stellar mass."

Two years later, Mirabel and Rodríguez compared the relativistic outflows along the jets in these microquasars to superluminal motions that had been discovered in quasars decades earlier, adding that microquasars may offer the best opportunity to gain a general understanding of relativistic ejections seen elsewhere in the Universe.[22] In this paper they compared the velocities of two radio components that appeared to recede from each other at velocities apparently exceeding the speed of light in the transient gamma-ray source GRS1915+105 near the Galactic Center.

The radio source Centaurus A exhibits such relativistic jets as seen in the X-ray image of Figure 4.2

Fig. 4.2 Giant Black Holes Appear to Power All Known Quasars. Shown here is an image of the source Centaurus A obtained by the Chandra X-ray Observatory. Quasars frequently exhibit powerful relativistic jets ejecting matter from the central portions of their host galaxies into extragalactic space. The processes involved in the black holes stripping matter from infalling stars and hurling this out into space are not yet satisfactorily understood. (Credit: Chandra X-Ray Observatory/NASA/Science Photo Library: NASA/SAO/R. Kraft et al.)

Such *superluminal velocities* had been predicted by Martin Rees at Cambridge University many years earlier, and represented a projection effect rather than a violation of relativity.[23]

Mirabel and Rodríguez further called GRS1915+105 "a scaled-up version of the famous stellar source of radio jets SS433," which until then had been believed to be unique in the Galaxy.

In a crowning paper of 1998, Mirabel and Rodríguez, collaborating with six other authors, noted, "In particular, it has been realized that since the characteristic dynamical times in the flow of matter onto a black hole are proportional to its mass, the events with intervals of minutes in a microquasar could correspond to analogous phenomena with duration of thousands of years in a quasar of 10^9 $M_\odot$."[24]

Three decades later Margon recalled the instrumentation that had revealed the striking outflows from SS 433:[25]

> Optical spectroscopy in the mid 1970s was just starting to make the transition away from the photographic emulsion as a detector, and those lucky enough to have access to the first generation of electronic detectors (... we're discussing image tubes here) had a huge advantage not just of sensitivity ... but of convenience. A photographic spectrum was never taken on whim, as it was followed by an extra two hours after dawn, struggling to stay awake in the smelly dark room while developing the plates after a long cold sleepless night outdoors, and then countless hours more on ancient measuring engines which Hubble, Baade, and Struve would have instantly recognized (even in 1980!), trying to extract the damn information. Conversely, with the then-new image tube scanners in use at Lick Observatory ... a spectrum on whim was trivial – flip a few switches and wait a few minutes, and then claim the trophy of an almost fully reduced spectrum. And it was at least almost on whim that I took my first spectrum of SS 433 even though Clark & Murdin had solved the problem already (pointed out the positional coincidence of the radio, X-ray and optical objects). Would I have done that if it had meant another one hour outside, plus time in the dark room and measuring afterwards? Quite possibly not.
>
> Perhaps my own greatest contribution was examining the early optical spectroscopic results of Clark & Murdin and deciding that there had to be far more to the story, ... how exotic the optical spectrum was. In this regard I was very lucky to have had classical training in stellar astronomy from Berkeley, as well as interest in high energy astrophysics.

Regarding the influence of theory: Once the realization of the existence of microquasars sank in, it dramatically shifted citations to a theoretical paper that had correctly predicted such effects and had been written many years earlier, in 1973, by two young Soviet astrophysicists, the 28-year old Nikolai Ivanovich Shakura and the 30-year old Rashid Alievich Sunyaev. For nearly a quarter of a century their paper had been systematically ignored.[26] Because the Shakura/Sunyaev paper for many years was seldom cited, it was not likely to have significantly influenced Murdin, Clark, or Margon and his colleagues.

Sometimes appreciation of genuine foresight comes slowly! And then, a decisive event breathes a new vitality into the search.

4.4   The Orbital Evolution of Binary Pulsars (1982)

In 1975, Russell A. Hulse, an unusually gifted graduate student, and his equally gifted thesis advisor, Joseph H. Taylor at the University of Massachusetts at Amherst, detected the radio emission from a new pulsar, designated PSR 1913+16 by its position on the sky. The interval between its pulses varied systematically between 58.967 and 59.045 milliseconds over a cycle of precisely 27908 ±7 seconds – just over 7 hours and 45 minutes. Obtaining approximately 200 independent observations at 5-minute intervals, Hulse and Taylor concluded that two neutron stars, one of them a radio pulsar, were orbiting their mutual center of mass with an orbital radius roughly equal to a solar radius, $R_\odot$, and an orbital eccentricity of $\epsilon = 0.615$.

The large variation in period told them that the orbital plane could not be perpendicular to the line of sight. An absence of mutual eclipses also meant that the orbital plane did not precisely contain the line of sight; and the unseen companion was unlikely to be a white dwarf whose relatively large diameter would have tended to produce an eclipse. Hulse and Taylor inferred:

> [T]he unseen companion is a compact object with mass comparable to that of the pulsar. In addition to the obvious potential for determining the masses of the pulsar and its companion, the discovery makes feasible a number of studies involving the physics of compact objects, the astrophysics of close binary systems, and special- and general-relativistic effects. [27]

Seven years later, Taylor and another young coworker, Joel M. Weisberg, had the additional data in hand to explore these possibilities in greater depth. These showed the masses of the two stars to be almost precisely identical and close to $1.4M_\odot$. Using these masses and the observed orbital period and eccentricity, general relativistic calculations predicted that the orbital period should have shortened at a rate of 2.403×10^{-12} parts per period over the roughly 6000 periods that had elapsed since the original observations of Hulse and Taylor. The actually observed decline in orbital period was $2.30 \pm 0.22 \times 10^{-12}$ parts per period, meaning that the observed and predicted periods were identical within observational uncertainties. Taylor and Weisberg concluded, "The excellent agreement provides compelling evidence for the existence of gravitational radiation, as well as a new and profound confirmation of the general theory of relativity."[28]

Though Taylor, Hulse, and Weisberg had not directly observed gravitational waves, their careful timing of the rate at which radio-wave pulses were emanating from one of the two neutron stars orbiting their common center of mass showed that the rate at which the pair was dissipating gravitational energy agreed precisely with the predictions of general relativity.

With the passage of nearly four decades since these seminal observations, several other binary pulsar systems have been discovered, and a variety of further tests of general relativity and potentially competing gravitational theories have been pursued. To date, no deviations from Einstein's general relativity have surfaced.[29]

4.5   Gravitational Waves from Black Hole Mergers (2016–17)

Although gravitational waves now appeared to exist, it took another 28 years for a direct detection of a gravitational wave pulse to be observed by the Advanced Laser Interferometer Gravitational-Wave Observatories, *Advanced LIGO*: On September 14, 2015 at 09:50:45 *universal time*, the observatories' two gravitational wave detectors, one at Hanford, Washington, the other at Livingston, Louisiana, mutually separated by a distance of ∼3000 km, provided a persuasive detection.

The signals were first detected at Livingston, and $6.9^{+0.5}_{-0.4}$ milliseconds later at Hanford – well within the maximal 10 milliseconds that the 3000 km distance between the two observatories would have permitted, but also sufficiently separated to permit a coarse sense of directions in the sky from which the gravitational waves could have arrived at Earth. The wave forms detected at the two observatories, shown in Figure 4.3, are essentially identical within statistical limits of detector noise and pulse arrival times. The combined signal-to-noise ratio assuring the fidelity of the detection was SNR ∼ 24, nearly five times higher than the standard SNR ∼ 5 level generally considered persuasive.

The event was assigned the label GW150914, providing the year, month, and day of detection.[30]

The theoretically predicted gravitational radiation emitted by mutually orbiting masses had been well determined by late in the twentieth century. The predictions of general relativity were particularly clear for gravitational wave emission from two mutually orbiting black holes. Observed in isolation, neither black hole permits energy to escape its surface. We can thus be sure that whatever gravitational energy they do radiate must emanate through gravitational interaction.

Three considerations come into play. The first is that general relativity predicts that a source consisting of masses M_1 and M_2 in circular orbit about their common center of mass radiates gravitational energy E at a rate

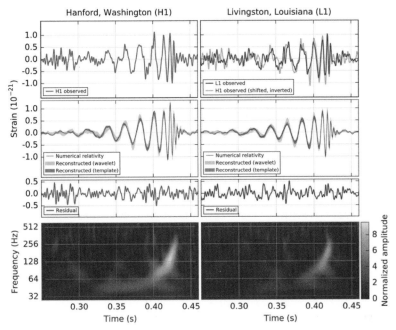

Fig. 4.3 The GW150914 Merger of Two Stellar-Mass Black Holes Reported by the LIGO Scientific Collaboration in the United States and the Virgo Collaboration in Europe. Plotted on the left are data from the Michelson interferometer at Hanford, Washington. At right are data from the interferometer at Livingston, Louisiana. The two top traces plot the strain h – the fractional change, positive or negative, induced in the lengths of the arms of the interferometer by the passage of the gravitational wave. The second set of plots shows the best fit to the data. The difference between these appears in the third set of data. The display at bottom provides the recorded frequencies of the gravitational waves, in units of hertz (Hz) – the number of cycles detected per second – as a function of time. Theoretical predictions of the wave patterns the merger of two stellar-mass black holes would produce enabled interpretation of the data obtained, not only to yield the individual masses of the merging black holes, but also the distance across the Universe at which the merger had occurred. (From "Observation of Gravitational Waves from a Binary Black Hole Merger," B. P. Abbott, et al., the LIGO Scientific Collaboration and Virgo Collaboration, *Physical Review Letters* 116, 061102 (February 12, 2016), Fig. 1, DOI:10.1103/PhysRevLett.116.061102.)

$$L_{GW} = -\frac{dE}{dt} = \frac{32}{5}\frac{G^4}{c^5}\frac{M(M_1M_2)^2}{a^5}.$$ (4.1)

Here, the gravitational wave luminosity, L_{GW}, is the system's orbital energy loss per unit time, $-dE/dt$; the total mass of the two stars is $M \equiv (M_1 + M_2)$; and a is the separation between masses.[31]

Second, the two masses must gradually spiral in toward each other to accommodate this energy loss. As the separation a between the two masses shrinks, they circle each other ever faster, in conformance with Kepler's third law, which states that the square of the orbital frequency, ω^2, is inversely proportional to the cube of the separation between the masses, a^3. The relation between kinetic energy per unit mass, which is always positive, and total energy of orbital motion, which is always negative, also tells us that $(a\omega)^2$ is twice the absolute value of the total energy per unit mass, $-E$, leading to the expressions

$$\omega^2 = \frac{GM}{a^3} = -\frac{2E}{a^2} \text{ , so that } \frac{1}{\omega}\frac{d\omega}{dt} = \frac{3}{2E}\frac{dE}{dt}, \qquad (4.2)$$

where the orbital frequency ω is measured in radians per second. Substituting this result in equation (4.1) we obtain

$$L_{GW} = \left(\frac{G(M_1 M_2)}{3a}\right)\left(\frac{1}{\omega}\frac{d\omega}{dt}\right). \qquad (4.3)$$

This classically derived relation, which applies at low velocities, holds until the very last instants where the orbital speed $a\omega$ begins to approach the speed of light.

We can now interpret the frequency parameters provided by the merger data offered by Figure 4.3. The oscillatory frequency f plotted at the bottom of the figure is $f \sim 150$ cycles per second at the highest amplitude just before disappearance of the wave.

Gravitational radiation is generated by the quadrupole component of the orbital motion, for which the frequency f is twice the orbital frequency, 75 cps. The corresponding angular frequency ω of the orbiting masses thus becomes $\omega \sim \pi f \sim 470$ radians per second.

The curves at the bottom of Figure 4.3 also show the rapid rise time in the oscillatory as well as the orbital frequency. We estimate this frequency's fractional rate of change with time as $(d\omega/dt)/\omega \sim 100$ s^{-1}. Inserting this value in equations (4.2) and (4.3), and assuming the two black hole masses to at least be roughly identical by setting $M_1 = M_2$, we can solve for the separation between masses at that epoch, $a \sim 342$ km, and the total mass to be $M = 1.32 \times 10^{35}$ gram $= 66 M_\odot$, and individual masses $\sim 33 M_\odot$ each.

These values, while clearly approximate, are in rough agreement with the LIGO/ Virgo conclusion that the masses respectively are roughly 36 and 29 $M_\odot$ with approximate uncertainties $\sim 10\%$ apiece.[32]

While the gradual change of period per unit time can be directly read off evolving wave trains, such as those shown in Figure 4.3, we still need to derive the distance of the source in order to determine its intrinsic luminosity. This can be determined through a peculiar characteristic of binary black hole formation, namely, that the two masses ultimately cannot approach their final merger along circular orbits.

An *Innermost Stable Circular Orbit, ISCO*, exists, beyond which the mutual gravitational attraction becomes so strong that the two masses no longer can sustain their orbital motion but dive into each other. This transition, marked by the sudden cessation of gravitational waves witnessed in Figure 4.3, occurs at a separation $a_{ISCO} \sim 6MG/c^2$, three times the *Schwarzschild radius* R_s for a black hole of final mass M. For two equal masses at separation a

$$L_{GW} = \frac{2}{5} \frac{G}{c^5} \left(\frac{M^2}{a^2} \right) (a\omega)^6 . \tag{4.4}$$

If we set $a = a_{ISCO}$, and take the velocity there to be half the speed of light, we obtain $a\omega \sim c/2$, and $M/a = c^2/6G$ as well as $M \sim c^3/12G\omega$, and arrive at a rough estimate of the luminosity at the surface defined by the innermost stable circular orbit, the highest luminosity the radiation can attain.

In terms of the *Planck luminosity* defined purely by the ratio of two fundamental constants, $L_{Planck} \equiv c^5/G$, this leads to[33]

$$\frac{L_{GW}}{L_{Planck}} \sim 2 \times 10^{-4} . \tag{4.5}$$

The strength of a gravitational wave arriving at Earth is determined by the alternation between fractional elongation and fractional contraction of a body with which the wave interacts. This fractional change in dimension is called the *relative strain* or simply the *strain* h. It is also referred to as the *amplitude* of the induced wave pattern. Figure 4.3 shows the oscillating amplitude of the strain observed for GW150914. The strain measured at Earth for GW150914, $h \sim 10^{-21}$, corresponds to the fractional change in the separation between two points, here given by the fraction of the projected length c/ω registered by the LIGO-Virgo interferometers.

The strain diminishes inversely with increasing distance from the gravitational merger. The *luminosity distance* d_L is defined as the distance at which the strain would have exhibited its maximum value roughly given by the derived value $L_{GW}/L_{Planck} \sim 2 \times 10^{-4}$.

For an orbital frequency $\omega \sim 470$, the *luminosity distance* d_L of the source was

$$d_L = \frac{(L_{GW}/L_{Planck})^{1/2} c}{h\omega} \sim 9 \times 10^{26} \, \text{cm} \sim 300 \, \text{Mpc} . \tag{4.6}$$

The square root of the luminosity ratios $(L_{GW}/L_{Planck})^{1/2}$ enters because the strain h involves a change in length whereas the luminosity ratios are integrated over a surface.

A more rigorous calculation than the rough estimate presented here shows that the total gravitational wave energy liberated in this first detection by Advanced LIGO amounted to a mass-loss of order $\Delta M \sim 3M_\odot$. This is equivalent to an energy loss $c^2 \Delta M \sim 5 \times 10^{54}$ erg, in the merger of the two stellar-mass black holes with a combined initial mass of $\sim 71M_\odot$.[34; 35]

Nearly two years after this initial discovery, a merger of another pair of black holes of similar mass and at a comparable distance led to a close-to-identical burst of gravitational radiation.[36]

The total energy emitted in gravitational radiation in each of these events, 10^{54} to 10^{55} erg, was roughly comparable to the total kinetic and electromagnetic energy liberated in one of the most powerful gamma-ray bursts ever detected, GRB 080916C observed on September 16, 2008.[37] The ultimate source of energy in these quite distinct appearing events was gravitational collapse to form a black hole with a final mass very roughly in the range of $50M_\odot$.

Supernova and hypernova explosions couple the gravitational energy released during a central black hole collapse to energy released in electromagnetic radiation and neutrinos. The collapse of two black holes lacks the means to provide this coupling directly, because neither of the mutually collapsing black holes is able to release electromagnetic energy or neutrinos on its own, and neither is likely to carry a significant electrical charge.

Figure 4.4, based solely on theoretical grounds, includes a binary stellar black hole merger at 300 Mpc, and its anticipated range of frequencies which overlap well with the span of 10 to 100 Hz observed by Advanced LIGO. The strain $h \sim 10^{-21}$ observed by Advanced LIGO is also in agreement with the amplitude or strain plotted in the figure.[38]

Gravitational waves appear to cover an energy range shown in Figure 4.4. That the range is finite follows from the finite range of masses in which black holes can form, roughly $3M_\odot \leq M \leq 10^{10}M_\odot$. Below the lower bound of this range, atomic matter resists catastrophic gravitational collapse. And if the Universe contained substantial numbers of bodies more massive than those of the most massive AGNs observed, these would by now have become apparent through gravitational lensing of background galaxies at rates we do not observe.

The ways in which gravitational energy is released by supermassive black holes in the central portions of galaxies are still not fully understood, though current models for these processes can be sketched along lines indicated immediately below. These suggest that the energy that supermassive black holes couple to their immediate surroundings ultimately gives rise to the same Fermi

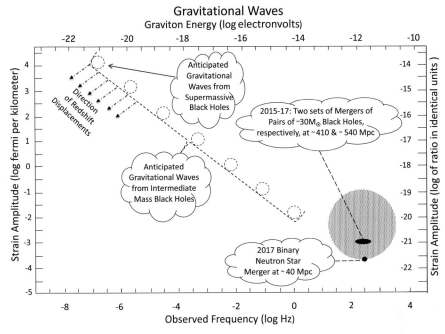

Fig. 4.4 The Gravitational Wave Landscape of Conceivable Observations – Covering the Range of Mergers of Pairs of Neutron Stars and Pairs of Black Holes of Different Masses. Note that the characteristic amplitude plotted on the ordinate is equivalent to the strain h induced at the typical distance R at which we may expect the sources listed to be found; the strain h is inversely proportional to that distance, h $\propto$ 1/R. The shaded region at lower right roughly outlines the sensitivity and gravitational wave frequency ranges in which the Advanced LIGO and Virgo cooperation was capable of detecting gravitational waves around the year 2017. Note the low energy of individual gravitons producing the gravitational wave signals the interferometers detected. A graviton frequency of 300 Hz has an energy of $h\nu \sim 1.24 \times 10^{-12}$ eV or $\sim 2 \times 10^{-24}$ erg. Solar photons in the visible range are about 10^{12} times more energetic. (Based, in part, on the work of G. H. Janssen, et al., *Proceedings of Advancing Astrophysics with the Square Kilometre Array (AASSKA14)*, June 9–13, 2014, Giardini Naxos, Italy; Proceedings of Science, POS(AASKA14)037.)

accelerations that supernovae and hypernovae generate. This is worth noting, because the processes ongoing in the massive central regions of galaxies now appear to be just as effective in producing highly energetic cosmic rays, gamma rays, and neutrinos as hypernovae may be.

The Potential Range of Gravitational Wave Observations

Equation (4.2) tells us that, when the product ωa approaches the speed of light c, the frequency ω should become inversely proportional to the total mass M. When ultimately detected, more massive black hole mergers are expected to roughly lie along a slope of -1 in Figure 4.4, though many of these sources may lie at high redshifts that will systematically lower the observed frequencies. For extremely high masses, the terminal gravitational wave merger frequency f_{GW} should be found to be a factor of $\sim 10^8$ lower than those for stellar black hole mergers, very roughly around $\sim 10^{-6}$ Hz for masses of $\sim 10^{10} M_\odot$, corresponding to the highest currently observed supermassive black hole masses.

Today, the earliest epochs at which supermassive black holes might have formed appear to correspond to redshifts $z \sim 6$, by which time the Universe will have reached an age of roughly a billion years.[39] Telescopes under construction today may soon be able to reveal these. Galaxies containing such large black holes may thus appear at $z \sim 6$, at a co-moving radial distance an order of magnitude more distant than where the $\sim 70 M_\odot$ binary black holes of GW 150914 appeared to have been located.

Equation (4.6) for a frequency 10^9 times lower, and d_L just 10 times more distant, would thus lead to a strain $\sim 10^8$ times higher, h $\sim 10^{-14}$, roughly as indicated in Figure 4.4.

A question of particular interest is the anticipated range of observable gravitational wave phenomena, i.e., the span of the gravitational wave phase space. Figure 4.3 shows that the first direct detection of gravitational waves by the Advanced LIGO collaboration involved oscillations on time scales of order milliseconds and corresponding frequencies around $\nu \sim 10$ to 10^3 Hz, with an energy per graviton of $h\nu \sim 10^{-13}$ to 10^{-11} eV. We should note, however, that the existence of gravitons has not yet been established. All that the direct observations of gravitational waves have told us to date is that these waves definitely exist. Whether they also are quantized, as electromagnetic waves are, in terms of photons, has not been established.

The merger of supermassive black holes could be expected to produce oscillations on time scales of order 100 years or frequencies measured in 10^{-9} Hz. The graviton energies would then be of order 10^{-23} eV – minute in comparison to the energies of photons conveying electromagnetic radiation across the Cosmos. But the total energy radiated in gravitational waves in such giant mergers would be a staggering $M_{rad}c^2$, where M_{rad} could be a sizeable fraction of a supermassive black hole mass, $10^8 M_\odot$, so that $M_{rad}c^2 \geq 10^{61}$ erg – comparable to the explosion of roughly $\sim 10^{10}$ supernovae. Although this high amount of energy would be spread over a time scale of order of a century, it still amounts to energy generation rates of the order of three supernovae each second. Most of this energy may simply escape into extragalactic space because the gravitational waves produced are not readily reabsorbed.

We know that accretion of matter onto a galaxy's central supermassive black hole appears responsible for much of the energy dissipated in the galaxy's nucleus. Accretion may not be directly related to the generation of gravitational waves. The vast jets in these galaxies, hurling large amounts of matter from their nuclei in directions perpendicular to their galaxies' planes of rotation, appear to be due to accretion disks of matter spiralling into the central supermassive black hole. Part of the in-spiralling matter ultimately falls into the supermassive black hole, spinning it up and increasing its mass, while another fraction of the disk's contents escapes this fate and instead is forcibly ejected out of the plane of the disk along each of the two oppositely directed jets.

4.6   The Merger of Neutron Stars (2017)

Soon after the discovery of mergers of stellar-mass black holes, binary mergers of neutron stars were similarly detected. The neutron star mergers are more informative in that they generate additional electromagnetic signals that black hole mergers do not. Because of their far shorter wavelengths, the location on the sky from which the electromagnetic signals arrive can be more accurately determined than the location at which the gravitational waves originate. Once the galaxy in which the neutron star merger occurred is thus identified, its distance is readily determined and compared to the rather cruder distance assessed from gravitational wave emission patterns alone.

The almost simultaneous arrival at Earth of the gravitational and electromagnetic signals announcing the merger of the neutron stars, after hundreds of million years of travel across the Universe, also confirms general relativity's prediction that gravitational waves, just as electromagnetic waves, travel at the speed of light.[40]

The electromagnetic signals from neutron star mergers can be spectroscopically analyzed to show that their merger is accompanied by the stripping of some of the dense outer layers of the neutron stars as their central cores begin to gravitationally merge to form a final black hole. This stripped-off material may be sufficiently dense to permit the rapid formation of neutron-rich chemical elements with nuclear masses well exceeding those of iron. Previously, the abundant formation of these massive chemical elements had not been satisfactorily explained by the only other plausible process that had been known – formation of heavy elements in supernova explosions. The prospects of direct spectroscopic detection of these elements in the ejecta of neutron star mergers raises hopes that the generation of these high-mass elements may soon be fully understood.

This information would not be provided by mergers of stellar black holes because no material escapes either black hole during their final merger. In contrast, the atomic matter detected after a neutron star merger would have been abrasively stripped off the surfaces of the merging neutron stars well before the final moments of their merger, and thereby escaped consumption into the merging black hole.

Cosmic gamma-ray burst detections had been known since 1973, when the military Vela program of the late 1960s, which had first recognized the burst's cosmic origin, became de-classified.[41] For some years, as well, astronomers had concluded that two different types of bursts might need to be distinguished – some lasting several minutes and others that at most lasted a few seconds.

Theoretical models suggested that the short bursts could come from neutron star mergers, but there was no independent evidence for this, until the event of 2017 showed the coincidence of a short gamma-ray burst accompanied by a spike in gravitational radiation clearly accompanying a merger creating a black hole. Some months later, radio observations revealed a structured jet reminiscent of relativistically ejected material frequently observed in quasars, as in Figure 4.2.[42]

4.7   Instrumental Perspectives

Each of the Advanced Laser Interferometer Gravitational-Wave Observatories, *Advanced LIGO*, consisted of a Michelson interferometer with orthogonal 4-kilometer-long arms. In each of the two arms, laser beams were reflected back and forth hundreds of times, to detect the tiny alternating expansions and contractions in the arms as a gravitational wave passed through, and thus to derive the relative delays between the two light beams to an accuracy of $\sim 10^{-17}$ cm $= 10^{-4}$ fermi. The *fermi* is a unit of length amounting to 10^{-13} cm, named in honor of Enrico Fermi. As Rainer Weiss, one of the leaders of the project, put it, this relative delay in the light beams traveling along the two interferometer arms amounted to something like one part in ten thousand of the diameter of the nucleus of a hydrogen atom – the proton.[43]

The roughly \$0.5B that the US National Science Foundation had spent was only a portion of the actual cost internationally. Over the decades physicists from many nations had been tackling the difficult design problems faced in the construction of gravitational wave sensing interferometers. Experience gained, for example in erecting the GEO600 interferometric gravitational wave detector in Hannover, Germany, by the Max Planck Institut für Gravitationsphysik in collaboration with UK partners supported by their Science and Technology Facilities Council, was essential for the GEO600 scientists collaborating with the Laser

Zentrum Hannover to build the lasers needed for Advanced LIGO.[44; 45] Similarly, Italian colleagues from the Virgo collaboration and Japanese physicists at the Kamioka Gravitational Wave Detector, KAGRA, contributed their expertise to the success of Advanced LIGO.

4.8   Supermassive Black Holes (2004)

As early as the mid-1980s, many astronomers suspected that quasars and other compact mass concentrations at the centers of galaxies represent massive black holes. That galactic nuclei might harbor such black holes had been widely discussed once the high redshift of quasars had become apparent from optical spectra obtained by Maarten Schmidt and others.[46] Initially it was not clear whether this redshift indicated that quasars were very distant, or very massive but nearer to us. Either criterion could produce a high redshift.

Many years passed before the faint emission spectrum of the galaxies harboring the quasars could be easily detected; the glare of the compact quasars was so much brighter than the diffuse light from the galaxies in which they were embedded that the host galaxies were difficult to detect. With considerable effort, ground-based observations had, by 1985, convincingly shown that the spectra of the hosts, vaguely apparent only as faint *fuzz* surrounding a quasar, exhibited the same redshift as their central quasar. This meant that the quasar redshifts predominantly arose through cosmic expansion rather than from a massive gravitational redshift.[47] This conclusion was further confirmed with more sensitive Hubble Space Telescope observations.[48]

The origin of supermassive black holes appears to be quite different from that of stellar-mass black holes. For a stellar mass to collapse to a sufficiently compact state to ever become a black hole, electrons surrounding atomic nuclei in the star's center must first be forced into the nuclei where they combine with nuclear protons to form neutron-enriched nuclei. If this forcing persists, eventually the entire stellar interior becomes a giant neutron-rich domain – a neutron star. Neutrons are Fermi particles. They can be compressed into a small volume only by increasing their momentum. This prevents further collapse unless the mass of the star exceeds a value believed to lie somewhere in excess of $2M_\odot$, where stellar matter no longer can resist collapse.

For supermassive black holes no such barriers exist, although a need arises for transporting heat away efficiently in the early stages of collapse. To date the mechanisms for achieving this heat transport have remained speculative.[49; 50]

Nevertheless, by the turn of the millennium, rotation curves of many nearby galaxies left little doubt that a number of them housed highly compact galactic nuclei with masses that could range to $\geq 10^9 M_\odot$. The massive body in the Milky

Way's nucleus has a far more modest mass, but still amounts to $\sim 4 \times 10^6 M_\odot$. The question was whether this and other compact nuclear sources might be black holes.

Two separate thrusts were undertaken. One was led by Andrea Ghez and her coworkers at the 10-meter Keck telescope site in Hawaii. The other, conducted by Reinhard Genzel, Frank Eisenhauer, and their group at the Max Planck Institute for Extraterrestrial Physics in Garching, Bavaria, used the European Southern Observatory's Very Large Telescope, VLT, in Chile.

Over a period of years, these two groups traced the motions of a set of stars in the immediate vicinity of the Milky Way radio source Sgr A*, which appeared to coincide with the gravitational center of this central distribution of stars. By obtaining both high-resolution spectra of these stars, and high angular determinations of their proper motions – yielding both their radial and transverse velocity components – both research groups were able to accurately determine the distance to the Galactic Center, $\sim 7.94 \pm 0.42$ kpc. These efforts yielded the dimensions of the stars' elliptical orbits around Sgr A*; the velocity components of the stars at each point in their orbits; and thus also the mass of Sgr A* $\sim 3.59 \pm 0.59 \times 10^6 M_\odot$, and a maximum dimension of this massive body as determined from the nearest approach to it by one of the stars of ~ 45 AU, where the star's velocity reached 12,000 km per second.[51; 52]

The careful analytic approach followed by both groups, respectively using the Keck and the ESO telescopes, gradually reached accuracies that left little doubt that the massive body at the Galactic Center, indeed, could only be a supermassive black hole – not as massive as some, but far more massive than any stellar black holes.

4.9   Gravitational Radiation from Supermassive Black Hole Mergers

As already emphasized, the time scales for mergers of supermassive black holes should be of the order of many months to years. The Universe should, by now, be awash with long-wavelength gravitational wave radiation emitted from such occasional mergers over the ages. As the waves traverse the Universe they should be pulling any masses they encounter back and forth. Millisecond pulsars emitting exceptionally stable radio pulse trains should exhibit velocity components approaching or receding from Earth in response to these tugs.

One can imagine Earth sitting in the midst of a random selection of such stably emitting pulsars within the Galaxy. Use of highly accurate clocks could then determine the radial line-of-sight components of the pulsars' undulating motions in response to the gravitational waves criss-crossing interstellar space.

Those waves would, of course, also move Earth as they encounter our planet. With the highly accurate clocks available by now, these distinct motions could be sorted out. A long-term program of radio observations using eight of the world's largest radio telescopes to keep track of such pulsar motions has now been launched by a consortium established to operate this *International Pulsar Timing Array*. Upper limits on the strains produced by these extremely slow gravitational waves already are in a reasonably relevant range of h $\leq 10^{-15}$.[53] An order of magnitude improvement down to a detection limit of $\sim 10^{-16}$ in the strain is expected in the next few years as the *Square Kilometer Array*, currently under construction in the Southern Hemisphere, comes on line. Detection of the anticipated long-wavelength gravitational waves should indicate the rate at which mergers of supermassive black holes may have occurred over the eons.[54]

4.10 Gravitational Wave Background Radiation

Given the possibility of gravitational radiation emanating from a variety of sources, some of which may already be known to emit much shorter bursts of electromagnetic radiation, such as fast radio bursts described later in Chapter 6, I deliberately left the range of gravitational wave frequencies in Figure 3.16 open-ended at high frequencies. Some very general bounds may nevertheless be cited. Once generated, these waves are not readily reabsorbed. They spread out through the entire Universe, and should appear isotropic.

We know very little about this radiation except that it should obey at least two bounds. The first is that the maximum energy of an individual graviton cannot exceed the Planck mass–energy, $\Delta E = m_p c^2$, which it radiates over a time span Δt given by the Planck time t_p. Earlier, in Chapter 2 we encountered both these quantities. Their product is

$$\Delta t \Delta E \sim t_p m_p c^2 = h/2\pi, \tag{4.7}$$

where h is Planck's constant. This satisfies the Heisenberg uncertainty principle.

The second bound is that the total energy entailed in gravitational radiation should not exceed any other cosmic energy reservoir with currently unassigned properties. The Fukugita–Peebles Table 1.2 assigns the cosmological constant an energy density $\Lambda c^2 \sim 6 \times 10^{-9}$ erg cm^{-3}, which could easily comprise the energy density of several different gravitationally active components, and presumably already includes the energy density of gravitational radiation globally accumulated over the eons.

Not having observed this background, to date, does not mean that it does not exist. In time, we should be able to detect it.

Notes

1 Die Grundlage der allgemeinen relativitätstheorie, A. Einstein, *Annalen der Physik* 49, 769–822, 1916

2 Über Gravitationswellen, A. Einstein, *Sitzungsberichte der Königlich Preussischen Akademie der Wissenschaften zu Berlin* VII, 154–67, 1918

3 On Continued Gravitational Contraction, J. R. Oppenheimer & H. S. Snyder, *Physical Review* 56, 455–59, 1939

4 On a Stationary System with Spherical Symmetry Consisting of Many Gravitating Masses, A. Einstein, *Annals of Mathematics* 40, 922–36, 1939

5 Gravitational Waves in General Relativity III. Exact Plane Waves, H. Bondi, F. A. E. Pirani, & I. Robinson, *Proceedings of the Royal Society of London A* 251, 519–33, 1959

6 Reality of the Cylindrical Gravitational Waves of Einstein and Rosen, J. Weber & J. A. Wheeler, *Reviews of Modern Physics* 29, 509, 1957

7 Detection and Generation of Gravitational Waves, J. Weber, *Physical Review* 117, 306–13, 1960

8 Gravitational Radiation, J. A. Tyson, *Annals of the New York Academy of Sciences* 224, 74–92, 1973

9 Accretion of Interstellar Matter by Massive Objects, E. E. Salpeter, *Astrophysical Journal* 146, 796–800, 1964

10 The Fate of a Star and the Evolution of Gravitational Energy Upon Accretion, Ya. B. Zel'dovich, *Soviet Physics Doklady* 9, 195, 1964

11 X-Ray Emission from the Direction of Scorpius, S. Bowyer, E. T. Byram, T. A. Chubb, & H. Friedman, *Astronomical Journal* 69, 135, 1963

12 Time Variation in Scorpius X-1 and Cygnus XR-1, J. W. Overbeck & H. D. Tannanbaum, *Astrophysical Journal* 153, 899–903, 1968

13 On the Binary Nature of CYG X-2, E. M. Burbidge, C. R, Lynds, & A. N. Stockton, *Astrophysical Journal* 150, L95–97, 1967

14 On Massive Neutron Cores, J. R. Oppenheimer & G. M. Volkoff, *Physical Review* 55, 374–81, 1939

15 New H-alpha Emission Stars in the Milky Way, C. B. Stephenson & N. Sanduleak, *Astrophysical Journal Supplement Series* 33, 459–69, 1977

16 An Unusual Emission-Line Star/X-Ray Source/Radio Star, Possibly Associated with an SNR, D. H. Clark & P. Murdin, *Nature* 276, 44–45, 1978

17 The Bizarre Spectrum of SS 433, B. Margon, et al., *Astrophysical Journal* 230, L41–45, 1979

18 The 164 and 13 Day Periods of SS 433: Confirmation of the Kinematic Model, B. D. Margon, S. A. Grandi, & R. A. Downes, *Astrophysical Journal* 241, 306–15, 1980

19 *The Quest for SS433*, D. H. Clark, Viking, New York, 1985, p. 131

20 Ibid., Clark, 1985, p. 195

21 A Double-Sided Radio Jet from the Compact Galactic Centre Annihilator 1E1740.7–2942, I. F. Mirabel, et al., *Nature* 358, 215–17, 1992

22 A Superluminal Source in the Galaxy, I. F. Mirabel & L. F. Rodríguez, *Nature* 371, 46–48, 1994

23 Appearance of Relativistically Expanding Radio Sources, M. J. Rees, *Nature* 211, 468–70, 1966

24 Accretion Instabilities and Jet Formation in GRS 1915+105, I. F. Mirabel, et al., *Astronomy & Astrophysics* 330, L9–12, 1998

25 Bruce Margon, *email* to Martin Harwit, April 6, 2008

26 Black Holes in Binary Systems. Observational Appearance, N. I. Shakura & R. A. Sunyaev, *Astronomy & Astrophysics* 24, 337–55, 1973

27 Discovery of a Pulsar in a Binary System, R. A. Hulse & J. H. Taylor, *Astrophysical Journal* 195, L51–53, 1975

28 A New Test of General Relativity: Gravitational Radiation and the Binary Pulsar SR 1913+16, J. H. Taylor & J. M. Weisberg, *Astrophysical Journal* 253, 908–20, 1982

29 Testing Relativistic Gravity with Radio Pulsars, N. Wex, *2014arXiv1402.5594W*

30 Observation of Gravitational Waves from a Binary Black Hole Merger, B. P. Abbott, et al., and the LIGO Scientific and Virgo Collaboration, *Physical Review Letters* 116, 061102

31 *The Classical Theory of Fields*, L. Landau & E. Lifshitz. Translated from Russian by M. Hammermesh, Addison-Wesley, Reading MA, 1951, p. 332

32 Ibid., Abbott, et al., 2016

33 The Basic Physics of the Binary Black Hole Merger GW150914: The LIGO Scientific and Virgo Collaborations, *Annalen der Physik* 529, No. 1–2, 1600209, 2017

34 Ibid., Abbott, et al., 2016

35 Astrophysical Implications of the Binary Black Hole Merger GW 150914, B. P. Abbott, et al., *Astrophysical Journal* 818, L22, 15 pp., 2016

36 GW170814: A Three-Detector Observation of Gravitational Waves from a Binary Black Hole Coalescence, B. P. Abbott, et al., *Physical Review Letters* 119, 141101, 2017

37 GRB 080916C and GRB 090510: The High-Energy Emission and the Afterglow, Wei-Hong Gao, et al., *Astrophysical Journal* 706, L33–36, November 20, 2009

38 Based on the article "Gravitational Wave Astronomy with the SKA," G. H, Janssen, et al. *Proceedings of Advancing Astrophysics with the Square Kilometre Array (AASSKA14)*, June 9–13, 2014, Giardini Naxos, Italy

39 Formation of Massive Black Holes in Rapidly Growing Pre-galactic Gas Clouds, J. H. Wise, et al., *Nature* 566, 85–88, 2019

40 For more detail, see "The Basic Physics of the Binary Black Hole Merger GW150914," LIGO Scientific and Virgo Collaborations, *Annalen der Physik (Berlin)* 529, No. 1–2, 1600209, 2017A. Appendix A of this paper provides a clear auxiliary explanation of the relation between orbital energy decay and gravitational wave radiation.

41 Observations of Gamma-Ray Bursts of Cosmic Origin, R. W. Klebesadel, I. B. Strong, & R. A. Olson, *Astrophysical Journal* 182, L85–88, 1973

42 Compact Radio Emission Indicates a Structured Jet Was Produced by a Binary Neutron Star Merger, G. Ghirlanda, *Science* 363, 968–71, 2019

43 The Story Teller, A. Cho, *Science* 353, 532–37, 2016

44 The History and Future of Gravitational Waves, K. Raynor-Evans & B. Schutz, *Astronomy & Geophysics* 60, 3.28–3.29, June 2019

45 See www.geo600.org

46 3C 273: A Star-Like Object with a Large Red-Shift, M. Schmidt, *Nature* 197, 1040, 1963

47 More Spectroscopy of the Fuzz around QSOs: Additional Evidence for Two Types of QSO, T. A. Boroson, S. E. Persson, & J. B. Oke, *Astrophysical Journal* 293, 120–31, 1985

48 The Apparently Normal Galaxy Hosts for two Luminous Quasars, J. N. Bahcall, S. Kirhakos, & D. P. Schneider, *Astrophysical Journal* 457, 557–64, 1996

49 Opacity Limit for Supermassive Black Hole Seeds, F. Becerra, F. Marinacci, K. Inayoshi, V. Bromm, & L. E. Hernquist, *arXiv:1702.03941v1 [astro-ph.GA]* February 13, 2017

50 Is There a Maxium Mass for Black Holes in Galactic Nuclei? K. Inayoshi & Z. Haiman, *Astrophysical Journal* 828, 110 (8 pp.), September 10, 2016

51 Stellar Orbits about the Galactic Center Black Hole, A. M. Ghez, et al., *Astrophysical Journal* 620, 744–57, 2005

52 A Geometric Determination of the Distance to the Galactic Center, F. Eisenhauer, et al., *Astrophysical Journal* 597, L121–24, 2003

53 The International Pulsar Timing Array: First Data Release, J. P. W. Verbiest, and a group of roughly 100 international colleagues, *Monthly Notices of the Royal Astronomical Society* 458, 1267–88, 2016

54 Pulsar Timing Array Based Search for Supermassive Black Hole Binaries in the Square Kilometer Array Era, Yan Wang & S. D. Mohanty, *Physical Review Letters* 118, 151104, 2017

5

Gravitational Lensing

In the last half of the twentieth century astronomers became aware that distant galaxies and galaxy clusters could act as giant gravitational lenses amplifying light from more remote sources, potentially revealing them as they had appeared eons ago at the dawn of time. It seemed that gravitational lensing might well become a wonder weapon in the arsenals of astronomy – a natural boon that would, at least occasionally, provide us a view of the distant Universe far clearer than any telescope we could ever hope to construct here on Earth!

In devising his general theory of relativity, Einstein had made clear that gravity, geometry, and time are intimately connected. Place a massive object randomly anywhere in an otherwise empty space and a ray of light that would otherwise have passed by along a straight-line trajectory now finds its direction altered by the gravitational tug of the mass. The light's arrival at its final destination is also delayed because time elapses more gradually near massive bodies than it does far away. Both of these consequences are part of *gravitational lensing*. As with a glass lens, gravity can concentrate light at a focus or merely displace it from its original trajectory.

The Universe is filled with myriad objects, some as massive as clusters of galaxies, innumerable others as small as asteroids. Each acts as a gravitational lens. The less massive bend trajectories less effectively, but tend to also be far more numerous and far harder to identify. Do we inevitably reach an impasse where accumulating deflections prevent the construction of ever-more accurate maps of the heavens or ever-higher fidelity timing?

Though gravitational lensing originally promised astronomers ways to see further than any affordable telescope would permit, it gradually became clear that it would also constitute an ultimate hindrance.

5.1   Gravitational Lensing to Increase Angular Resolution

The highest angular resolving power R_α astronomical instruments can offer is limited by the longest baseline D an interferometric array may span, or the widest aperture D a telescope provides.

The longest baselines attainable with ground-based interferometers are limited by Earth's diameter. Interferometers launched into space and orbiting the Sun can overcome this limit but are more expensive. If the angular resolution provided by existing observatories does not sufficiently resolve remote stars, quasars, or galaxies, naturally occurring gravitational lenses can sometimes help.

To see how this option arose, we may turn to a letter Einstein received in the mid-1930s from R. W. Mandl, an inquiring amateur who asked about the amplification of light from a distant star focused by a foreground star nearer to an observer.

In his reply, published in 1936, Einstein wrote that an observer at the focal point of Mandl's proposed configuration would see light from the more distant star forming a ring encircling the foreground star. Einstein concluded, "Of course, there is no hope of observing this phenomenon directly." An observer would be very unlikely to ever find himself located precisely along an extension of the straight line drawn through the centers of the two stars. Besides, Einstein asserted, the angular resolving power of astronomical instruments would never suffice to resolve that ring of light from the glare of light emitted by the lensing foreground star.[1]

Undeterred, the ever-prescient Swiss astronomer Fritz Zwicky, shown in Figure 5.1, took another look at the problem the following year. Working at Caltech, where he was studying the nature of galaxies, he pointed out that gravitational lensing of a remote galaxy by a foreground galaxy along the same line of sight would be far more readily observed than the gravitational lensing of one star by another. An observer would see the distant galaxy appearing as a ring around the foreground galaxy. The existence of such rings could further confirm Einstein's general theory of relativity.

Zwicky noted that the apparent total luminosity of this ring might be as much as a hundred times the luminosity of the un-lensed background galaxy, enabling one "to see nebulae [Zwicky's term for galaxies] at distances greater than those ordinarily reached by even the greatest telescopes … Observations on the deflection of light around nebulae may provide the most direct determination of nebular masses."[2] As an afterthought, Zwicky published a second letter in which he estimated the chances of finding

Fig. 5.1 Portrait of Fritz Zwicky 1898–1974. Zwicky predicted that massive galaxies would be found to act as gravitational lenses magnifying the images of more remote stars or galaxies. As mentioned in Section 1.16, he also was the first astronomer to suggest that some form of dark matter must exist to account for the apparent stability of clusters of galaxies whose members were exhibiting velocities sufficiently high to escape unless each cluster was far more massive than the summed mass of its individual galaxies. With Walter Baade of the Mount Wilson Observatory, Zwicky also was first to recognize the existence of rare supernovae many orders of magnitude more energetic than the ordinary novae abundant in all spiral galaxies. The two astronomers then also suggested that supernovae might well be the sources of cosmic rays. (Courtesy of the Archives, California Institute of Technology.)

such a lensed galaxy and correctly concluded, "Provided that our present estimates of the masses of *cluster nebulae* are correct, the probability that nebulae which act as gravitational lenses will be found becomes practically a certainty."[3]

Figure 5.2 shows how correct Zwicky's forecast was. The faint elongated streaks encircling the foreground cluster of galaxies are gravitationally lensed images of a background source.

Several dozen or more well-isolated, strongly-lensing galaxies are now known to be strewn across the sky. We can make use of at least some of these to magnify remote regions of the Universe to view the earliest generations of stars and galaxies as they were just beginning to form.

Fig. 5.2 The Galaxy Cluster 0024+1654. A number of elongated tangential features appearing as fragments of a longer continuous arc can be discerned primarily at upper left and also around lower right. These appear to be images of a single galaxy gravitationally lensed by the foreground galaxy cluster. Several other galaxies in the background appear to also be lensed in this image obtained by the Hubble Space Telescope in November 2004. (Credit: W. N. Colley & E. Turner at Princeton University; J. A. Tyson at Bell Labs and Lucent Technologies; and NASA/ESA.)

5.2   The Foundations of Gravitational Lensing

In 1964, more than two decades after Zwicky's predictions, Sjur Refsdal, at the University of Oslo in Norway, re-examined gravitational lensing more quantitatively. Aged 29 at the time, he was particularly interested in the gravitational lensing of "a supernova lying far behind, and close to, the line of sight through a distant galaxy." Refsdal noted a plane defined by three points

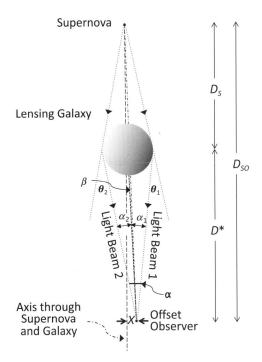

Fig. 5.3 Gravitational Lensing of a Remote Supernova by a Galaxy Near the Line of Sight from the Supernova to an Observer. The finely dotted lines show trajectories 1 and 2 of radiation, respectively arriving at the observer along the shorter and longer light paths. The observer is displaced by a distance X from a line of sight running from the supernova through the center of mass of the lensing galaxy. If the lensing galaxy were replaced by an equally massive, but far more compact mass, the observer would see the supernova at an angular separation β from this center of mass. The observer's two gravitationally deflected images arrive, respectively, from directions α_1 and α_2 displaced from that center. Both angles lie in a plane defined by the supernova, the galaxy's center, and the observer's own position. The total angle separating the two images is $\alpha \equiv (\alpha_1 + \alpha_2)$. The corresponding gravitational deflection angles of radiation passing the galaxy at right and at left are θ_1 and θ_2.

in Figure 5.3 – a remote supernova, a nearer galaxy acting as a gravitational lens, and an observer located off the projected axis joining the supernova and the galaxy's center of mass. The light from the supernova may then follow two different paths to the observer. Instead of the ring seen by an on-axis observer, the observer will see two images in this plane, one on each side of the focusing galaxy. This can only happen if the light passes by the galaxy along either the shortest path to the observer or the longest. Two papers, both written in 1964, presented Refsdal's findings.[4; 5]

The light path of the less-bent trajectory, Beam 1 in the figure, is shorter. Light along this path reaches the observer first. Light following the longer trajectory, Beam 2, lags. One can verify that these two images indeed come from the supernova by checking, as Refsdal noted, for "a difference Δt in the time of light travel for these two paths, which can amount to a couple of months or more, and may be measurable." As the luminosity of the supernova evolves with time, its evolutionary pattern observed in Beam 1 systematically repeats in Beam 2 with a constant time lag.

Not surprisingly, Refsdal found that the angle θ by which the radiation from the distant supernova is bent on closest approach to the galaxy is just the angle Einstein had derived and predicted in 1916 for the deflection of light from a distant star passing close by the Sun. For a solar mass $M_\odot$ this deflection at closest approach r_{min} was

$$\theta \sim \frac{4GM_\odot}{c^2 r_{min}}. \tag{5.1}$$

Here c is the speed of light, and G the gravitational constant.[a]

Refsdal now emphasized that the flux per unit solid angle remains unaltered through gravitational lensing, and that the increased flux reaching an observer by virtue of the gravitational lens results from the magnified solid angle across which the supernova's radiation arrives. This magnification along the respective light paths may be represented by symbols μ_1 and μ_2.[6]

Defining an angle $\alpha \equiv (\alpha_1 + \alpha_2)$, often blocked by the intervening lensing galaxy, and a second angle $\beta \equiv (\alpha_1 - \alpha_2)$, between a line of sight from the observer to the center of the lensing galaxy and the line of sight to the supernova – even if blocked by the galaxy – Refsdal derived the magnifications along the two beams, respectively, as

$$\mu_1 = \frac{1}{4}\left(\frac{\beta}{\alpha} + \frac{\alpha}{\beta} + 2\right); \mu_2 = \frac{1}{4}\left(\frac{\beta}{\alpha} + \frac{\alpha}{\beta} - 2\right); \mu_T = \frac{1}{2}\left(\frac{\beta}{\alpha} + \frac{\alpha}{\beta}\right), \tag{5.2}$$

where $\mu_T \equiv \mu_1 + \mu_2$ is the total magnification observed.

For observers centered along the axis of symmetry, where $\beta = 0$, these expressions diverged because they assumed the supernova to be a point source.

---

[a] The gravitational deflection of radiation passing the Sun barely above its limb should amount to an angle $\theta \sim 1.75$ arcseconds – twice the angle that Newtonian dynamics predicted. When observationally verified in 1921, Einstein became a celebrity and general relativity began gaining adherents.

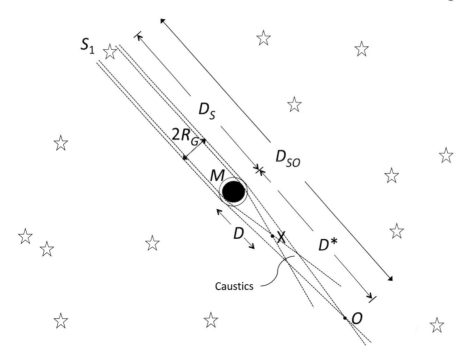

Fig. 5.4 Light from a remote star S_1 is shown incident on the mass M at center. The rays from the remote star are deflected along a series of nested conical surfaces converging less rapidly for rays gravitationally deflected at increasing radial distances R_G from M. These mutually nesting conical surfaces produce caustics rather than a single well-defined focus. Light from any of the myriad other stars, a few of which are symbolically included in this figure, would follow similar axially symmetric surface trajectories about lines joining the central star to any of these others. The point labeled X is one location at which radiation from star S_1 is focused by the mass M. An observer located at O sees a different set of rays from S_1 focused by M.

For a more realistic supernova angular radius u in the absence of the lensing galaxy, Refsdal derived a magnification $\mu_T = \alpha_0/u$, where α_0 is the value of α along the axis of symmetry.

Galactic Star and Planetary System Studies

Within our Milky Way light from a remote star, S_1 in Figure 5.4, may similarly be amplified through gravitational lensing by a foreground star, S. For an on-axis observer O, Einstein's total deflection angle at the star shown in Figures 5.3 and 5.4 then consists of the sum $\theta = R_G/D_S + R_G/D^*$, where R_G is the distance at which light reaching the observer passes M. Using Einstein's value for θ then allows us to solve for the radial distance R_G from the mass M at which the light cone begins to converge toward O:

$$R_G = \left[\left(\frac{4GM}{c^2} \right) \frac{D_S D^*}{D_{SO}} \right]^{1/2}. \tag{5.3}$$

Refsdal provided an example of a star of Solar mass $M = M_\odot$ at a distance $D^* = 10$ pc deflecting light from a more remote star of similar mass at distance $D_{SO} = 100$ pc. For this set of parameters, the light reaching O is deflected at a distance of $R_G \sim 57$ solar radii $R_\odot$ from the mass M. Light deflected at closer radial distances converges too rapidly and crosses at a point X far closer to M before again diverging.

For Refsdal's sizeable deflecting radius R_G, the magnification is significant, $\mu_T \sim 1100$, largely thanks to the high radial distance from M at which light is deflected. Such high values of R_G also assure a relatively high probability that significant numbers of remote Galactic stars and their planetary systems will be lensed.

5.3   Gravitational Lensing in Planetary Studies

Predictions for a planned US Wide-Field Infrared Survey Telescope provisorily designated WFIRST-AFTA suggest that a survey by a telescope having an aperture of 2.4 meters, identical to that of the Hubble Space Telescope, would find large numbers of lensed sources. Scanning a $\sim$2.8 square degrees patch of the sky around the Galaxy's central bulge at near-infrared wavelengths, the telescope would detect some 37,000 lensed sources including stars and planets orbiting stars or freely floating in interstellar space. The anticipated angular resolution of this telescope would be of the order of 0.14 arcseconds, but the precision to which the central position of radiating sources might be determined would be as accurate as $\sim 10^{-4}$ arcseconds.[7]

A proposal for this telescope submitted in 2015 by two leading astrophysicists, David Spergel at Princeton and the late Neil Gehrels at NASA's Goddard Space Flight Center, represented the input of more than 50 co-authors from a spate of separate US institutions.[8] The proposal dwelled at considerable length on gravitational lensing by individual stars and planets leading to deflection, imaging, or blocking of radiation termed *microlensing*, a set of effects readily observed within our own Galaxy.

Microlensing by planets bound to their parent stars can readily be inferred when a foreground star microlenses a more distant star. Occasionally this exhibits not only a broad magnification peak as it lenses that star but also a

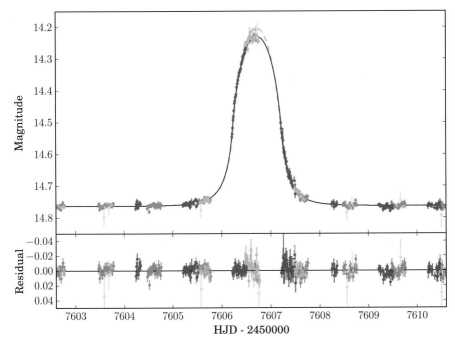

Fig. 5.5 On August 6, 2016, a distant free-floating planet in the mass range of
Neptune or Saturn microlensed the light from a Galactic star many kiloparsecs
further away, amplifying the light reaching Earth in an event designated
OGLE-2016-BLG-1540. The lensing lasted less than 8 hours as the planet's motion
traversed the remote star's projected disk. (From "A Neptune-Mass Planet Candidate
Discovered by Microlensing Surveys," Przemek Mróz, et al., *Astronomical Journal* 155,
121, March 2018, with permission of the authors and the American Astronomical
Society.)

slightly offset narrower peak as a planet at some lateral displacement from its
parent star lenses that remote star as well.[9] Free-floating interstellar planets
similarly can microlens background stars, and can likewise be recognized by the
single narrow peaks of their light curves. In all such traversals, the shape and
timing of the observed light curve will be identical in all wavelength ranges –
the defining hallmark of all gravitational lensing. Usually an isolated planet's
traversal of a remote star's disk lasts less than two days.

A Neptune-mass free-floating planet microlensing a more remote star in the
Galactic bulge exhibited short-time lensing, lasting only 0.320 ± 0.003 days,
just under 8 hours. Its light curve is shown in Figure 5.5.[10] Microlensing
by free-floating interstellar planet-sized bodies appears to be as common as
microlensing by planets orbiting stars. Spergel and Gehrels estimated that freely
floating giant planets "may outnumber the stars in our galaxy by two to one."[11]

The planets appear to have been ejected from their system of origin through interaction with sister planets which survived the encounter and became more strongly bound. Energetics suggest that lower-mass planets, comets, and asteroids could have been ejected into interstellar space in appreciable numbers as well. One such asteroid, now named Oumuamua, entered, crossed, and left the Solar System in 2017. It was highly elongated, with dimensions approximately $1 \times 0.1 \times 0.1$ km. Its appearance taken together with its tumbling motion provided some possible hints on its origins.[12]

Planets at large orbital separations from their parent stars can be difficult to discover. Their orbital periods are long, sometimes lasting several decades, making their differential tug on their parent star difficult to detect. Their reflected starlight and thermal radiation both are faint and similarly hard to detect. Microlensing of a remote star by such foreground planets may then be the simplest way of obtaining at least a census of how abundant such planets may be. By 2018, roughly 50 planetary microlensing events of this type had been observed through ground-based detections. Because detection of microlensing from space could be considerably easier, missions for this purpose are being planned.[13; 14]

Equation (5.1) specifies the angular deflection of light by a spherical mass M, taken in Figure 5.4 to be a star. Let us consider part of this deflection exerted just as the radiation is reaching M from remote source S_1. A complementing part of the deflection is further imposed as the radiation recedes from M. If the lensing star is much closer to the observer than the lensed source, virtually the full deflection comes from this second phase, leading to a convergence onto the axis of symmetry at an angle

$$\theta = \frac{4MG}{c^2 R_G}. \tag{5.4}$$

Here R_G is the radial distance of the gravitationally deflected light ray at its closest approach to M.

5.4   Gravitational Time Delay

In devising his general theory of relativity in 1915, Einstein predicted that time should pass more slowly in the immediate vicinity of a star than at considerable distances from it. In 1968 a group led by Irwin I. Shapiro working at MIT in Cambridge, Massachusetts, at the time, employed a radar technique to measure a delay in radiation passing close to the surface of the Sun.

The radar beam passing by the Sun both on its way to Venus, and after bouncing off that planet and returning to Earth, again via a close approach to the Solar surface, produced a delay on this roundtrip of the order of 200 microseconds.[15; 16]

The delay Δt on individual passages of radiation past a star is largely determined by the radiation's closest approach to the star:

$$\Delta t \sim \left(\frac{2MG}{c^3} \right) \ln \left(\frac{R_{max}}{R} \right), \tag{5.5}$$

where R can range from R_G to R_{max}. For $R_G \ll R_{max}$ this reduces to $\Delta t \sim MG/c^3$.

For a stellar mass $M_\odot$ we can insert values $R_G \sim 4 \times 10^{13}$ and $R_{max} \sim 1.5 \times 10^{18}$ cm, respectively, for a distance of closest approach and the most remote distance from the star still lying within its prime sphere of influence. With these values we obtain a delay $\Delta t \sim 5$ μs for the mean time delay obtained by integrating over the entire weighted range of R values. Within a cylindrical volume of cross section 1 pc^2 stretching from Earth to S_1 and beyond, we may expect roughly ten thousand solar-mass stars in the Galactic plane to reside along a distance of 10 kpc.

If all the stars were of precisely identical mass and were arrayed along a precisely constructed grid, this delay would cause little confusion. But we do not know the precise masses of the individual stars along such a line of sight, nor the masses of some of the planets or dark matter freely floating in interstellar space, or orbiting the observed stars. So, uncertainties at levels of microseconds to seconds, entirely due to gravitational uncertainties, need to be anticipated as built into any absolute timing sequences currently sought.

This time delay accumulates linearly with distance traversed. For stars 10 kpc away, the total expected time delay would be about 50 milliseconds.

Today, all these perturbations may still be considered to be negligibly small. But as instrumental resolving powers continue to increase by two or three orders of magnitude, levels will be reached at which gravitational deflections by progressively smaller masses will increasingly intrude. They may not be dominant in any given wavelength or energy region, but they will be present in any and all energy domains and constitute an irrepressible source of confusion as every astronomical body, no matter how small its mass, contributes cumulatively to the total lensing mass $M = \sum_{i=1}^{N} M_i$ in all locales.

5.5   Cosmological Investigations

In a cosmological context, Refsdal went a step further to calculate the relative time delay for light arriving from a remote supernova along its respective trajectories around a lensing galaxy. His important insight was that the Hubble constant H could be calculated if the angles α_1, α_2, and α shown in Figure 5.3 were measured, as well as the redshifts z_s of the supernova and z_g of the galaxy. He specifically assumed that redshift was directly proportional to distance, and that the time delay Δt between the time of arrival along the two beams could be accurately determined – presumably if the delay could be identified from a supernova's evolving luminosity tracked in the respective beams as a function of time.

The Hubble constant could then be derived from the ratio

$$H = \frac{z_s z_g \alpha (\alpha_1 - \alpha_2)}{\Delta t (z_s - z_g)}. \tag{5.6}$$

Refsdal's realization with this calculation was that, whereas the first of his two papers had provided the relative lengths of the two lensed light paths expressed in terms of the angles shown in Figure 5.3, he would need to find some measure of absolute length, to provide his analysis physical significance. He found this in the time delay, Δt, of an identifiable signal emitted at the source, arriving at the observer along the two distinct optical paths. That delay multiplied by the speed of light provided a physical path length, $c\Delta t$, from which he could readily derive a value for the Hubble constant in the observed galaxy's immediate vicinity.

5.6   Quasar $0957 + 561$ (1979)

In 1979, Dennis Walsh of the University of Manchester, Ray Weymann at the University of Arizona, and Bob Carswell at the Institute of Astronomy in Cambridge discovered two closely spaced, identical appearing luminous sources they identified by their celestial coordinates as $0957 + 561$.[17]

Among the known radio galaxies listed at the time, the source $0957 + 561$ revealed itself to have a close pair of apparently associated blue stellar objects, A and B, with identical redshifts $z = 1.405$, though A and B appeared separated by as much as ~6 arcseconds. As the three astronomers pointed out, optical observations of the two sources appeared to have essentially identical redshifts, magnitudes, and colors. Spectroscopy of the twin images also showed their spectral features closely mirroring each other.[18] To the three researchers the conclusion that the two observed sources were simply the lensed images of one and the same distant quasar appeared inescapable.

The discovery was widely appreciated and stimulated a range of theoretical studies as well as further observations.

Over the past several years, the possibility that the Hubble constant might not have a unique value for specified redshifts but could vary from locale to locale in the Universe has been raised. If so, it is not clear whether equation (5.6) would remain an entirely trustworthy guide, among other possible reasons because Refsdal had specifically assumed his redshifts z_g and z_s to be measures of the distances to the gravitational lens and the remote source, and that those distances were *not* arbitrarily changing from one cosmic locale to another.

5.7   Quasar 0957 + 561 Revisited

By 1997, the two images A and B of the lensed quasar 0957 + 561 were judged to lie respectively at angular separations $\alpha_1 \sim 5.2''$ and $\alpha_2 \sim 1.034''$ from the observer's line of sight to the lensing galaxy's center. By then the galaxy's redshift was also known to be $z_g = 0.36$, for which the *scale factor* – the transverse separation perpendicular to the line of sight – is 5.078 kpc per arcsecond. The line-of-sight separation between the quasar and the lensing galaxy judged by their respective redshifts then was $(D_{SO} - D_L) \sim 0.721$ Gpc.[19]

That year, 1997, the time delay originally predicted by Refsdal's second paper of 1964 was finally detected as well. The brightness variations of the quasar observed in one of the images, labeled A, occurred about 420 days ahead of those in the other image, B.[20]

Over time, a wide range of mutually inconsistent values for the Hubble constant, H, have been determined for the system 0957 + 561. For other lensed galaxies H also has not converged onto a singe value.[21; 22] Further efforts will be needed to determine how well such lenses can lead to reliable values for the Hubble constant in different cosmic realms.

By 1997, a combination of the highest available angular resolving power at radio frequencies, and the best optical images yielded by the Hubble Space Telescope at the time, showed a group led by Gary Bernstein, at the University of Michigan at Ann Arbor, a profusion of small arcs, blobs, and other features permeating the landscape around the twin quasar. Figure 5.6 documents the many previously unknown additional features that now appeared to have entered the discussion and would have to be resolved and understood.[23]

The messy appearance of the region around 0957 + 561 might however have been anticipated by the main thrust of a series of papers to which we now turn.

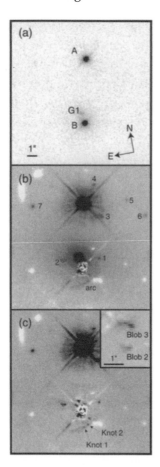

Fig. 5.6 Detailed optical images of the gravitational lensing QSOs 0957+561 A, B, obtained in 1997 by a team led by Gary Bernstein at the University of Michigan at Ann Arbor. The second and third images from the top were attempts to highlight weak features otherwise undetected. One was the apparent location of an optical center of the primary lensing galaxy coinciding, within 10 milli-arcseconds, with a previously reported Very Long Baseline Radio Interferometry source. Others, the authors suggested, might be "multiple images of each other." White features seen in the central and bottom panels are negative *ghosts* resulting from the image subtractions that revealed these fainter identified sources. The authors hoped that such insights would lead to a more reliable value of the Hubble constant, H, than equation (5.6) would otherwise permit. (From "Improved Parameters and New Lensed Features for Q 0957 + 4561 from WFPC2 Imaging," G. Bernstein, P. Fischer, J. A. Tyson, & G. Rhee, *Astrophysical Journal* 483, L79, 1997, with permission of the authors and the American Astronomical Society.)

5.8   Two Significant Advances

Two decades after Refsdal's original work, his approach was further fleshed out by a pair of perceptive theoretical studies. The first was a paper published in 1985 by Peter Schneider of the Max Planck Institute for Astrophysics in Munich, who showed that Refsdal's approach to gravitational lensing could be reformulated to readily fit into general relativistic models of the Universe.[24] This led to notable simplifications and greater confidence that the equations of gravitational lensing rested on a broader theoretical foundation.

The second, published in 1986, was an article by Roger Blandford and Ramesh Narayan at the California Institute of Technology, which introduced an entirely new aspect of gravitational lensing – *caustics* and *optical catastrophes*.[25] This was unwelcome news, indeed, though, as we will see below, its full significance was not immediately appreciated!

5.9   Caustics and Catastrophe Theory

The penetrating examination of gravitational lensing, undertaken by Blandford and Narayan in their paper of 1986, had warned of an alarming feature that appeared not to have been heeded earlier. A 1980 paper by Michael V. Berry and C. Upstill at the University of Bristol in the United Kingdom had ominously predicted:

> As light propagates, it is common for focusing to occur. In the most elementary picture all the rays pass through a focal point. It is exceptional, however, for this to correspond to reality: what happens typically is that the rays envelop a caustic surface in space. The light intensity is greatest near caustics, and they dominate optical images … [Caustics] have been awkward to fit into optical theory … [but] … [t]his unsatisfactory state of affairs has been changed dramatically by the injection into optics of a new branch of mathematics called catastrophe theory … The perfect point sources so sought after in optical technology is in this mathematical sense highly unstable, a fact borne out by the difficulty of producing it. Thus catastrophe optics is complementary to traditional optics, where the emphasis is on solving ray or wave equations in highly symmetric situations.[26]

Blandford and Narayan thus took on the task of introducing the concepts of *caustics* and *catastrophe optics* to the astronomical community and, more specifically, to show their potential dominance in gravitationally lensed images because cosmic mass distributions are generally far more complex than the optical lenses used in constructing astronomical telescopes.

A simple analogy may help: The distribution of mass across clusters of stars or clouds of gas and dust in typical galaxies gives rise to unpredictable deviations in the trajectories of radiations akin to the scattering of light passing through speckled glass panes. On the smallest angular scales, it is these gravitational catastrophes that may well constitute the ultimate limitations on the angular resolution and thus the imaging information transmissible across the Universe. For essentially everything we wish to learn about the Universe involves remote clusters, galaxies, stars, planets, or asteroids, all lying beyond some distribution of foreground masses, whose gravitational attraction inherently deflects and distorts bypassing radiation. The degree of deflection may be large or small; we may respectively refer to it as "lensing," "weak lensing," or "microlensing"; but it is always there – sometimes as an apparent blessing, but ultimately also an irreducible nuisance.

Because all *gravitational lensing* effects distort electromagnetic radiation and gravitational waves, across all energy ranges, and have the same effect also on neutrinos and highly energetic cosmic rays, such distortions are universal, as Figure 5.4 indicates. Once gravitational deflection has set in, even telescopes of superb angular resolving power cannot reverse these distortions to yield corrected shapes and dimensions of individual galaxies or the separation between them.

The complexities of the arcs and blobs shown in Figure 5.6 should, therefore, not have appeared anomalous. The predictions Blandford and Narayan had made more than a decade earlier, in 1986, had simply not caught the astronomical community's full attention until Bernstein and his co-authors had undertaken their careful analysis.

More conclusive catastrophes, however, emerged only another two decades later, when the gravitationally lensed image of a remote supernova was observed split into four distinct parts, as we will later note in Figure 5.7. First, however, we need to return to Refsdal's prediction that gravitational lensing could provide a measure of the Hubble constant.

5.10   Hubble Constant Inconsistencies

Over the past decade, two well-established methods for determining the Hubble constant H have raised concerns because their respective measures of the cosmic expansion rate, which had usually been assumed to be identical throughout the Cosmos at any well-defined epoch, appeared to differ by a small percentage. One of the measures has arisen from the thorough analysis of the cosmic microwave background radiation; the other from a step-by-step set of calibrations known as the *cosmic distance ladder*.

The more traditional distance ladder method begins with a determination of the distances of the nearest stars using simple triangulation methods, with Earth's orbit about the Sun providing a useful baseline. Once a sufficient number of luminous stars with well-calibrated luminosities and readily identified features are in hand, the distances of progressively more remote stars within this group can be derived from their apparent brightness, and occasionally compared to far more luminous but extremely rare events such as supernova explosions within the same galaxies as the well-calibrated stars. Some of these supernovae in turn exhibit mutually identical luminosities so that their apparent brightness can tell us how far away they and their immediate surroundings are. The observed redshift at that distance then provides the rate of expansion at the same distance, which in one of the most carefully compiled studies corresponds to a Hubble constant, $H = 73.8 \pm 1.1$ km s^{-1} Mpc^{-1}.

The alternative method comes from an in-depth investigation of the cosmic microwave background radiation based on an assumed theoretical model of the Universe. This is based on an understanding that the cosmic expansion obeys general relativity and that the Universe contains three types of mass–energy – dark energy, dark matter, and baryonic matter and radiation. The last of these comprises atomic and subatomic matter, and electromagnetic as well as other recognized radiations interacting with this matter. Three other factors also enter the model: The opacity of the Universe to the transmission of electromagnetic radiation that would otherwise reach us unimpaired, called the *optical depth* of the Universe, τ; the geometry that channels the radiation our way, called the *scalar spectral index*, which comes very close to being the simplest line-of-sight transmission; and the *angular acoustic scale*, which reflects the lumpiness of the Universe. Once these six parameters are in hand, general relativity permits a straightforward derivation of the rate at which the Universe expands, the Hubble constant H. The value the Planck mission favors is $H = 67.4 \pm 0.5$ km s^{-1} Mpc^{-1}.

The discrepancy between the values of the Hubble constant determined by these two independent methods amounts to only a percentage, but this still is roughly five times higher than their combined anticipated errors should allow, suggesting that either one or both methods are overlooking a real cosmological difference, or else are making some failed assumption. If so, Refsdal's lensing method might confirm the reality of the discrepant results or refute them through determination of H in widely separated cosmic realms occupied by distinct gravitational lenses. Each lens could provide an independent Hubble constant appropriate to its particular location within the Cosmos if H indeed varied from place to place.

This possible application of lensed observations was recently addressed by Christofer Kochanek at Ohio State University, who investigated the question in depth. He found that, although observation along Refsdal's proposed lines might yield independently acceptable values for the Hubble constant, a number of inherent problems would systematically introduce uncertainties of order 10% in the values of H Refsdal's method would yield. No amount of cross-checking would reduce these differences. Refsdal's method would thus not be able to resolve current tensions between the two rival means of determining the Hubble constant today, because their measures of H differed by no more than about 3%.[27] We will return to these uncertainties in Section 5.15.

In line with Kochanek's assessments, direct measurements of the Hubble constant by Refsdal's technique have indeed yielded values for the constant that mutually differ by about 10% for different sets of observed lensed sources.[28]

5.11   Supernova Refsdal

A dramatic discovery, of the year 2017, began to outline how useful or difficult gravitational lensing might turn out to be for attempts to directly observe the earliest stars ever formed. A paper spearheaded by Patrick S. Kelly at the University of California at Berkeley announced that the catastrophes anticipated by Blandford and Narayan three decades earlier had been directly confirmed.[29]

On April 29 of the previous year, 2016, the Hubble Space Telescope had begun a long series of observations of a remote supernova whose redshift $z = 1.49$ indicated the explosion of the star roughly 9.4 billion years ago. These observations had been enabled through gravitational lensing by an elliptical foreground galaxy at redshift $z = 0.54$ designated MACS J1149.6+2223 (Figure 5.7). The lens had split the radiation from the supernova into four individual images arranged in an *Einstein cross* – four individual images appearing to surround the elliptical galaxy cross-wise. In honor of Sjur Refsdal, who had first predicted the possibility of studying remote supernovae by means of a gravitationally lensing galaxy, the supernova was named *Supernova Refsdal*. It lies in a spiral galaxy, seen virtually face on and catalogued as Sp1149. The galaxy has an approximate stellar mass $M \sim 5 \times 10^9 M_\odot$ and a ratio of stellar to dark matter mass estimated at 1:5, comparable to galaxies in the immediate surroundings of the Milky Way today.

To study the evolution of the supernova over a number of years, a sequence of several dozen Hubble Space Telescope observations was obtained at intervals separated by roughy 10 days. Part way through the sequence a faint source appeared at the periphery of Sp1149. Lens magnifications of compact sources in

Fig. 5.7 Hubble Space Telescope image of the large cluster of galaxies at redshift $z = 0.542$ designated MACS J1149.5+2223. One of the cluster's member galaxies shown in the expanded inset is lensing the more distant Supernova Refsdal at redshift $z = 1.49$. The supernova's four images are indicated by the four arrows. (Hubble Space Telescope gallery "Galaxy Cluster MACSJ1149.5+223 and a supernova four times over." (Credit: NASA, ESA, S. Rodney (Johns Hopkins University, USA) and the FrontierSN team; T. Treu (University of California Los Angeles, USA), P. Kelly (University of California Berkeley, USA), and the GLASS team; J. Lotz (STScI) and the Frontier Fields team; M. Postman (STScI) and the CLASH team; and Z. Levay (STScI).)

the catastrophic range can lead to amplification of their radiation by factors of several hundred or even in the thousands, depending on the precise location of the source within the catastrophic range. In contrast, light from different parts of an extended background source tend to be amplified by different amounts, leading to a scrambled mix of amplification and damping that maintains the integrated flux across the source's extended span virtually constant.

If a compact region or an isolated star moves laterally across the sky, traversing the catastrophic region, its motion can lead to an amplification of its emission, followed by a decline as the star emerges from the highly amplified domain. Crossing a convoluted catastrophic region may also produce a series of high-amplification phases, making it difficult to determine whether the successively amplified radiation emanates from a single isolated star, or from different members of an aggregate of stars of comparable magnitude, such as those of a young star cluster – all systematically traversing the same highly lensed region.

The slightly different locations from which the observed emission from the catastrophic domain appeared to emanate might be interpreted either way. The angular resolution of the Hubble Space Telescope, slightly better than $\sim$0.1 arcseconds, did not suffice to permit an unambiguous conclusion.[30]

5.12   Searching Back for Primordial Star Formation

The redshift of the galaxy Sp1149 in which the catastrophically lensed region was observed was $z \sim 1.49$, telling us that the Universe was roughly 4.3 billion years old when the galaxy's light was emitted. We might ask what it would take to observe such a region at a higher redshift, say, $z \sim 12$ when the age of the Universe was approximately 370 million years. We do not yet know whether stars had begun forming by that time, but the temperature of the Universe at that epoch would have been $(z + 1) = 13$ times higher than the microwave background temperature observed today, or roughly 35 K. This is about as high as the temperature in star-forming regions in the Galaxy today.[31]

Either way, a redshift of $z \sim 12$ would at least roughly correspond to an epoch at which the first stars could have formed. At that early epoch and distance an angular resolution of ~ 0.1 arcseconds, roughly comparable to that of the Hubble Space Telescope's, would resolve a region of order 375 pc, and thus be inadequate for the study of isolated star-forming regions.

We would need to have an angular resolution able to discern regions no more than only 0.3 pc apart, roughly comparable to the separation between stars we see in today's galaxies. This would require an angular resolving power of $\sim$100 micro-arcseconds, a thousand-fold improvement over the Hubble Space Telescope but, under favorable conditions, already available today with the *GRAVITY* instrument at the European Southern Observatory. However, this complex instrument operates at a wavelength of about 2 μm rather than the more appropriate $\sim$10 μm wavelength region into which the radiation from some of the earliest-formed stars would be redshifted.[32]

The $\sim$10 μm wavelength falls in the range covered by the 6-meter James Webb Space Telescope expected to be launched around 2021. Although this telescope's highest angular resolving power in most of the 10 μm wavelength range would only barely resolve the 375 pc region mentioned above, gravitational lenses revealing catastrophically magnified regions might nevertheless permit informative observations on some of these early star-forming regions.

Long-baseline interferometers launched into space or deployed on the Moon might likewise readily provide the requisite angular resolving power within a century – as long as catastrophically magnified regions didn't also introduce nuisance effects. Experience in building such instruments on Earth, as we have by now done at both visible and radio wavelengths, should make construction of interferometers operating in the mid-infrared regime somewhat easier, even if those instruments would have to conduct observations from space, well beyond Earth's strongly absorbing atmosphere.

The only remaining question is whether obtaining images, spectra, and possibly high-time-resolution data on sources at redshifts $z \sim 12$ is likely to fulfill all astronomical needs.

Today, it appears likely that earlier than redshift $z \sim 12$, no compact radiating sources may have ever existed. Certainly, from what we know by now, at epochs preceding $z \sim 12$, the atomic makeup of the Universe will have comprised mainly hydrogen and helium. Atoms of carbon, oxygen, or more massive elements would not yet have formed. Because of this, planets generally containing a considerable quantity of heavier elements could not have been born either until well after a first generation of massive stars had formed, run the course of their lives for a few million years, ejected the heavy elements formed in their interior, and given rise to a new generation of stars which might then have been accompanied by the formation of planetary systems. Accordingly, instruments capable of carrying out high-angular-resolution observations at redshifts $z \sim 12$, with sufficient spectral resolving power to sense radiation emitted in prestellar collapse, are likely to yield some of their most valuable information in studies of early star, galaxy, and quasar formation.

The blank spaces in Figure 6.8 may provide a reasonable estimate to the limits we will ever need to consider for improving instrumental angular resolving power for studies of the most remote, earliest galaxies.

We may think of observers wishing to resolve individual stars in the first galaxies ever formed. These galaxies now known to have been quite small might contain perhaps a hundred million stars. The redshifted ultraviolet radiation from some of these stars might already be discernible with instrumentation expected to be available soon. Instrumental angular resolving power improvements by a further factor of 10, over and above current instrumental capabilities, could be relatively straightforward to achieve.

5.13   Gravitational Microlensing

Gravitational lensing, usually by individual stars and potentially by substellar bodies, leads to distortions termed *microlensing*, readily observed within our own Galaxy.

In the mid-1990s several teams of astronomers undertook an ambitious search for massive astronomical bodies that might or might not exist. This was no idle venture: Newton's laws of gravitational attraction tell us that, in equilibrium, the combined kinetic energies of all the stars orbiting the center of a galaxy must be bound to the galaxy by a (negative) potential energy due to the combined mass of this selfsame assembly of stars.

A decades-long puzzle of astrophysics had been, and continues to be, that a galaxy's kinetic energy of rotation always appears to be far higher than the potential energy accounted for by the gravitational attraction of the galaxy's observed stars. This has led to the suggestion that some form of *dark matter* – so called because it fails to emit observable radiation – might account for the required potential energy to keep the galaxy's stars gravitationally bound.

Dark matter could exist in many forms, none of them familiar. During the 1990s a popular guess was that the dark matter might be concentrated in large compact chunks of ordinary matter in the spherical halo surrounding the central core of many galaxies, including ours. These massive compact objects possibly could be stellar black holes or cold, Jupiter-sized planets, or any compact bodies that do not radiate electromagnetic energy effectively and therefore are not readily discerned.

In 1986, Bohdan Paczyński, a highly gifted Polish astrophysicist visiting Princeton University in the United States, had chosen to remain in the United States following increased political repression in Poland. He became intrigued by the work of Vera Rubin and Kent Ford, who had discovered and documented the kinetic energy discrepancy in galaxies suggesting the possible existence of dark matter. The dynamics of the stars studied by Rubin and Ford suggested that the dark matter responsible for the observed stellar motions should be forming a spherical "halo" around the Galactic center. Paczyński proposed that the presence of a population of postulated *massive compact halo objects* might potentially be discovered through their gravitational deflection of starlight.[33]

If the Galaxy's halo contained such massive dark objects their motion transverse to the line of sight to a more distant galaxy could modulate – gravitationally lens – the light of individual stars in that galaxy. As seen by an observer at Earth, the light from the remote star would first appear to brighten as the dark object's trajectory passed close to the star's line of sight and focused the star's radiation. The star would later appear to dim as the foreground mass moved out of the line of sight.

Paczyński predicted that the light from stars in the Large Magellanic Cloud, LMC, a satellite galaxy to our Milky Way, might exhibit this lensing effect. If compact dark matter masses existed in the Galaxy's halo, the brightening and dimming of LMC stars should be observable. Of course, ordinary stars in the halo could produce similar lensing effects, but these stars could be readily identified by the visible light they emit, and would not be mistaken for dark matter.

Although Paczyński had correctly predicted that only one halo object in a million would, at any given time, be in the right position to gravitationally lens an LMC star, a number of astronomers felt that even this rare occurrence might be verified provided a project could be launched that potentially monitored hundreds of millions of these objects, including the myriad ordinary stars populating the Galactic halo.

Three major surveys were launched. One of these, initiated by Paczyński himself, was the *Optical Gravitational Lensing Experiment, OGLE*, led by Andrzej Udalski of Warsaw University Observatory.   In the United States, Charles R. Alcock began to assemble a team of like-minded colleagues to observationally test Paczyński's prediction in a survey called the *Massive Compact Halo Object, MACHO, Project*.[34] And a French-led survey, *Expérience de Recherche d'Objets Sombres, EROS*, was undertaken at the European Southern Observatory at La Silla, Chile. These surveys worked in friendly competition, reaching consensus on many fronts while puzzled by a few differences. The OGLE consortium conducted a number of surveys to look for potential transits – occultations of a remote star by a foreground object. Their survey would include a search involving more than a hundred thousand stars.

Aside from occultations by dark matter, other forms of variability also were observed. One exceptionally well-documented identification was that for the source OGLE-TR-56b.[35] There, a Sun-like parent star of mass $M = 1.04M_\odot$ is orbited by a planet with mass $0.9M_{Jupiter}$ and a period of only 1.2 days, meaning that its orbital semi-major axis is only $a = 0.023$ AU. This close to the star, the planet's surface temperature must be as high as $\sim$1900 K.[36]

The MACHO consortium's study of the microlensing of Large Magellanic Cloud (LMC) stars by compact masses among foreground Galactic halo stars comprised 9×10^9 individual photometric measurements. In one of their early papers, the team noted that, by 1997, their compilation of data had become "the largest survey of astronomical variability in history."[37]

They pointed out that, because their survey had been so comprehensive, they had not been able to merely disregard known types of variable stars, whose variability might otherwise have been mistaken for the brightening of a gravitationally lensed star. They had actually been forced to look at any observed variability regardless of whether it emanated from sources already known; it thus had included many types of variabilities never previously registered.

The team's criteria for announcing the observation of a gravitational lensing event thus were exceptionally thorough. They wrote they had needed to "learn how to perform event selection from the data; it was not possible to develop meaningful selection criteria independent of the data."

Members of the EROS-2 survey, by 2007, compared the results of their many independent efforts to conclude, "In the mass range from 0.6×10^{-7} to $15M_\odot$ [dark matter objects] were ruled out as the primary occupants of the Milky Way Halo."[38]

This conclusion differed somewhat from that of the MACHO team, which could not rule out that some objects in the mass range of $0.5M_\odot$ might account for a small contribution by massive compact objects. Nevertheless, a major conclusion was unambiguous: Massive compact halo objects did not appear to constitute a major component of the Galactic halo, at least not among objects that were in the cited stellar, brown dwarf, or planetary-mass ranges. Some other origin for the apparent existence of dark matter would need to be found.

To date no satisfactory model of dark matter has been identified![39]

The difficulties encountered in all of these observations may be appreciated in terms of the series of regions examined in Figure 5.8. They illustrate some of the complexities that needed to be taken into account to correctly interpret such observations.

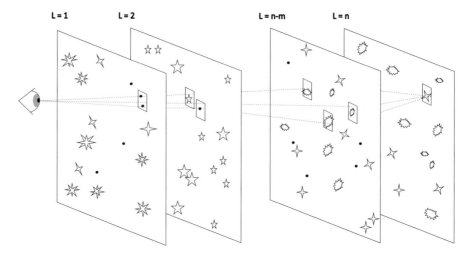

Fig. 5.8 Multiple microlensing by stars in different layers. At great distances from us, gravitational deflection or lensing by myriad stars in intervening galaxies deflect and distort the light from some of the earliest-formed stars and quasars, often in coherent ways displacing the locations and stretching the forms of the lensed background sources. Nearer by, individual stars in intervening galaxies microlense more distant stars. Within the Galaxy, foreground stars and even isolated planets can magnify more distant sources or deflect the trajectories of photons and all other radiations or particles passing by.

5.14   Quantitative Limits from Gravitational Microlensing

Gravitational microlensing by Galactic foreground stars generally is significant only when the angular separation θ of a more remote source from the foreground lensing star diminishes to less than the radius of that star's Einstein radius, θ_E, the radius of the ring of light that Einstein had predicted would surround a foreground star focusing the light from a remote, precisely co-aligned background star. The Einstein angular radius is

$$\theta_E = \left(\frac{4GMD_S}{c^2 D^* D_{SO}} \right)^{1/2} \leq 3 \left(\frac{M}{M_\odot} \right)^{1/2} \left(\frac{\text{kpc}}{D^*} \right)^{1/2} \text{ milli-arcseconds,} \qquad (5.7)$$

where the symbols denote the same quantities as in equation (5.3).[40]

The mild inequality at the right of expression (5.7) arises because the ratio D_S/D_{SO} is always less than 1, but usually a significant fraction of 1. For a gravitational lens of mass of $M = 10M_\odot$, at a distance 10 kpc, θ_E is $\sim$ 3 milli-arcseconds (mas). For $\theta_E/\theta \gg 1$, the amplification of the background source rises to $A_s(\theta) \sim (\theta_E/\theta)$.

Along sight lines toward the Galactic Center observed by the MACHO Project in its first year of operations, the amplifications A of 45 microlensing events of a background star by a foreground star ranged from typical values of $A \geq 2$ up to $A \sim 70$ in one highly lensed object. Typical durations of lensing were of the order of 15–50 days, with a few lenses prevailing for more than 150 days.[41]

The selection criteria used by the MACHO team in reporting these first-year results were considerably refined over time, leading to a somewhat higher number of clearly defined lensing events by the time the Large Magellanic Cloud observations were undertaken.

By 2001, the MACHO team excluded, at a 95% confidence level, the existence of dark matter objects in the 0.3 to 30.0$M_\odot$ mass range from contributing a mass of more than $4 \times 10^{11} M_\odot$ to the Galactic halo. Combined with earlier results, this meant that objects with masses under 30$M_\odot$ could not make up the entire dark matter halo if the halo had this assumed total mass. Masses lower than 10$M_\odot$ contributed less than 40% of such a hefty dark matter component.[42]

One significant limitation to both the Galactic halo and the Large Magellanic Cloud surveys was "crowding" – the automatic rejection of any events that were close to regions where stars along the line of sight to a prospective gravitational lens appeared unresolvable from the lens. This is worth mentioning because a limited range of improved microlensing studies could be achieved through increased angular resolving power able to disentangle crowded regions.

The Large Magellanic Cloud studies indicated that the gravitational mass of the Galaxy's halo can be accounted for by the total stellar mass of potential lensing bodies the MACHO survey identified. The most likely mass for typical lensing stars was in the range of ~0.05 to $1M_\odot$. Bodies with masses in the range ~10^{-4} to $3 \times 10^{-2}M_\odot$ contributed less than a total of $10^{11}M_\odot$ to the overall gravitational Galactic halo mass of ~$4 \times 10^{11}M_\odot$ estimated to lie within 50 kpc from the Galaxy's center.

5.15   Complexities of Gravitational Lensing

We may consider gravitational lensing as taking place in successive layers of deflectors at different distances from us. Figure 5.8 presents this in a simplified schema. Actual complexities require simulations through computational models. Figure 5.8 sketches gravitational deflection of radiation by galaxies, stars, or even masses as low as those of planets freely floating along a line of sight to a more remote source within the Galaxy. For simplicity, we may consider these trajectory-deflecting masses as confined to a number of hypothetical surfaces at different distances from the observer.

Consider light emitted isotropically from a star in the layer labeled L = n. Some of the rays are gravitationally deflected by three distinct galaxies that happen to all lie in some layer L = (n − m). Two of these deflected rays pass nearby a planet, indicated by a dot in layer L = 2. A third ray is deflected on passage by a star in the same layer. From there on, all three rays pass close by two planets in layer L = 1 before reaching the observer.

The three trajectories followed by the light may have been emitted by the original star at epochs separated by hundreds of days. The deflecting bodies in the different layers are likely to be continually in motion. The observer will be uncertain about whether the arriving radiation was emitted by a single star, or many, nor will the location or locations from which the radiation originated be clear.

The situation indeed is catastrophic!

A paper by S. S. Tie and Christopher S. Kochanek at the Ohio State University in Columbus, Ohio, clarifies many of the complexities that universal lensing imposes on observations of the remote Universe.

The two authors' primary concerns were often-assumed oversimplified expectations about the structure of foreground galaxies lensing variable background sources such as supernovae or quasars. They wrote:[43]

> The time delays of gravitationally lensed quasars are generally believed to be unique numbers whose measurement is limited only by the quality of the light curves and the models of the contaminating contribution of gravitational microlensing to the light curves.

> This belief is incorrect – gravitational microlensing also produces changes in the actual time delays on the [approximate number of] day(s) of light-crossing time scale of the emission region....
> Microlensing changes time delays on the scale of the light crossing time of the [quasar's] accretion disk, which has a typical scale of light days. These microlensing induced time delays will then slowly change as the accretion disk moves relative to the stars doing the microlensing.

To illustrate this point, Tie and Kochanek concentrated on two time delays normally not incorporated, for example, in evaluations of the Hubble constants based on Refsdal's equation (5.6).

The first of these delays is a term of the form $D_S \phi_0(\theta)$ that needs to be added to the time delays already incorporated in equation (5.5). Here D_S, the distance from the deflector to the source, has to take into account any anticipated tilt angle $\phi_0(\theta)$ of the lensing galaxy to the line of sight from the observer. This is needed because different portions of the lens lie at correspondingly different distances from the observer. A second added delay addresses a similar concern about the difference in distance of various points within the source – the quasar or remote galaxy observed – relative to the lensing galaxy. It takes the form $[(1 + z_s)\Delta z]/c$, where Δz is the proper redshift separation of the two sources contributing to an image of the source, and the term in parentheses is the time-dilated source redshift.

Figure 5.9 indicates the computer-modeled point to point intensity fluctuations Tie and Kochanek found for regions of a quasar at considerable distances from its central black hole lensed as well as microlensed by a foreground galaxy dominated by a randomly distributed 20% fractional mass of $0.3 M_\odot$ stars. The intricacy of the lensing patterns is striking as are the brief relative delay times computed. The computations do not include possible motions of the microlensing stars relative to one another in the lensing galaxy. Confirmation of the significance of extragalactic microlensing emerges also in a paper published slightly later by Jose M. Diego et al., exhibiting computer-generated images similar to those of Figure 5.9.[44]

If we take the circle highlighted by an arrow at the upper left of Figure 5.9 to indicate the angular resolving power at which the spatial noise due to microlensing by individual stars was adequately simulated, then a spatial domain of order $\sim 4 \times 10^{15}$ cm in the lensing galaxy, say at a nominal distance of 1 Gpc $\equiv 3 \times 10^{27}$ cm, equivalent to an angular resolution of $\sim$0.3 micro-arcseconds, would provide essentially all the observational data reliably available to map the quasar seen through the foreground galaxy.

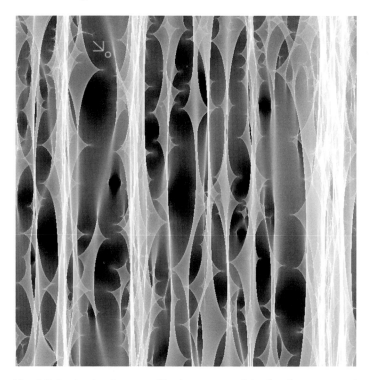

Fig. 5.9 A microlensing magnification pattern of the four image quasar lens RXJ 1131-1231, a quasar of mass $(1.3 \pm 0.3) \times 10^8 M_\odot$, at redshift $z_s = 0.658$, whose lensed image is observed split into four separate images, similar to the four lensed images observed for Supernova Refsdal shown in Figure 5.7. The magnification increases from darker/de-magnified to brighter/magnified regions. The map simulates source intensities measured by an observer at redshift $z_0 = 0$ due to amplification in an intervening lensing galaxy at redshift $z_\ell = 0.295$ with a stellar mass fraction 0.2 consisting solely of $0.3 M_\odot$ stars, each star independently acting as a microlens. The quasar accretion disk has a radius $R_0 = 7.34 \times 10^{14}$ cm, amounting to a light travel radius of 0.28 days. The map dimensions are 8192×8192 pixels covering a region with outer scale 5.02×10^{17} cm and pixel size 6.12×10^{13} cm. The circle pointed at by an arrow in the upper left corner shows the expected size of the quasar accretion disk for this system, assumed to be many times the pixel scale. (From "Microlensing Makes Lensed Quasar Time Delays Significantly Time Variable," S. S. Tie & C. S. Kochanek, *Monthly Notices of the Royal Astronomical Society* 473, 80, 2018, with permissions of the authors and Oxford University Press.)

Instruments available today are already approaching such capabilities. At the optical/near-infrared wavelengths involved, the highest angular resolution achieved to date has been $\sim$10 micro-arcseconds, $R = 5 \times 10^{-11}$, obtained with the "Phase Referencing Optical Interferometry for the Very Large Telescope Interferometer GRAVITY," at the European Southern Observatory.[45]

For the particular purpose of fully exploring the gravitational lensing of remote galaxies, our current capabilities already are within a factor of ∼30 of the resolving power that would fully resolve all the indicated foreground noise due to Tie and Kochanek's simulated microlensing. For stars moving at a lensing galaxy's rotational speed, ∼300 km/s, individual stars would move some ∼10^{15} cm per year. Over a time span of 4 years, stars would traverse distances roughly comparable to the separation of one of Tie and Kochanek's domains from another. So the foreground galaxy's microlensing would significantly change on such time scales further complicating the deciphering of details of a background galaxy's or quasar's structure.

Specifically, for a lensing star of Solar mass $M_\odot$ at a distance D^* of 1 Gpc from the observer and in transverse motion to the line of sight at $v \sim 300$ km/s, or one-thousandth the speed of light, c, the elapsed time for a fully resolved occultation is of order $t_e \sim 1.4 \times 10^9$ seconds, or roughly 40 years:

$$t_e \sim 40 \left(\frac{c/10^3}{v} \right) \left[\left(\frac{M}{M_\odot} \right) \left(\frac{D^*}{1\text{Gpc}} \right) \right]^{1/2} \text{ years.} \tag{5.8}$$

An observer will detect significant blinking from the galaxy's innumerable stars randomly turning on and off with individual periods of order tens of years.

5.16   Weak Lensing by Dark Matter

Radiation reaching us from great distances and early epochs passes through a gravitationally undulating extragalactic terrain, pulling it back and forth, potentially splitting its wavefront into parts arriving downstream from different directions with distinct time delays and redshifts.

The gravitational lensing originally depicted by Zwicky and Refsdal, which occasionally provides benefits of increased angular resolving power, is generally referred to, today, as *strong lensing*. More often images of remote galaxies are subjected only to *weak lensing*, which typically magnifies the radiation received from these sources while also distorting their shapes by amounts of order 1%. Effects this small can only be determined statistically through correlations observed in the shapes of galaxies lying within the celestial field of view.

The number of galaxies in the Universe is of order 10^{11}. Each galaxy or cluster of galaxies partially magnifies and partially distorts the shapes of others at greater distances. The magnitude of this effect can be statistically estimated.

Just as in strong lensing, weak lensing magnifies the image of regions directly beyond a foreground galaxy or cluster. Magnification spreads the apparent separation among background galaxies, while also making the fainter galaxies among them visible. A *tidal* drag exerted by massive lateral motions within the foreground lens can also *shear* the appearance of the background scene.[46]

The long axes of elliptic galaxies are expected to be randomly oriented in the plane of the celestial sphere. The shear due to weak lensing distorts these orientations, making them appear correlated and providing additional information on the foreground lens.

Galactic stars and interstellar matter are embedded in an ambient halo of dark matter. The mass of this dark matter in any foreground galaxy can be assessed by mapping the distribution of galaxies lying beyond it. Usually, this will reveal deviations among these background sources from a purely random arrangement. By attributing these deviations to weak lensing by the foreground galaxy, it becomes possible to assess that galaxy's total mass. Eliminating the mass of the galaxy's stars and gas from this total, then indicates the fractional mass of dark matter present.

The several galaxy clusters comprising the components of Abell 1758 at redshift $z = 0.278$ shown in Figure 5.10 act as gravitational lenses distorting the shapes of more remote galaxies in this fashion. The distortions are tiny and only permit rather crude, but currently the most direct, measures of the masses of the various components.[47] The northern components, A1758NW and A1758NE, respectively exhibit lensing masses of $\sim 8 \times 10^{14}$ and $\sim 5.5 \times 10^{14} M_\odot$ with uncertainties of order $\pm 25\%$. The southern component has a mass of $\sim 5 \times 10^{14} M_\odot$ with similar uncertainties.

Hydrodynamic estimates of these galaxies' masses, obtained from X-ray and radio spectra of the extended X-ray emitting regions identified in Figure 5.10, lead to similar derived masses, and comparable estimated errors, but depend on assumptions on the adopted hydrostatic models. Despite their own uncertainties, masses derived from weak lensing may therefore appear more directly obtained.[48]

A major study undertaken by David H. Weinberg and coworkers in 2013 pointed out that the images of typical galaxies deviate by a systematic elongation of $\epsilon \sim 20\%$ from appearing perfectly round.[49] Accordingly, if we wish to determine whether the galaxies in a particular patch of the sky are systematically elongated along a particular direction by, say, a further $\Delta\epsilon = 1\%$ through gravitational lensing, we would need to overcome the statistically expected random deviation of any set of these galaxies, which declines in proportion to ϵ^2.

Statistically, a patch containing 400 randomly oriented galaxies would exhibit a deviation from circular symmetry amounting to 1%. If we wish to measure the actual elongation produced by gravitational lensing to an accuracy of 1%, over and above this random 1% elongation, we would need to examine a patch of sky another factor of 10^4 larger, containing $\geq 4 \times 10^6$ galaxies.

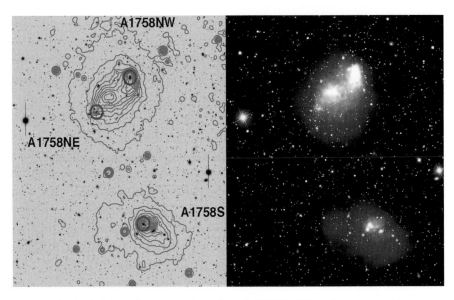

Fig. 5.10 Galaxy clusters catalogued as Abell 1758. Although we are unable to
detect dark matter directly because it neither absorbs nor emits radiation as far as
we know, we do detect its weak lensing – its gravitational bending – of radiation
emitted by more remote sources. The diffuse outer realms and bright central regions
of pairs of merging clusters of galaxies, respectively displayed at top and bottom in
both the photograph at right and the analytic display of surface brightness at left,
comprise regions of X-ray emitting hot gas embedded in a far more massive halo of
dark matter enveloping both the hot gas and stars in the colliding galaxies. The
more compact spots emitting radiation throughout each image, as well as within
the X-ray emitting regions, are due to stars spread throughout this part of the sky.
Regions of identical surface brightness are indicated by *isophotes*, the roughly
concentric closed curves in the diagram on the left, where the three regions named
in the text are also labeled. (The X-ray map on the left is from "The Merger History
of the Complex Cluster Abell 1758: A Combined Weak Lensing and Spectroscopic
View," R. Monteiro-Oliveira, et al., *Monthly Notices of the Royal Astronomical Society* 466,
2614–32, 2017, with permission of the lead author and Oxford University Press. The
X-ray image at right, showing diffuse intensely heated gas in the clusters superposed
on an optical image of the region was obtained with the Chandra X-ray Observatory,
reproduced with permission of NASA/CXC/SAO/G.Schellenberger, et al.; optical data
are from the Sloan Digital Sky Survey.)

As the paper by Weinberg and his co-authors pointed out, with angular
resolving powers and sensitivities available in the visible domain at current
ground-based telescopes, the celestial sphere exhibits ∼15 resolved galaxies per
square arcminute. So, the induced elongation due to gravitational lensing would
require examination of a patch of sky amounting to ∼75 square degrees. This
is a sizeable fraction of the sky, whose entire extent covers just ∼40,000 square

degrees, or ~530 non-overlapping areas 75 square degrees in size, each often appearing to have a quite different configuration of foreground galaxies. For observations conducted from space, the number of galaxies resolved per unit area would increase by another factor of order ~3, and the number of galaxies observable over the entire sky would reach ~10^{10}, of the order of 10% of all the galaxies the Universe contains. The angular resolving power required for this is $R_\alpha \sim 10^6$. Plans for the ground-based *Vera C. Rubin Observatory*, currently underway, and two planned space missions, the *Euclid* dark energy space mission led by the European Space Agency, ESA, and the *Wide-Field Infrared Survey Telescope, WFIRST*, now renamed the *Nancy Grace Roman Space Telescope*, to be led by NASA. Each would plan to observe ~10^9 galaxies, but to succeed would have to reduce systematic observing errors by an order of magnitude over current (2019) capabilities, which the authors consider "a daunting task…. There are unfortunately many causes of these systematic errors, and most of the effort of the weak lensing community has been devoted to defeating them."[50]

Weinberg and his co-authors asked themselves whether there is a future beyond this stage, "both in terms of science motivation and technical capability? It seems unlikely that there would be a follow-on experiment that consists of simply a super-size *LSST* (now renamed the *Vera C. Rubin Observatory*), *Euclid*, or *WFIRST*, particularly because these experiments will come within a factor of a few of the *cosmic variance limit* at several tens of galaxies per square arc minute." They surmised, "A more distant future would have to involve new technology and a new science case not subject to the usual limitations." They did not amplify on just what they meant by this, but suggest looking "for lensing by primordial gravitational waves … using highly-redshifted 21 cm radiation as the source." But they admit "We have now entered the speculative realm of … science and technology, where our ability to forecast the future is of limited reliability. We conclude our discussion of weak lensing here."[51]

A statistical study of this *weak lensing* effect due to the combined gravitational tug of myriad individual galaxies acting independently has by now been undertaken by H. Hildebrandt and coworkers, with current observational capabilities resolving angles of order 0.35–0.7 arcseconds, amounting to resolving powers of ~10^6. Their findings to date are in line with the estimates of Weinberg and coworkers.[52]

As the study by Weinberg's group indicated, statistically weak lensing provides a useful tool, as long as solely statistical information is needed to investigate cosmological properties. For the study of individual galaxies at high redshifts, however, weak lensing becomes a limitation that hinders reliable assessment of inherent shape and size.

It thus appears that, at least for observations on the first galaxies ever formed, an angular resolution capability of 10^{-3} arcseconds or angular resolving powers

$\sim 10^8$ – two orders of magnitude higher than the limits imposed by gravitational lensing – could well suffice to extract as much information as practical in overcoming unpredictable naturally occurring weak lensing distortions. As we will see in Figure 6.8, for some types of observations this level of performance has already been achieved at the highest angular resolving powers attained in dedicated studies at optical and radio frequencies.

The largest survey of galaxy–galaxy lensing, to date, in which as many as 660,000 galaxies in a celestial region covering 1321 square degrees were involved, indicated in its first year of investigations that galaxy clustering and weak lensing by them provide significant information about the structure and evolution of the Universe over extensive periods. The results of the survey are consistent with many of the conclusions of the Planck Survey described in the next section.[53; 54]

5.17   Gravitational Deflection of the Microwave Background

The cosmic microwave background radiation provides us a direct view of the Universe when it first became transparent to radiation, roughly 375,000 years after the start of time. Ever since, these photons have journeyed through the Universe to reach us. Along the way their trajectories were distorted by the gravitational pull of accumulating concentrations of matter, as indicated in Figure 5.11. This lensing effect can still be traced because it systematically remapped primordial fluctuations, so that radiation now observed in a celestial direction $\hat{n}$ actually represents the unlensed anisotropy from a primordial direction $\hat{n} - \nabla\phi(\hat{n})$. Here $\phi(\hat{n})$ is a *lensing potential* defined by the integrated effect of the mass concentrations along the paths traversed by the observed radiation reaching Earth.[55] The lensing potential thus is an integrated measure of the mass distributions the radiation has encountered since it first decoupled from matter.

When mapped over the celestial sphere, the *lensing potential* $C_L^{\phi\phi}$ undulates from one angular location ϕ to another, exhibiting modulations on angular scales $\phi \sim 2\pi/L$ radians or $\sim 180/L$ degrees across the sky, as shown in Figure 5.12. The quantity $[L(L+1)]^2 C_L^{\phi\phi}/2\pi$ on the figure's ordinate is the flux per steradian expected within a solid angle whose span is defined by the angular frequency L on the abscissa. A small value of $L(L+1)$ thus comprises a large solid angle. This lensing of the microwave background radiation exhibits a number of distinct features.

The remarkably small but capable Planck telescope had a beam width respectively of 7 and 5 arcminutes in its highest frequency bands at 143 and 217 GHz. On small angular scales, where $L \geq 1000$, angular variations shown in Figure 5.12 fall within the noise ranges indicated by the rectangular boxes. Over larger angles and lower L values, angular displacements of the radiation

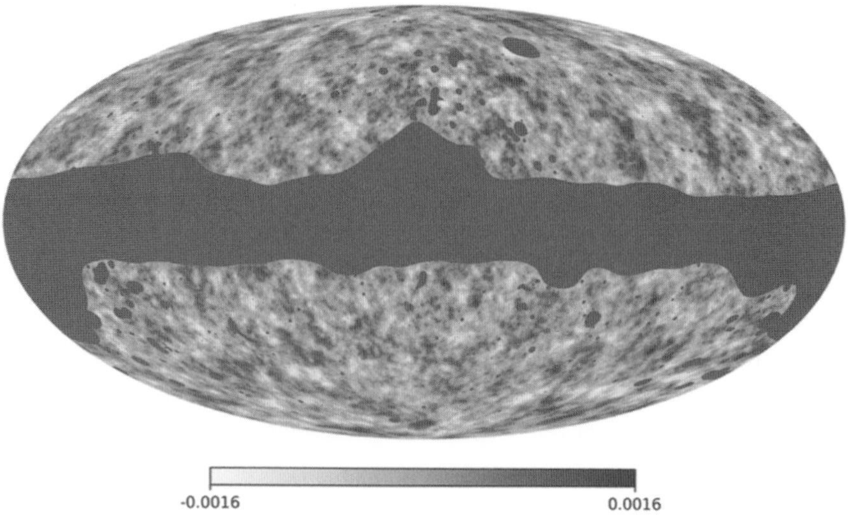

-0.0016   0.0016

Fig. 5.11 Weak Lensing by Myriad Foreground Galaxies Distorts the Map of Realms Emitting the Microwave Background Radiation. The root mean square displacement of points on the map is $\langle d^2 \rangle^{1/2} = 2.46 \pm 0.04$ arcminutes, about 7×10^{-4} radians, roughly a 25th of a degree on the celestial sphere. (From "Planck Final Results VIII," *Astronomy & Astrophysics*, in press, 2020, with permission of the Planck Collaboration.)

become distinctly higher, peaking in the range around $L \sim 35$, and an angular scale around $\sim 5°$.

Integrated over all angles, the typical displacement of radiation from its direction of origin amounts to $\langle d^2 \rangle^{1/2} = 2.46 \pm 0.04$ arcminutes, about a 25th of a degree on the celestial sphere. These displacements are confined to features whose angular scales exceed ~ 20 arcminutes. Over smaller separations, ≤ 10 arcminutes, the relative displacements between radiating regions are far lower, yielding such detailed background maps as shown in Figure 2.7.

To a radio-astronomy community that is now building instruments such as the Event Horizon interferometric array, capable of measuring relative displacements of the order of 25 micro-arcseconds in nearby galaxies, these numbers may appear staggeringly high. They also are an order of magnitude higher than the gravitational arcsecond displacements of remote supernovae or quasars lensed by foreground galaxies. However, these large-scale shifts at large angular scales are consistent with data on the appearance of remote galaxies. Weak lensing distorting the true shapes of galaxies also distorts our maps of the microwave background. The same distortions apply to electromagnetic radiation at all frequencies, because gravitational lensing is frequency independent.

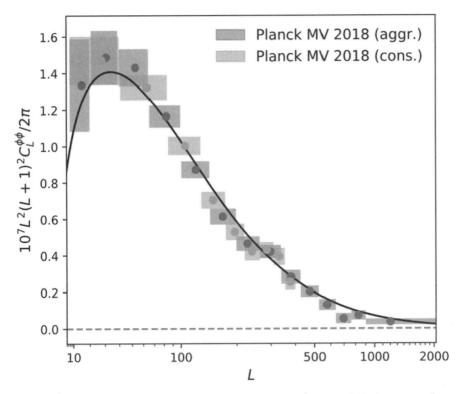

Fig. 5.12 Weak lensing by foreground galaxies induces mutual displacements of spatial regions contributing to the microwave background radiation. This effect is most pronounced among regions separated by angles $180°/L \sim 5°$, as witnessed by the peak around $L \sim 35$. Although the root mean square displacement of points on the map is about a 25th of a degree on the celestial sphere, the curve indicates that for widely separated regions the angular displacement of one region relative to others is several times higher. On the smallest angular scales, below 10 arcminutes, the displacement of one region relative to others is comparable to or less than other uncertainties. Data marked aggressive (aggr.) cover a wider range of radiation frequency components L than those labeled conservative (cons.). Dots indicate centers of their respective data bins. (From "Planck Final Results VIII," *Astronomy & Astrophysics*, in press, 2020, with permission of the Planck Collaboration.)

5.18   Summary

Gravitational lensing was initially welcomed as a way of occasionally providing the means to view some of the most remote or some of the most compact sources that telescopes we could reasonably afford to construct would be unable to reveal. Later, weak lensing techniques enabled us to determine the amount of dark matter present in foreground galaxies by means of the distortions of more remote galaxies the dark matter was lensing. Both were unique techniques that gravitational lensing enabled.

A major message this chapter attempted to convey is how strongly gravity and geometry are intertwined. We cannot tell precisely where a messenger originated unless we know the trajectory it followed. But we cannot determine that trajectory with precision without adequate knowledge of how mass is distributed along and surrounding its trajectory. Nor can we precisely determine the time delays a set of messengers may suffer in following that route.

The Planck project demonstrated how such uncertainties affect the accuracy with which we can define the angular distribution of radiation arriving from the cosmic microwave background. The tools the mission provided seemed fully adequate to assess these displacements.

The work of David H. Weinberg and his co-investigators suggests that weak lensing by foreground galaxies would already constrain realistic depictions of mass distributions across remote cosmic regions even at resolving powers no better than 10^{-3} arcseconds.

Tie and Kochanek remarked that the spatial noise due to microlensing by individual stars in a gravitationally lensing galaxy at a distance of 1 Gpc, mapping more remote sources, would exhibit structure resolved on scales of 10^{-12} radians or 0.2 microseconds of arc – only a factor of ∼50 better than currently provided by the GRAVITY instrument. As expected, the spatial noise would be accompanied by noise in the lensing time delay.

Nearer to home, the MACHO team found that the 9×10^9 individual photometric measurements they had undertaken in observations of stars in the Large Magellanic Cloud had helped them detect several new sources of stellar variability unrelated to gravitational lensing by massive objects. Their first year of observations of the Galactic Center also revealed some 45 microlensing events of background stars lensed by foreground objects, requiring alignments along our sight line at levels of order 3 milliseconds of arc, well within GRAVITY's resolving powers. Some of these lensing events lasted more than 100 days, providing adequate time for careful measurements if deemed interesting.

By 2018 ground-based observations had also revealed 50 lensings of background stars by foreground planets. These observations generally lasted a few hours and could more readily have been carried out from space. Because the number of planets detached from their stellar systems and now freely floating in space could equal or exceed the number of planets still orbiting their parent stars, significantly more of these events should soon be discovered. Presumably, more sensitive observations would ultimately also reveal even higher numbers of stars faintly lensed by freely floating interstellar asteroids or comets.

Such randomly occurring events may ultimately become sources of unavoidable transient noise in sensitive extragalactic, or stellar, or planetary studies. Whether improved instrumental capabilities will then help or defy attempts to obtain progressively more informative data should become clearer over the next few years. Chances are we may soon reach limits where improved instrumental performance will no longer help because the sources involved emit radiation at levels well below random photon noise at the wavelengths involved.

No matter how deeply we search, high-angular-resolution studies will remain essential as long as all observations are carried out locally without embarking on space travel. At the 10 micro-arcsecond performance level of the GRAVITY instrument, an efficient coronagraph might succeed in singling out some of the more massive planets within a 30 pc range. At distances of order 3 pc, or 9×10^{18} cm, such an instrumental configuration could partially resolve planetary patches roughly 5×10^8 cm or 5000 km across – sufficient to resolve an Earth-like planet's diameter, and potentially a terrestrial-size ocean, though not enough to resolve cities like New York or comparable macroscopic signs of potentially organized life.

In the millimeter wavelength range, where the ratio of energy received from star and planet would be greatly reduced, the angular resolution of the Event Horizon Telescope, an interferometric array consisting of eight or more radio telescopes spanning Earth's globe, might play a similar role with its 25 micro-arcsecond resolving power at a wavelength of 1.3 mm.[56]

High angular resolving power, nevertheless, may not be the simplest way to resolve surface features on nearby habitable planets. We already know that simpler ways can be offered by the rotation of planets, which successively bring different surface features into an observer's field of view in a repeated pattern which, under favorable conditions, might be reconstructed into surface maps. Spectroscopic analyses could then further resolve emissions from different latitudes velocity-shifted by distinct fractions of a planet's rotational velocity as a function of distance from the planet's equator.

Stray radiation from a planet's parent star may be reduced by well-designed coronagraphs. Gathering a sufficient number of photons to obtain reliable observations, however, may still require the construction of large telescopes, as we will find in Section 6.15. Such sequences of repeated observations may replace high-resolution angular requirements by potentially less demanding timing studies, though these may entail difficulties of their own. The optimum way to map a planet's surface may thus depend on a range of circumstances.

Although angular and time resolution capabilities may be degraded through gravitational lensing and some of its catastrophic consequences, spectral resolution is not affected. This is because gravitational lensing affects all types of radiations and all species of cosmic-ray particles identically, regardless of their energy. Images and time sequences may therefore be distorted by lensing, but their spectra will remain unchanged.

The extent of these different sources of noise should become clearer over the next few decades, and will likely reveal just where improvements in angular or temporal resolving powers at levels exceeding those already available are no longer going to help even if we can afford to build larger aperture telescopes or interferometers to collect information more rapidly.

We may then have reached a point where further improvements in observational capabilities no longer appear worthwhile, on either astrophysical or economic grounds. Gravitational lensing, deflection, and time delay are so universal that they appear certain to dominate that final decision.

Notes

1 Lens-Like Action of a Star by the Deviation of Light in the Gravitational Field, A. Einstein, *Science* 84, 506–7, 1936

2 Nebulae as Gravitational Lenses, F. Zwicky, *Physical Review* 51, 290, 1937

3 On the Probability of Detecting Nebulae Which Act as Gravitational Lenses, F. Zwicky, *Physical Review* 51, 679, 1937

4 The Gravitational Lens Effect, S. Refsdal, *Monthly Notices of the Royal Astronomical Society* 128, 295–306, 1964

5 On the Possibility of Determining Hubble's Parameter and the Masses of Galaxies from the Gravitational Lens Effect, S. Refsdal, *Monthly Notices of the Royal Astronomical Society* 128, 307–10 , 1964

6 *Gravitational Lenses*, P. Schneider, J. Ehlers, & E. E. Falco, Springer Verlag, Berlin and New York, 1992

7 Wide-Field Infrared Survey Telescope – Astrophysics Focused Telescope Assets WFIRST-AFTA 2015, co-chaired by D. Spergel & N. Gehrels, *Report by the Science Definition Team (SDT) and WFIRST Study Office*, pp. 52–53 and 72, March 10, 2015

8 Ibid., Wide-Field Infrared Survey Telescope

9 Ibid., Wide-Field Infrared Survey Telescope, p. 46

10 A Neptune-Mass Free-Floating Planet Candidate Discovered by Microlensing Surveys, Przemek Mróz, et al., *Astronomical Journal* 155, 121, March 2018

11 Ibid., Wide-Field Infrared Survey Telescope, p. 46

12 Why Is Interstellar Object I1/2017 U1(Oumuamua) Rocky, Tumbling and Possibly Very Prolate? J. I. Katz, *Monthly Notices of the Royal Astronomical Society* 478, Issue 1, L95–98, 2018

13 Ibid., Wide-Field Infrared Survey Telescope, p. 46

14 UKIRT-2017-BLG-001Lb: A Giant Planet Detected through the Dust, Y. Shvartzvald, et al., *Astrophysical Journal Letters* 857, 18, April 10, 2018

15 Fourth Test of General Relativity, I. I. Shapiro, *Physical Review Letters* 13, 789–91, 1964

16 Fourth Test of General Relativity: Preliminary Results, I. I. Shapiro, et al., *Physical Review Letters* 20, 1265–69, 1968

17 0957+561 A, B: Twin Quasistellar Objects or Gravitational Lens? D. Walsh, R. F., Carswell, & R. J. Weymann, *Nature* 279, 381–84, 1979

18 Multiple-Mirror Telescope Observations of the Twin QSOs 0957+561 A, B, R. J. Weymann, et al., *Astrophysical Journal* 233, L43–46, 1979

19 A Cosmology Calculator for the World Wide Web, E. L. Wright, *Publications of the Astronomical Society of the Pacific* 118, Issue 850, 1711–15, 2006.

http://www.astro.ucla.edu/ wright/ACC.html

20 A Robust Determination of the Time Delay in 0957+561A,B and a Measurement of the Global Value of Hubble's Constant, T. Kundić, et al., *Astrophysical Journal* 482, 75, 1997

21 Keck Spectroscopy of the Gravitational Lens System PG 1115+080: Redshifts of the Lensing Galaxies, T. Kundić, et al., *Astronomical Journal* 114, 507–10, 1997

22 Analytic Time Delays and H_0 Estimates for Gravitational Lenses, H. J, Witt, S. Mao, & C. R. Keeton, *Astrophysical Journal* 544, 98–103, 2000

23 Improved Parameters and New Lensed Features for Q0957+561 from WFPC2 Imaging, G. Bernstein, P. Fischer, J. A. Tyson, & G. Rhee, *Astrophysical Journal* 483, L79, 1997

24 A New Formulation of Gravitational Lens Theory, Time-Delay, and Fermat's Principle, P. Schneider, *Astronomy & Astrophysics* 143, 413–20, 1985

25 Fermat's Principle, Caustics, and the Classification of Gravitational Lens Images, R. Blandford & R. Narayan, *Astrophysical Journal* 310, 568–82, 1986

26 Catastrophe Optics: Morphologies of Caustics and their Diffraction Patterns, M. V. Berry & C. Upstill, in *Progress in Optics, XVIII* edited by E. Wolf, North-Holland, 1980, pp. 257–346, see pp. 259–60

27 Over-constrained Gravitational Lens Models and the Hubble Constant, C. S. Kochanek, *Monthly Notices of the Royal Astronomical Society* 493, Issue 2, 1725–35, April 2020

28 Improved Constraints on the Gravitational Lens Q0957+561.II Strong Lensing, R. Fadely, et al., *Astrophysical Journal* 711, 246–67, March 1, 2010, see p. 247

29 Multiple Images of a Highly Magnified Supernova Formed by an Early-Type Cluster Galaxy Lens, P. L. Kelly, et al., *Science* 347, 1123–26, 2015

30 Understanding Caustic Crossings in Giant Arcs: Characteristic Scales, Event Rates, and Constraints on Compact Dark Matter, M. Oguri, et al., *Physical Review D* 97, 023518, 2018

31 http://www.astro.ucla.edu/~wright/ACC.html

32 *arXiv:1705.02345v1 [astro-ph.IM]* May 5, 2017

33 Gravitational Microlensing by the Galaxy's Halo, B. Paczyński, *Astrophysical Journal* 304, 1–4, 1986

34 The M. Project: 45 Candidate Microlensing Events from the First-Year Galactic Bulge Data, C. Alcock, et al., *Astrophysical Journal* 479, 119–46, 1997

35 Konacki, M., et al., *astro-ph/0301210*, 2003

36 Sasselov, D. D., *astro-ph 0303403*, 2003

37 The MACHO Project Large Magellanic Cloud Microlensing Results from the First Two Years and the Nature of the Galactic Dark Halo, C. Alcock, et al., *Astrophysical Journal* 486, 697–726, September 10, 1997

38 Limits on the Macho Content of the Galactic Halo from the EROS-2 Survey of the Magellanic Clouds, P. Tisserand, et al., *Astronomy & Astrophysics* 469, 387–404, 2007

39 Dark-Matter Hunt Comes up Empty, Elisabeth Gibney, *Nature* 551, 153–54, 2017

40 Bohdan Paczyński, Cosmic Dark Matter, and Gravitational Microlensing, C. Alcock, in *The Variable Universe: A Celebration of Bohdan Paczyński*, Astronomical Society of the Pacific Series, volume 403, edited by K. Z. Stanek, pp. 71–85, 2009

41 The Macho Project: 45 Candidate Microlensing Events from the First-Year Galactic Bulge Data, C. Alcock, et al., *Astrophysical Journal* 479, 119–46, 1997

42 MACHO Project Limits on Black Hole Dark Matter in the 1–30$M_\odot$ Range, C. Alcock, et al., *Astrophysical Journal* 550, L169–72, 2001

43 Microlensing Makes Lensed Quasar Time Delays Significantly Time Variable, S. S. Tie & C. S. Kochanek, *Monthly Notices of the Royal Astronomical Society* 473, 80, 2018

44 Dark Matter under the Microscope: Constraining Compact Dark Matter with Caustic Crossing Events, Jose M. Diego, et al., *Astrophysical Journal* 857, 25, 2018

45 First Light for GRAVITY: Phase Referencing Optical Interferometry for the Very Large Telescope Interferometer, GRAVITY Collaboration, *Astronomy & Astrophysics* 602, A94, 2017, see Figure 19

46 Ibid., Schneider, et al., 1992, p. 328

47 The Merger History of the Complex Cluster Abell 1758: Combined Weak Lensing and Spectroscopic View, R. Monteiro-Oliveira, *Monthly Notices of the Royal Astronomical Society* 466, 2614–32, 2017

48 Forming One of the Most Massive Objects in the Universe: The Quadruple Merger in Abell 1758, G. Schellenberger, et al., *Astrophysical Journal* 882, 59, 2019

49 Observational Probes of Cosmic Acceleration, D. H. Weinberg, et al., *Physics Reports* 530, 87–255, 2013, see in particular pp. 134 ff

50 Ibid., Weinberg, et al. Figure 17, p. 149

51 Ibid., Weinberg, et al., p. 166

52 KiDS-450: Cosmological Parameter Constraints from Tomographic Weak Gravitational Lensing, H. Hildebrandt, et al., *Monthly Notices of the Royal Astronomical Society* 465, 1454, 2017

53 Dark Energy Survey Year 1 Results: Galaxy-Galaxy Lensing, J. Prat, et al., *Physical Review D* 98, 042005, 2018

54 Dark Energy Survey Year 1 Results: Cosmological Constraints from Galaxy Clustering and Weak Lensing, T. M. C. Abbott, et al., *Physical Review D* 98, 043526, 2018

55 Planck 2015 Results XV. Gravitational Lensing, Planck Collaboration, *Astronomy & Astrophysics* 594, A15, 2016

56 First M87 Event Horizon Telescope Results: The Shadow of the Supermassive Black Hole, The Event Horizon Telescope Collaboration, *Astrophysical Journal Letters* 875, Letters 2019, April 10, a journal issue comprising a series of six Letters, L1–L6

Part III

Parameters Specifying Individual Messengers

6

The Ranges of Messenger Parameters

6.1 Three Searches for Planets Circling Other Stars: Why Three?

Peter van de Kamp

In 1938, Peter van de Kamp, a respected, Dutch–American astronomer born in Amsterdam in 1901, began a systematic study of the motions of Barnard's star, the nearest star to the Sun visible from the Northern Hemisphere.[1] Over the next 43 years van de Kamp photographed the position of the star with great precision some 1200 times, through his Sproul Observatory telescope in Swarthmore, Pennsylvania. Analysis of these data led him to conclude that the star exhibited a slight periodic displacement across the sky due to the gravitational tug of two orbiting planets. He calculated their respective masses to be 0.7 and 0.5 times that of Jupiter with orbital periods of roughly 12 and 20 years. Both planets moved in elliptical orbits inclined at about 11° to the line of sight. These were difficult measurements carried out at the very extreme capabilities of the Sproul Observatory.[2]

We now know that neither planet exists. A life's work dedicated to an important question had tragically failed and yielded nothing, partly perhaps because of an over-reliance on repetition of observations without equally dedicated attempts to seriously improve instrumental capabilities. Van de Kamp may have been misled by discrepancies introduced over the years through refurbishments of the Sproul telescope.

Work at the extreme edge of an instrument's capabilities tends to emphasize both random and systematic instrumental errors that can be misinterpreted. To be certain of a discovery, the observations on which it is based must stand out well above such potential errors, a task which the Sproul Observatory apparently could not fulfill. Today's most difficult observations tend to be carried out with instruments whose capabilities are deliberately improved whenever possible to avoid the fate van de Kamp's work encountered.[a]

Aleksander Wolszczan and Dale A. Frail

The first planets reliably detected outside the Solar System were discovered by two radio astronomers.[4; 5] The timing accuracy of these observations was an unprecedented 3 microseconds over a three-year span of observations.

In the winter of 1990, the Arecibo Observatory, the world's largest radio telescope at the time, was being repaired and could make observations only of sources that passed directly overhead within a limited angular field of view. This made available large amounts of observing time to programs that could actually benefit from these restrictions.

Arecibo was outfitted with fast-sampling pulsar equipment ideal for pulsar searches and timing observations. To determine whether or not a pulsar might be orbited by one or more planets, Aleksander Wolszczan and Dale A. Frail embarked on a program of systematically observing millisecond pulsars at high latitudes above the Galactic plane where timing irregularities due to patches of ionized gas occupying the plane would be minimized.

Early in February 1990, the two astronomers noticed that the pulsar PSR1257+12, a rapidly rotating neutron star emitting a sharp pulse with great regularity every 6.2 milliseconds, exhibited a slight periodic anomaly. The star appeared to be pulled back and forth along the line of sight, reaching a speed of 0.7 meters per second toward us – roughly the speed of a person walking – before slowing down and accelerating away until it reached the same maximum speed receding from us.

Wolszczan and Frail traced this cyclic motion to a superposition of two periodicities, and inferred that the neutron star must be orbited by two planets, one having a mass at least 2.8 times, the other 3.4 times the Earth's mass. The planets orbit the star in close-to-circular orbits with respective periods of 98.2 and 66.6 days, first tugging the star gravitationally along the line of sight toward us, and then propelling it back in the opposite direction.[6] Two years later, a further analysis showed Wolszczan the existence of a third planet of significantly lower mass orbiting the pulsar with a period of 25.3 days.[7]

As Wolszczan later wrote, he had benefitted from opportunities to try unconventional observations at the Arecibo Observatory during a rare period

---

[a] Recent observations suggest that Barnard's star might have a companion with a minimum mass roughly three times that of Earth in an orbit with a period of 233 days, clearly quite different from van de Kamp's findings.[3]

reserved for major repairs. But it had also involved "just a bit of pure old good luck."[8] Luck often does play a role in astonishing achievements, but Wolszczan's inspired idea of trying new ways of using available Arecibo instruments for a quite fantastic search, and his steadfastness in guiding his set of painstaking observations through to success exhibit some of the finest traits a scientist can possess.

Neutron stars, however, are rare, resulting from the gravitational collapse of evolved stars not sufficiently massive to explode as supernovae. Most astronomers were more interested in whether, and how frequently, the far-more-abundant, lower-mass main sequence stars like the Sun were orbited by planets. For this, a different technique seemed indicated.

Timing studies of large numbers of other pulsars, including those observed by Wolszczan and Frail, and two decades earlier by Hulse and Taylor, had abundantly confirmed that the gravitational tug of Jupiter as it orbits the Sun leads to a reflex motion of the Sun amounting to a maximal velocity of 13 meters per second. This is a reflex velocity expected also for many other low-mass stars orbited by massive planets.

The task facing an observer thus appeared to be the detection of the speed of a star as it alternately approached and receded from the Sun. This had to take into account the roughly 13-meter per second variation induced in the Sun's own velocity through space, mainly due to the orbital motion of Jupiter, but to a lesser extent also those of the other Solar System planets. These relative speeds of the Sun about the Solar System's center of mass – its barycenter – had to be given due weight.

Michel Mayor and Didier Queloz

A difficulty of such observations is that the cyclic motions of stars orbited by planets tend to have periodicities of the order of months or years, so that detection of planets through their reactions on the parent star they orbit generally requires investigations stable in their performance across periods lasting months or years. For a long time this was difficult to achieve.

At an October 6, 1995 workshop in Florence on "Cool Stars, Stellar Systems, and the Sun," Michel Mayor and Didier Queloz of the Geneva Observatory in Switzerland created a sensation when they reported the discovery of a Jupiter-mass companion to the star 51 Pegasi![9] 51 Pegasi closely resembles the Sun. The planet orbiting the star, which Mayor and Queloz had discovered through repeated observations of its parent star's changing line-of-sight velocity during the previous 18 months, was puzzling. It had a mass in the range of 0.5–2 Jupiter masses, but its orbital period was merely ~4 days and its orbital radius only ~0.05 astronomical units (AU), roughly one-twentieth the orbital radius of Earth's motion about the Sun. This small an orbital radius came as a surprise since planets up to then had been assumed to form and survive in stable orbits only if they resided at far greater distances from their parent stars.

The discovery of this Jupiter-mass planet hinged on the exquisitely high spectral resolution and highly stable spectrometric techniques that Mayor and Queloz had developed, permitting them to discern the small velocity changes exhibited by 51 Pegasi. This was optical spectroscopy at its very highest spectral resolving power, denoted by the small triangular symbol shown later in this chapter, in Figure 6.7.

Coincident with the efforts of Mayor and Queloz, Geoffrey W. Marcy and R. Paul Butler at the University of California in Berkeley had been employing a gaseous iodine cell whose sharp spectral lines were superposed on a star's spectrum to provide a precise wavelength standard. This permitted them to measure instantaneous velocities of bright stars with an accuracy of 3 meters per second on any given night, implying a spectral resolving power, $R \sim 10^8$.[10]

Following the October 6 announcement by Mayor and Queloz, it took Marcy and Butler only a few days to confirm the existence of the newly found planet, a finding further supported by astronomers working at four other observatories. On October 25, this group, by now numbering 15 astronomers, jointly issued an astronomical telegram through the International Astronomical Union alerting the scientific community to this momentous finding.[11] Since then, thousands of other stars orbited by planets have been discovered in our Solar System's vicinity, and many hundreds of astronomers have switched fields to study how these planetary systems may have formed and evolved.

Why Three Distinct Attempts?

The tools available to astronomers generally fall into three distinct classes.

A telescope may enable us to measure the relative positions of stars, planets, or also galaxies on the sky in terms of the angles separating them. The ability to conduct this measurement precisely is expressed as the *angular resolving power* of the telescope. Peter van de Kamp's attempts to measure a regularly changing angular separation between Barnard's star and its immediate neighbors on the celestial sphere were based on his assessments of his telescope's angular resolving power.

Alternatively, a highly stable apparatus may permit us to accurately keep track of a regularly timed signal emitted by an astronomical source. By the

1990s, highly stable clocks had become available at Arecibo to accurately measure patterns of temporal change in radio signals emitted by pulsars. The extreme regularity with which pulsars emitted their sequences of pulses initially surprised astronomers and soon were recognized as potentially providing new insights. Wolszczan and Frail recognized that a regular sequence of approach and recession by a pulsar in a planetary system might yield observable period-icities in the pulsar's pulse rates as it orbited its planetary system's center of mass. They succeeded in detecting this pattern.

Third, a quite different type of instrument may allow us to measure the spectrum radiated by known atoms and ions in an observed star to determine whether it may systematically differ from the spectra emitted by the same atoms and ions here on Earth. By 1995, spectral resolving powers had reached levels at which small but systematic line-of-sight velocity changes in the spectra of stars could be discerned and similarly related to their Doppler shift. Mayor and Queloz perfected spectral resolving power to the point where these tiny repeated spectroscopic shifts could be reliably discerned over periods of months, opening a floodgate of further discoveries of planets around other stars.

A final capability of an astronomical instrument may be its ability to measure the polarization of an arriving astronomical signal. This usually is only used to highlight additional peculiarities in conjunction with one of the other three capabilities.

6.2   The Traits of Arriving Messengers

The information transmitted by any given class of messengers – pho-tons, neutrinos, gravitational waves, cosmic-ray particles, or any other forms of energy – may be specified by citing five distinct parameters an observation may yield.[12; 13] Each of these usually requires a dedicated instrument capable of yielding the desired data. It may provide:

• the energy, or equivalently the spectral frequency or wavelength of the messenger.

• the angular resolution the transmitted data can yield – the level of detail an anticipated image or celestial map will provide.

• the time resolution with which a sequence of observed events can be recorded; in some astronomical observations such events might require records stretching over centuries; in others, critical changes can occur within microsec-onds.

• the spectral resolution the data can provide: In observations restricted to measuring total energy, the integrated energy transmitted by all arriving messengers needs to be included. For chemical analyses derived from the spectra of molecules a spectral resolving of one part in 10^8 of the spectral frequency may be required. For planetary mass studies, periodic velocity changes of the parent star along the line of sight may be recognized only at spectral resolving powers of the order of one part in 10^9.

• finally, the degree and direction of polarization of the radiation may be recoverable. Often the degree of polarization or its wavelength dependence is informative and may require added care in specifying the fractional polarization detected, and whether both linear and circular polarization were measured to determine a degree of elliptical polarization.

These five parameters specify the *design requirements* for instruments and observatories constructed to detect and measure one or more of the *traits* any specific messenger can display.

A single arriving messenger can reveal nothing more than four *traits*: its precise energy or equivalently its spectral frequency; the precise direction from which it arrived; the precise time at which it arrived; and the spin or polarization it may display. This is all we can hope to learn about an individual messenger; and not even all of these traits can be simultaneously determined with high precision. Heisenberg's uncertainty principle teaches that a precise determination of an individual messenger's arrival time will interfere with a precise determination of its energy.

Frequently, an observer may also be interested in the level of instrumental sensitivity required. This used to be of particular concern in early years when the detectors constituting the heart of an instrument often generated undesirable *noise*, limiting the levels at which faint celestial signals might be discerned. Today, this tends to be of somewhat lesser concern; modern instruments increasingly, though not yet universally, are beginning to be equipped with detectors generating less noise than introduced by natural disturbances or the random arrival times of messengers from faint sources.

High sensitivity, nevertheless, is often useful. To detect faint signals, telescopes with extremely large apertures – light-collecting areas – may then be required, or else a faint source may need to be simultaneously observed through several smaller telescopes whose outputs are combined to yield greater sensitivity.

Figure 6.1 provides a simplified preview of the range of imaging capabilities instruments sensitive to electromagnetic radiation at different wavelengths make available today. Those capabilities often reflect an accumulation of design requirements the astronomical community may have prioritized at various times to advance ongoing research. The scale at the bottom indicates the wavelength at which an observation is undertaken. The scale at the top translates this wavelength into photon energy measured in units of electronvolts, eV.

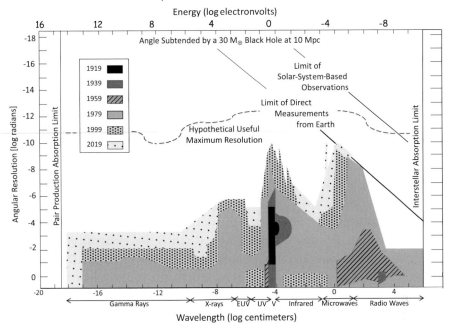

Fig. 6.1 Angular resolving powers of astronomical instruments capable of detecting electromagnetic radiation. Distinct levels of shading indicate the rapidly increasing capabilities attained over the century from 1919 to 2019. They depict progress in our ability to map the sky at different wavelengths. Bounds to transmission at the left and right are imposed by nature, as Figure 1.5 and Table 1.1 earlier explained. No similarly clear bounds appear in this figure at top but, as we will see later in the chapter, appear to be emerging due to ubiquitous gravitational deflections. A shaded region ranging upward in this figure indicates an ability to obtain high-resolution images at wavelengths shown on the scale at bottom. This stretches from the longest observable radio waves at extreme right, through visible and ultraviolet radiation around the middle of the scale, to X-rays and the highest-energy gamma rays at extreme left. The wavy dashed line meandering horizontally across the figure suggests that we may ultimately encounter bounds to high angular resolving power inherently imposed on photons traversing vast stretches of space. Diagonal lines at upper right emphasize the highest angular resolving power enabled by telescopes or spatial interferometers whose dimensions, respectively, are comparable to Earth's diameter or the diameter of our planet's orbital motion within the Solar System.

Long wavelengths correspond to low photon energies; short wavelengths to high energies. The scale at left indicates the angular resolution that may be required for a particular observation. All three of these scales are logarithmic, meaning that each notch on a scale increases or diminishes wavelength, energy, or angular resolution by a factor of 10. Negative values on each scale show fractions of centimeters in wavelength at bottom, and fractions of a radian, an angle of about 57°, on the scale at left. The figure documents the growth of observational capabilities over successive 20-year intervals identified through progressively lighter shading.

Observations at a given wavelength and angular resolving power are currently possible only if the intersection of wavelength and resolving power fall into a shaded region of the figure. If not, the observation will not become possible until the range of instrumental capabilities further expands – as it steadily has over the past century.

The darkest shaded areas on this plot show the limited range of angular resolution that was available for astronomical observations around 1919. At that time, astronomy was largely constrained to observations at optical wavelengths, with minimal extension into the near infrared through the use of heat-sensing detectors. The angular resolution at visible wavelengths at mountain top observatories was about one arcsecond, $\sim 5 \times 10^{-6}$ radians. This high an angular resolving power, $R_a = 2 \times 10^5$ radians^{-1}, the reciprocal of the angular resolution, was available only under the most favorable atmospheric conditions at observatories on high mountains.

By 1939, observing capabilities, as shown by somewhat lighter shading, had been extended further into the infrared domain, and to higher as well as lower angular resolving powers at optical wavelengths. And, as indicated by the small striped patch at wavelength $\sim 1.5 \times 10^3$ cm at lower right, Karl Jansky at the Bell Telephone Laboratories in Holmdel, New Jersey, had also discovered radio waves emanating from the center of the Galaxy.

Low angular resolving power implies an ability to detect small changes in surface brightness over large regions in the sky. The figure's border at bottom defines the lowest bound on angular resolution, a solid angle of 4π. This defines a capability of detecting fully isotropic radiation – a bland distribution of radiation uniformly arriving from across the sky. Shading at the very bottom of the figure provides insight on the epochs at which isotropic radiation at different wavelengths was first discerned. Attaining this capability can be especially difficult. It generally requires means for convincingly distinguishing an observed signal from noise inadvertently generated within an observer's own instrument.

Figure 6.1 refers solely to angular resolution capabilities, and is simplified in order to introduce some of the basic concepts, which more comprehensive plots will later also include. The ultimate aim of these diagrams is to visually display the full range of spectral, temporal, and angular resolving powers available to astronomers today, at any and all electromagnetic wavelengths the Cosmos transmits.

In principle, it is easy to assemble similar plots exhibiting the spectral, angular, and temporal resolution of instruments currently available for neutrino, graviton, or high-energy particle detection across available energy ranges. The resolving powers these other messengers currently offer are far more modest than those routinely available, by now, for detecting photons.

6.3   Instruments for Deciphering the Messengers

Each messenger can reveal a different perspective on the Cosmos. The properties of distinct messengers determine the range of observations the Universe will ultimately permit us to fruitfully undertake.

Where we do obtain the information the messengers convey, interpreting the significance of any given message can be difficult. Often, the messengers reaching us represent just a tiny fraction of all that could have arrived. A vast majority of them may have been intercepted in gaseous clouds they would have needed to traverse to reach us. Cosmic-ray particles may have been deflected by magnetic fields along their trajectories, displacing their direction of arrival so that it no longer points back to their source of origin. Our tools also need to determine the amount of energy each messenger conveys, and record its time of arrival, though those properties too may be of doubtful value if passage through interstellar or intergalactic space altered the messenger's energy or delayed its arrival. Whenever we are unable to faithfully reconstruct the time and place of a messenger's origin, or its energy and polarization at launch, our astrophysical insight wanes.

The number of tools a basic toolkit will require to faithfully detect, potentially clean, and finally reconstruct all the usably transmitted information reaching us, depends on our correctly recognizing the various means by which the data the messengers convey may have been erased or distorted along their trek.

An important further realization by now is that, far from being arbitrary, each class of messenger the Universe routes our way, be it photons, neutrinos, gravitons, cosmic rays, or particulate matter, is confined to a peculiar energy range. Outside that range it may either be generated nowhere, as Figure 3.16 informs us, or be rapidly intercepted and altered or destroyed. This means that ultimately some types of instruments we already have in hand are likely to detect absolutely nothing in searches across the celestial sphere. Nature simply may not generate the messengers sought, or else the messengers may never reach us intact.

Here, it is not a lack of technical expertise that holds us back, but rather the inability of the Universe to generate or transmit the information sought.

Our understanding, today, of the dominant astrophysical processes at work in the Cosmos provides us with at least a rough grasp of the scope of a complete cosmic toolkit we may ultimately require to observe all reliably trans-mitted messengers. This understanding admittedly remains limited because we know too little about two dominant cosmic energy repositories, *dark matter* and *dark energy*. Both appear relatively inert except for the gravitational forces they exert. If so, they may ultimately not require the construction of further tools.

6.4   The Phase Space of Observations and Discovery Space

If we scan the sky tonight it may appear slightly changed from a few nights earlier. The color and magnitude of a remote supernova may have evolved; the tail of a comet approaching the Sun may have lengthened; the steady beat of radio waves emitted by a pulsar thousands of light years away may suddenly have been disrupted by an irregular *glitch*.

Despite such daily changes an immense equanimity prevails. A thousand years hence, a carefully kept log of a night's observations will show a similar evolution of some remote supernova, a changed length of some other comet's tail, an irregular beat displayed by some pulsar – none at the same positions in the sky as a millennium earlier, but nevertheless exhibiting a largely immutable face of the Cosmos.

In this way, various realms of the Universe may periodically undergo extremely rapid changes or periods of extremely low activity. Some of these phases may display finely structured detail, whereas others show up on far coarser scales; and others still might display only extremely narrow or exceptionally broad spectroscopic features, or unusually high or low levels of polarization. Each of these phases is best detected by an instrument specifically honed to discern these traits at the wavelengths or in energy ranges characteristically emitted by the source.

Accordingly, we can array the set of all conceivable instruments in a five-dimensional *phase space*, whose orthogonal axes indicate (i) the wavelength or energy range in which the instrument excels, (ii) the range of angular scales of celestial features it best resolves, (iii) the durations of temporal changes it optimally records, (iv) the spectral resolving power it most readily discerns, and (v) the polarization states it is designed to reveal.

Whereas Figure 6.2 provides more insight on a phase space filter's com-plexity, Figure 6.1 displays a more realistic projection of this five-dimensional display onto the two-dimensional page of this book. It displays the ranges of angular resolving powers at different electromagnetic wavelengths attained over successive 20-year periods, and the steady growth of instrumental capabilities earlier in the twentieth century. Later in this chapter we will encounter more

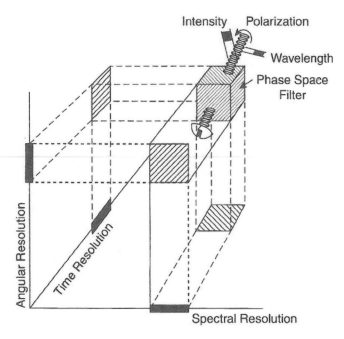

Fig. 6.2 A phase space filter combines the properties of several different devices. Three of the properties that characterize it are its definition of required angular, spectral, and time resolving powers. These traits specify the band of spatial frequencies, time variations, and spectral features transmitted by the filter. The filter also has three further properties. As I was unable to draw a six-dimensional space, these three are represented by the threaded bolt mounted at the location already designating the selected angular, spectral, and time resolution. The separation between windings on the bolt represents the wavelength the filter transmits. The bolt's diameter represents the intensity of the radiation encompassed by an observation – the number of individual messengers (photons or particles, etc.) at the given wavelength or energy transiting through the filter in unit time. The pitch angle of the windings, left- or right-handed, represents the type of polarization transmitted by the filter. Later, in Chapter 8 and Table 8.4, we will find these same six traits useful in identifying the principal distinctions between major astronomical phenomena.

comprehensive complementing Figures 6.6, 6.7, and 6.8. At 20-year intervals, each records a history of steadily increasing astronomical resolving power for electromagnetic radiation over the past century.

An important feature of all these plots is the prevalence of blank spaces highlighting instrumental capabilities we still lack. A select portion of these blank spaces currently of particular interest is often referred to as *discovery space*. As used by observational astronomers, discovery space comprises all astronomical capabilities that have to date *never* been implemented for lack of the requisite tools – though our knowledge of the Universe presumably would advance if those tools were available.

We can divide Figure 6.1 into three general realms: (i) those for which we have already constructed and successfully implemented tools revealing interesting cosmic features; (ii) those that could readily detect angular features of given dimensions if only the Universe transmitted information on features of those angular dimensions; (iii) intervening realms in which the Universe does transmit information, but for which we have not yet built instruments capable of fully decoding the information flow. These final realms constitute discovery space. They are the regions of Figure 6.1 depicting instrumental capabilities with which novel, previously unattainable astronomical discoveries are likely to be made and where the implementation of novel tools is most likely to reap rewards.

Understanding the bounds of discovery space is thus an essential element of astronomical studies. It defines the extent to which the Universe has *not yet* been as fully explored as should ultimately become possible.

Interaction of cosmic messengers with the medium they traverse generally limits the information they are able to transmit. These limits, in turn, prevent our learning all we might have hoped. Implementing the toolkit's full set of tools should eventually reveal to us just where nature's ultimate limits lie that we will never be able to transgress.

The conquest of discovery space will be complete once we assemble and implement the entire *observational toolkit* essential for tapping any and all information the Universe is intrinsically capable of transmitting. We may then still lack observations of phenomena the Universe actually displays, but it will not be for lack of the requisite instruments. Rather, it will indicate that we have exhausted all the information the Universe faithfully transmits.

As Figure 6.3 emphasizes, instrumental capabilities become useful only if the Universe generates, and is capable of transmitting, the information sought. Unless both these requirements are fulfilled, building instruments to detect such features will prove futile.

Later in this chapter we will focus on an assessment of messengers the Universe both generates and reliably transmits. This will permit us to reasonably estimate the cost of assembling a minimal set of observational tools for detecting those messengers and deciphering their messages.

To date, the astronomical tools we have been assembling have fallen into place remarkably quickly and at relatively low cost, suggesting that the conquest of discovery space should continue to remain affordable. But we do not yet know whether *dark matter* or *dark energy* may introduce novel messengers whose detection may prove expensive.

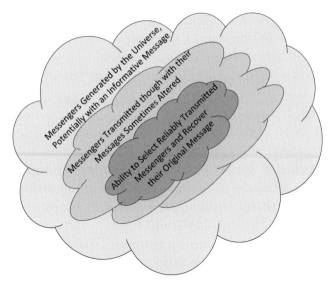

Fig. 6.3 The Universe generates considerably more messengers than it transmits with discernible fidelity. We do not yet have all the observational capabilities in hand today to detect all the cosmic messengers reliably transmitting information our way. But we appear to be approaching bounds beyond which further perfection of tools may no longer help.

Mastering all potential discovery tools, even with dark matter and dark energy in mind, will not necessarily assure our fully understanding the Universe. These means alone will likely not suffice to reveal all we would like to know. But unless we do assemble the required minimum kit of tools to detect and clearly discern all cosmic messengers reliably streaming our way, we will certainly not learn everything about the Cosmos we potentially could.

6.5 The Conquest of Discovery Space

The Universe does not reveal its workings readily. As Chapter 2 explained, most epochs of cosmic history have systematically erased their existence, bequeathing us just a few lone messengers from which to reconstruct the past.

Some of the features of the portion of phase space reflected in Figure 6.1 are worth highlighting. Near the figure's left border, high-energy photons strongly interact with the ubiquitous microwave background radiation. This bound is indicated by the vertical line marked "Pair Production Absorption Limit." As Chapter 3 showed, we cannot detect electromagnetic radiation at these or higher energies from galaxies more remote than roughly 300 million light years or 100 Mpc.

The *conquest of discovery space* will thus ultimately entail identification of all cosmological limits on transmitted information; define the tools needed to enable practicable observations; and estimate how long it may take and what it will cost to acquire the requisite instruments. These circumstances thus foreshadow an economics of cosmological inquiry which should display corresponding finite bounds as well.

The rationale for adopting this instrument-based phase space is that the purpose of astronomical instruments is to capture and analyze information the Universe transmits. A natural description of an instrument's capabilities thus could begin by listing the phase space properties of the *messengers* – the photons, neutrinos, gravitons, or cosmic-ray particles – conveying the desired information.

If the messenger is an X-ray photon, an astronomical instrument naturally must be sensitive to X-rays. If the aim of an observation is to extract a high-resolution image, the instrument's capabilities need to include high angular resolving power. The higher the required angular resolution of the instrument, the higher the shaded portions displayed in Figure 6.1 would have to extend. We can see that the shaded parts for the wavelength range labeled *X-rays* does not extend up as high in the figure as the shaded portions in the visible range labeled *V*, or in the *microwave* domain. This shows that, at least for now, astronomical X-ray instruments do not provide images as highly resolved as those that observations at visible or microwave frequencies could deliver.

Later, we will encounter similar comparisons of available instrumental capabilities for achieving high fidelity in the timing of rapidly changing cosmic events, for obtaining exquisite spectra to identify the chemical constituents of gaseous interstellar clouds, or for detecting whether any of these radiations arriving at Earth are linearly or circularly polarized – a trait that can inform us of the presence or absence of magnetic fields embedded in a radiating source or along the intervening line of sight to the source. As we venture further in discussing instrumental capabilities, we will find that resolving time sequences in cosmic processes, or obtaining spectroscopic resolution to determine chemical processes at work, will similarly be bounded. For each of these capabilities, the basic laws of nature will render instrumental improvements beyond certain bounding limits futile, simply because cosmic processes do not generate or else transmit information reliably beyond those bounds.

In Chapter 3 we defined the range of energies across which the Universe transmits each different type of messenger – electromagnetic radiation, cosmic rays, neutrinos, and gravitational waves. The present chapter pursues this topic further to define which messengers can be relied on for providing useful information through precise observations of their time of arrival, the portion of the sky from which they arrived, the spectral energy distribution of the family of messengers to which they belong, and the degree to which the arriving messengers are polarized. Any or all of these properties may be limited by inherent features of the Universe and by astrophysical processes, many of which we already recognize and understand well today.

6.6   Transmission of Information

Imaging, Timing, Spectroscopic, and Polarization Studies

Any astronomical exploration can be defined by the six capabilities cited in Figure 6.2. These specify the capacity to clearly image celestial scenes, accurately convey changes over time, correctly discern the amount of radiation and its spectrum transmitted in distinct energy ranges, and display the polarization of the radiation received.

A high spatial, temporal, or spectral resolution requires analyses of fine detail. A correspondingly low angular, timing, or energy resolution may require less attention to detail, but is often more difficult to achieve because the quality of the calibrations entailed depends on the availability of comparison sources whose strengths have been reliably established. Often these are hard to find or may not even exist. Finally, electromagnetic radiation also may exhibit polarization alerting us to distinct mechanisms enabling some astronomical sources to radiate.

Most of the information the Universe has revealed to us to date has been conveyed by photons, electromagnetic waves. The past few decades have brought three other types of messengers to the fore, cosmic rays, neutrinos, and gravitational waves. Our tools for extracting the information these more recently discovered messengers may be conveying are not yet fully developed, and so this chapter begins by dwelling on the distinct classes of information photons transmit. Later, we will turn to the three more recently recognized messengers to examine the types of signals they might uniquely convey once we are able to fully decipher them.

The Phase Space of Observational Capabilities

The information conveyed by each distinct set of messengers is derived from the traits they display in the phase space just encountered in Section 6.4. Some messengers, like the radio waves arriving from pulsars, display a highly stable sequence of beats, when averaged over time spans of a few minutes. We measure these messengers' directions of arrival, the interval between beats, and potentially also the frequency of anomalies to gain insight on the identity of the messengers' sources of origin. The wavelengths of the arriving photons, their energy spectra, and their polarization provide supplementary information. These observational traits are not entirely unrelated, and so it is useful to first examine how restrictions on our ability to accurately resolve the arrival times of signals can affect attempts to obtain highly resolved spectra or clear images.

6.7   High Time Resolution and Fast Radio Bursts (2007)

In 2007, Duncan Lorimer, his former PhD thesis advisor Matthew Bailes, and three colleagues announced the detection of a single, fast radio burst, FRB. It lasted less, possibly much less, than 5 milliseconds, launched from some source no one has ever identified reliably since.[14] The signal was hard to believe. Science is based on observations that any sufficiently equipped scientist can verify. What can one do with a single observation of a phenomenon never previously noted that appears to have vanished forever with no other similar effects spotted elsewhere again?

The burst had originally been detected six years earlier, on August 24, 2001, at the 64-meter Parkes Radio Telescope in Australia, during observations carried out by Lorimer, at the time working on his PhD thesis. These were part of a multibeam survey consisting of 209 different telescope pointings, each lasting 2.3 hours, during which the multibeam receiver collected signals from 13 different locations on the sky, sampling each signal once every millisecond over a range of 96 separate frequency channels spanning a frequency band 288 MHz wide centered on 1.4 GHz.

The observation yielded a flood of data that would require laborious analyses. Thus, it is not surprising that it took some six years to discover this single unanticipated pulse buried deep in the all-but-overwhelming data trove. Lorimer only came upon it during a painstaking sift through the data after finishing his thesis and moving on to West Virginia University in Morgantown, West Virginia.

Standing out in this unique signal was a feature made clear by the individual records in the 96 distinct frequency channels. As seen in Figure 6.4, the pulses recorded in the highest frequency ranges around 1.5 GHz had systematically arrived in a well-ordered sequence, up to 400 milliseconds ahead of the lowest frequency pulse components below 1.3 GHz . The arrival time of the pulse components shown in the figure increased quadratically with diminishing frequency in the curved swath crossing the figure from upper left to lower right. This spacing in the sequence indicated that the beam had passed through a strikingly high *column density* of electrons on its way to the Parkes telescope – far

rapidly that the merged body, though sufficiently massive, cannot collapse to form a stellar black hole until its angular momentum sufficiently declines.[20]

On December 27, 2004, a giant outburst of gamma rays and X-rays impacted Earth. Emitted by a magnetar designated SGR 1806-20 at a distance of almost 50,000 light years – roughly twice the distance of the Galactic Center – it briefly distorted our ionosphere, endangering the fleet of spacecraft circling Earth. The gamma-ray sensing SWIFT satellite registered an enormous initial pulse of radiation, with continuing ringing at intervals of 7.56 seconds, corresponding to the magnetar's rotation period, for close to 5 minutes thereafter. Had such an outburst reached us from anywhere much closer, the damage to the fleet of spacecraft linking our planet's entire economy could have been catastrophic. A variety of after-effects continued for months.[21]

Whether or not the conjectured connection between magnetars and FRBs holds up may be determined by a network of observatories that has begun to track these fast bursts wherever they originate. The network will also record other rare events that might be brief but consequential.

6.8 Restrictions on Information Transmitted by Electromagnetic Waves

A variety of inherent bounds limit the information that the Universe transmits. Some of these bounds arise from the way the contents of the Universe interact with matter or radiation in well-defined energy ranges. Among these are limits that ionized gases in interstellar space place on the transmission of low-frequency radio waves; that interstellar dust clouds place on the transmission of visible light; or the distortions that interstellar magnetic fields exert on the trajectories of charged cosmic-ray particles, preventing us from locating their regions of origin. All are examples of loss of information different types of messengers encounter on trajectories through the interstellar medium.

In the radio frequency domain these limitations are particularly clear. The interstellar medium does not transmit electromagnetic waves at frequencies lower than the *plasma frequency*. If the distance to a pulsar is well established, measurement of the pulse arrival delay permits recovery of the plasma frequency, and thereby the mean number density of electrons along the path from the pulsar. Information is nevertheless lost because the interstellar medium is patchy and irregular.

6.9 Temporal Resolving Power

More broadly confounding constraints exist as well: No instrument can resolve signals in frequency bands that would exceed the messengers' inherent

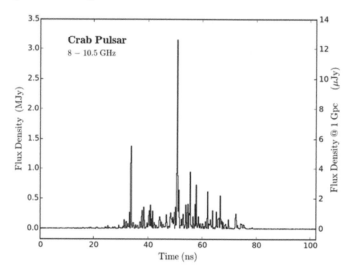

Fig. 6.5 Portion of a single pulse from the Crab pulsar showing sub-nanosecond structure consistent with a shot-pulse model for the radiation. The largest shot pulse exceeds 3 MJy and several others exceed 0.5 MJy. The right-hand axis indicates the flux-density scale if the source were at a distance of 1 Gpc. (From James M. Cordes & Ira Wasserman, *Monthly Notices of the Royal Astronomical Society* 457, 232–70, 2016, based on data from T. Hankins. Courtesy of Jim Cordes, Ira Wasserman, and the journal's publishers.)

or *carrier* frequencies. Other limits prevent any combination of instruments from simultaneously providing high-timing, high-spectral resolution, and high spatial resolution observations of conceivably interesting phenomena through one and the same transmission. This joint rate of information transmission cannot exceed the peak rate at which any individual timing, spectral, or angular information can be transmitted. A high-resolution observation optimizing any one of these traits may then lead to a combined loss of resolving power in the two others.

The Crab Nebula pulsar is a rapidly spinning *neutron star* formed in the explosion of a supernova that gave rise to the high-velocity outflows now forming the Crab Nebula. At times, pulsars emit brief exceptionally energetic outbursts. Figure 6.5 shows the arrival of such a burst from the Crab, detected in an 8 to 10.5 GHz radio frequency band. Some of the individual spikes seen in the figure and referred to as *nanoshots* last no more than $\Delta t \sim 0.4 \times 10^{-9}$ seconds apiece.[22] The Crab pulsar also emits similar nanoshots in the 6 to 8.5 GHz band.

James M. Cordes and Ira Wasserman at Cornell University have pointed out that the Crab pulsar nanoshots and extragalactic FRBs might be related physical phenomena that simply differ in scale.[23]

A point of particular interest is the extreme narrowness of the Crab pulsar's arriving pulses, ≤ 0.4 nanoseconds, in the 2.5 GHz bandwidth of the 8 to 10.5 GHz frequency band in which Hankins and Eilek were observing. They exhibit the sharpest possible pulses any electromagnetic waves detected in such a narrow bandwidth can in principle display.

We do not know whether the narrow individual pulses represent the fragments of a primary pulse splintered through scattering along its trajectory through the interstellar medium, or whether the pulsars emitted the sequential nanoshots inherently. Nor can we tell whether the width of the pulses already was as wide as ~0.4 nanoseconds at emission, or whether the observed nanoshots may have originated at the pulsar with even narrower pulse shapes, which the limited 2.5 GHz bandwidth of the receivers Hankins and Eilek used were unable to resolve.

Very roughly, the amount of information that can be obtained in one second from electromagnetic waves transmitted with modulating frequencies in a range from v to $v + \Delta v$ can be thought of in terms of the number of times per second a zero-electric-field state is encountered as the field strength systematically oscillates from crests to troughs and back from troughs to crests.

Electromagnetic waves exhibit two distinct states of polarization, each of which may, in principle, convey as many as W distinct messages per second, by itself, if an observer employs detectors able to distinguish between the two states of polarization. The number of zero crossings encountered for distinct waves in this frequency range can thus range from $2v$ to $2(v + \Delta v)$, enabling us to distinguish $2\Delta v$ distinct waves within an interval of one second. Since each of these distinct waves can transmit a different message, we speak of a message transmission rate $2W = 2\Delta v$, where W denotes the *bandwidth* defined as $W \equiv \Delta v$. In brief, the maximum amount of information an electromagnetic wave can transmit in one second – the maximum number of distinct messages that can be encoded within bandwidth W – is twice that bandwidth: $2W$. Fourier transform theory, and its closely related Heisenberg uncertainty principle, tell us that $\Delta t \Delta v$, the product of time resolution Δt and instrumental bandwidth Δv, can never be less than of order unity.[24] The bandwidth $\Delta v \sim 2.5$ GHz of the apparatus that Hankins and Eilek were using at frequencies of ~10.5 GHz already was operating within a factor of at most ~4 of the fastest frequency response possible for resolving the pulses, a spectral resolving power corresponding to a time resolution Δt well under an almost incredibly short billionth of a second. The only way to obtain higher temporal resolving powers will be to observe at higher electromagnetic frequencies, where observing in frequency bands a factor of 10 to 100 wider will permit corresponding improvements by those factors.

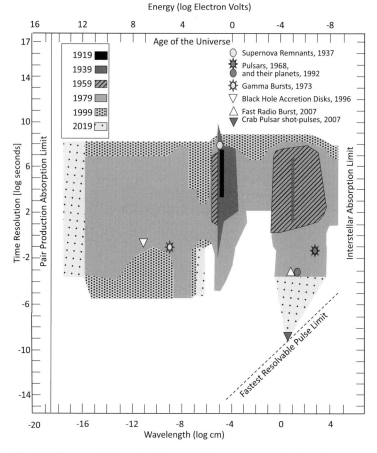

Fig. 6.6 The twentieth and early twenty-first century evolution of temporal resolving power in electromagnetic observations of astronomical sources and the discoveries they brought about. The dashed line at lower right indicates the highest (fastest) time resolution electromagnetic waves of shortest wavelengths, highest frequencies, can convey. The small superposed symbols display how enhanced timing capabilities enabled novel discoveries over the course of a century, both at extremely high and extremely low temporal resolving power and a broadening wavelength span. The legend "Age of the Universe" reminds us that changes on time scales far exceeding the age of the Universe cannot readily be detected. The bound marked "Pair Production Absorption Limit" denotes the energy at which photons annihilate in collisions with microwave background photons to form electron/positron pairs, a process limiting the energetic photons' transmission across intergalactic space.

Figure 6.6 shows a data point at wavelength 3 cm and time resolution $\sim 0.4 \times 10^{-10}$ s, at bottom right, corresponding to the Hankins/Eilek observation. Through this the drawn dashed line sloping downward to the left reaches a temporal resolving power of $\sim 0.4 \times 10^{-15}$ s around the visual frequency

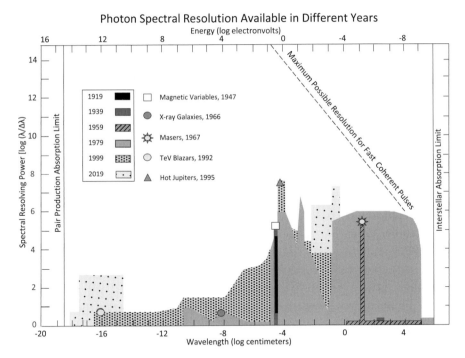

Fig. 6.7 The twentieth and early twenty-first century evolution of spectral resolving power in electromagnetic observations of astronomical sources and the discoveries these advances enabled. The vertical line at the right, marked "Interstellar Absorption Limit" designates the wavelength domain beyond which partially ionized interstellar gases strongly absorb radio waves.

domain, employing apparatus with correspondingly conceivable high bandwidths around 2.5×10^{15} Hz. This line roughly indicates the highest temporal resolving powers ideal instruments will ever attain at the respective electromagnetic frequencies.

The 8 to 10.5 GHz band Hankins and Eilek employed places the wavelength of radiation in the 3 cm wavelength region, a radio domain in which some of the clearest transmissions of messengers have consistently been attained. Conservatively, it thus appears that, in this domain at least, there is at very best less than a factor of 10 improvement we could in principle ever hope to achieve, either in time resolution or in spectral resolving power. For, as we earlier saw, the highly precise, fast-time resolution that enabled Wolszczan and Frail to discover the first extrasolar planets, registered in Figure 6.6, and the extreme spectral resolution, marked in Figure 6.7, that Mayor and Queloz had employed in their successful quest for such planets, were not entirely distinct lines of attack.

6.10   The Relation between Temporal and Spectral Resolving Power

We tend to think of instruments providing precise timing as highly accurate clocks, and those enabling high-resolution spectroscopy as concise ways of distinguishing distinct wavelengths or spectral frequencies. In terms of their design and appearance, these two types of devices could hardly seem more distinct. But, as Dennis Gabor, the Hungarian-born British engineer and inventor of holography, first pointed out in an article published in 1945, the lines of investigation these two quite different instruments provide are closely linked.[25]

We can think of the spectrometer of Mayor and Queloz as counting the number of electric field crossings in a time interval τ, at selected observed spectral frequencies, say $\nu_1, \nu_2, \ldots \nu_n$, of the parent star 51 Pegasi. Correspondingly, Wolszczan and Frail were counting numbers of pulsar peaks arriving per unit time, and noting those variations over many months in observations of Pulsar PSR1257+12.

Let us see by how much the spectral resolving power in the next-most-developed spectral range, the visual range, might still be improved. Here, the wavelengths involved lie around 6000 Å, or 6×10^{-5} cm – at a frequency 5×10^4 higher than in the 3 cm range, and over a bandwidth wider by a correspondingly high factor, and resolving power of order $R_\nu \equiv \nu/\Delta\nu \sim 1.25 \times 10^{14}$, equivalent to a passage of time of 4 cycles or 8×10^{-15} seconds in a measurement lasting one second.

In Figure 6.6 note that, over the longest time intervals, we currently resolve processes developing at optical frequencies as gradually as over periods of order 300 years, or 10^{10} seconds, for some of which we already may have reliable observational data dating back several centuries.

In Figure 6.7, current optical spectral resolving powers roughly reach 10^8. We would need to raise this to $R_\nu \sim 10^{14}$ to take full advantage of the available frequency band. This seems like an achievable target to set for the century ahead, given how much the past century has already advanced our spectroscopic capabilities.

6.11   Spectral Resolving Power

Figure 6.5 shows that, despite their narrow widths, we readily recognize that some of the arriving pulses exhibit flux densities several times higher than others in their immediate vicinity. The spectrometer examining this series of pulses evidently had a spectral resolving power sufficiently high to resolve the individual pulse heights.

Resolving the spectrum of radiation arriving from an astronomical source always requires sampling the incoming radiation's amplitude in its given frequency band at a rate sufficiently frequent to satisfy the *Nyquist sampling theorem*. For a signal bandwidth W, sampling the signal's amplitude at least at a rate $2W$, the *Nyquist rate*, can reproduce its frequency spectrum fully. Sampling the radiation's amplitude less frequently than this rate only permits a more approximate analysis of the signal's spectrum.

Radiation in the optical wavelength band accordingly requires a sampling rate also in the optical frequency range, unless there are good reasons to believe that the arriving optical wave trains were clearly damped to such a high extent that sampling at a considerably lower rate would not only suffice but be more economical.

If a radiating source emits pulses of radiation that are considerably damped – sufficiently broadened – the bandwidth of the sampling apparatus needed to optimally resolve the emitted spectrum may be considerably lower than the spectral frequency of the emitted radiation. For spectroscopic analyses of astronomical sources whose radiation arrives broadened through thermal or turbulent random motions of radiating atoms or molecules along the line of sight, this is standard practice. Only if the emitted radiation is highly coherent, such as radiation produced by a cosmic maser, or some other means distinctly out of thermal equilibrium – and if such a beam were, in addition, modulated at extremely rapid rates – would the sampling frequency ever approach the electromagnetic frequency of an arriving wave train. This is the highest resolving power that would ever be required for deriving all the information conveyed by an arriving astronomical electromagnetic wave.

For many, if not most, astronomical purposes such high resolving powers may ultimately never be needed. At the highest conceivable optical resolving powers, the accuracy with which line-of-sight velocities could be resolved would be $\sim 6 \times 10^{-4}$ cm/second, or roughly 2 cm/hour. In comparison, random thermal velocities in interstellar gas clouds tend to be measured in kilometers per second, as might random stellar velocities.

However, astronomical sources do often surprise us!

If astronomical sources exist that, for example, generate *optical frequency combs*, by now readily generated in laboratories, extremely high spectral or temporal resolving powers would be required to detect them. Optical combs comprise myriad simultaneously generated pulses, whose pulse frequencies

$$f, f + \Delta f, f + 2\Delta f, \ldots f + n\Delta f \ldots$$

mutually differ by highly stable pulse frequency differences Δf.[26] Recognizing the existence of such frequency combs would help in compensating for small delays in the rates at which pulses with different frequencies arrive, so that the originally transmitted signals might be recovered. This could be helpful both for reconstructing naturally generated astrophysical signals, and in searches for extraterrestrial intelligence, SETI.

Optical frequency combs are by now readily generated in the laboratory for a variety of purposes, among them for high-speed optical telecommunication, and might be of use in communication across astronomical distances.[27]

6.12   Angular Resolving Power

In Chapter 3 we dwelled mainly on the energies of transmitted messengers. While important, these represent only a fraction of the information the messengers usually are able to convey. With well-designed instruments, we readily determine the location on the celestial sphere from which an arriving photon reaches us. The precision with which we can determine the direction of arrival and the angular spread of the emitting region depends on the angular resolving power of our instruments.

With instruments of adequately high angular resolving power we can also recognize whether two or more arriving photons were emitted by a single source, or whether two or more stars were independently transmitting radiation reaching us from closely adjacent patches on the sky. High angular resolution enables us to distinguish between these alternatives and can provide detailed images and maps of the sky.

While improved angular resolving power is helpful in this way, it does not satisfy all astronomical needs. Radiation arriving from all directions in the sky with uniform surface brightness may be detected solely with instruments specifically designed for that purpose. Often the difficulty is not in detecting the arriving radiation; it rather lies with ascribing a cosmic origin, instead of, say, incidental atmospheric emission to a detected signal, or perhaps recognizing that the signal is actually generated within the astronomical instrument itself – much to the observer's chagrin.

Figure 6.8 shows that, without the distinct capabilities of both high- and lowangular resolving powers, across the entire photon energy/wavelength range transmitted by the Universe, many of the discoveries shown would have eluded us.

The discovery of the faint microwave background fluctuations, indicated by the square symbol marked "CMB Fluctuations, 2011," could not have come about much earlier than the year 2011 when instruments enabling detection of a patchy diffuse cosmic background radiation became available. The symbol's location in the microwave wavelength range of the diagram, with a

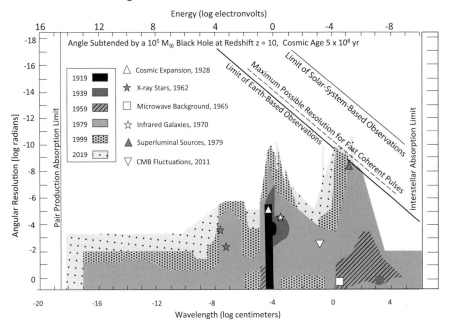

Fig. 6.8 The twentieth and early twenty-first century evolution of angular resolving power in electromagnetic observations of astronomical sources, highlighting discoveries these advances enabled. Note that the most significant advances in angular resolving power since 1999 were implemented in the shortest-wavelength, highest-energy ranges, at ultraviolet frequencies and in the far infrared. This is partly due to attempts in these wavelength ranges to catch up with the rapid earlier developments of radio and optical techniques originally developed for industrial or military applications.

spatial resolving power of order of minutes of arc, indicates the instrumental capabilities that were required to discover the fluctuations. The background shading at the symbol's location indicates that this capability became available sometime between 1979 and 1999, when plans for launching the ambitious Wilkinson Microwave Anisotropy Probe (WMAP) were reaching maturity for mapping these fluctuations across the entire sky. WMAP's successor mission, Planck, pursued these technological advances even further. The portrayal of time-keeping capabilities across the entire electromagnetic domain leads to a comparable display. Figure 6.6 covers the same wavelength range as Figure 6.8 but exhibits the pace at which higher, as well as lower, temporal resolving powers were being implemented.

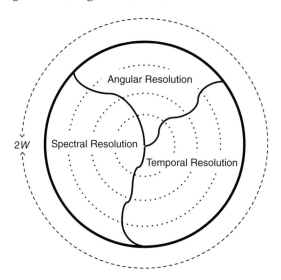

Fig. 6.9 The highest rate at which information can be conveyed by electromagnetic messengers is determined by twice the bandwidth of the transmitted radiation, $2W$, here represented by the circumferences of the various circles shown. For each such bandwidth, an observer needs to constrain the total bandwidth required by joint angular, spectral, and temporal resolving power requirements to fit the available $2W$ budget. If the observations had to include polarization studies, these various segments would need to be expanded, shrunk or subdivided to still accommodate the total data rate within this $2W$ limit.

6.13   Joint Angular, Spectral, and Temporal Resolving Powers

We have already seen that spectral and temporal resolving powers are closely related. We now need to ask whether similar constraints also hold for angular resolving powers?

The angular resolution α measured in radians, or angular resolving power $R_\alpha \equiv \alpha^{-1}$ obtained with a telescope of aperture D, or equivalently with an interferometer with baseline D, is $\alpha \sim \lambda/D$, where $\lambda \equiv c/\nu$ is the wavelength of electromagnetic radiation of frequency ν, and c is the speed of light. The angular resolution R_α obtained with a telescope or spatial interferometer arrayed with its axis of symmetry pointing directly at the source is then $R_\alpha \sim D/\lambda$.[28] We then have a relation between angular resolution, spectral resolution, and temporal resolving power available in observations lasting t seconds. The time resolution is $R_t \sim t/\Delta t$, where Δt is the time elapsed between passage of successive wave crests, λ/c, and $D = ct$. Consequently, the angular resolution $R_\alpha = D/\lambda = ct/c\Delta t = R_t$.

Similarly, the spectral resolving power R_ν, obtained with a Michelson interferometer, for which the baseline D is along, rather than perpendicular to the incoming beam, but can be of the same length D, is now given by the Michelson interferometer's baseline D divided by the wavelength $\lambda = c/\nu$. This yields $R_\nu = D/\lambda = D\nu/c = ct/c\Delta t = R_t$.

For instruments spanning comparable baselines, the angular, timing, and spectral resolving powers can thus be equal.

If the spectral resolution is set at its highest available in the electromagnetic frequency range, in order to best resolve a coherent source of radiation, the angular resolving power thus made available will also be maximized. However, as indicated in Figure 6.9, the total rate of information transmitted in one second cannot exceed twice the bandwidth, $2W$. Accordingly, the combined angular, spectral, and temporal information that can be extracted in observations spanning a limited observing period t remain bounded by the total frequency bandwidth of whatever instrument is in use.

6.14   Polarization

Even when we have determined the location of an emitting region on the sky, and obtained its image, time variability, and spectrum, the information received may be analyzed further to determine whether its emitted radiation is polarized. To date, astronomical capabilities for detecting polarization have provided particularly significant information mainly in the radio domain, where they have contributed invaluable data on the number of freely floating electrons in interstellar and extragalactic space and indications of the magnetic fields interwoven in these plasmas.

Polarized radio waves often also alert us to the possibility that the waves were generated through *synchrotron emission* by highly energetic relativistic electrons spiralling in a strong magnetic field at velocities approaching the speed of light. A radio wave's direction of polarization generally changes as it travels through interstellar space, where a magnetic field component along the direction of propagation may rotate its direction of polarization, twisting it at a rate determined by the integrated product of the number density of electrons n_e and the strength of the magnetic field component $B_\parallel$ along the line of sight. This *Faraday rotation*, the rate at which the direction of polarization changes, also depends on the square of the radiation's wavelength, λ^2. Knowing the wavelength and the angle through which the direction of polarization has rotated provides a measure of the mean product of electron number density and line-of-sight magnetic field strength $\langle n_e B_\parallel \rangle$ along the path the radiation has traversed.

In the visible and infrared wavelength ranges, linear polarization studies have provided insight on the mutual alignment of elongated dust grains permeating cool interstellar clouds. Polarization observations across broad wavelength domains have also provided novel insights on the existence of highly energetic streams of particles in jets collectively moving at *relativistic velocities*.

Astronomical instruments sensitive to the polarization of arriving X-rays or gamma radiation do not yet exist, though their potential appears promising.[29]

6.15   Exoplanetary Systems

Let us return to our initial discussion of exoplanetary systems to see how the evolution of instrumental capabilities has provided new insight.

By 2017, efforts were in place to detect not only the gravitational tug of massive planets orbiting a central star at short radial distances, but also for detecting smaller planets at distances from their parent star compatible with the existence of liquid water on the planet's surface – a potential requirement for the sustenance of life. The spectral resolving power required for such a search can be roughly estimated.

For a lone planet of mass m circling a common center of mass with its parent star of mass M, the planet and star maintain respective distances r and R from their center of mass at rest, related by $mr = MR$. Their respective tangential velocities v and V then are provided by ratios $v/V = r/R$, which locates the center of mass precisely along a line of sight separating them. If both the star and the planet follow roughly circular orbits, then their orbital velocities are related by $MV \sim mv$.

For a star with a mass comparable to that of the Sun, the requirement for the existence of water on a planet's surface requires the planet to lie at a distance comparable to the distance of Earth from the Sun. There, the planet's orbital velocity is of order $v = 30$ km/s. The respective masses of Earth and Sun are 6×10^{27} and 2×10^{33} gram, leading to a corresponding orbital velocity component of the parent star, $V \sim 9$ cm/s. The requirements for a spectrometer capable of detecting the reflex velocity imposed on a star by a habitable planet would therefore be an ability to detect and measure the star's orbital velocity components to accuracies of order $V \sim 10$ cm/s, over epochs of several years. As a fraction of the speed of light, $c = 3 \times 10^{10}$ cm/s, this corresponds to a spectral resolving power $\mathcal{R}_s \equiv c/V \sim 3 \times 10^9$.[30]

The habitable planet and its parent star's properties assumed here are fairly typical. In order to retain water vapor, H_2O, in its atmosphere, the planet's mass has to be sufficiently high to retain atomic hydrogen, H, which the parent star's ultraviolet radiation may occasionally dissociate from the planet's atmospheric molecules. The planet Mars appears to have lost this retention, though it lies further away from the Sun and is thus cooler. The escape velocity of hydrogen atoms from Earth is ~11 km/s. For Mars it is only ~5 km/s, meaning that gases could more readily escape if the upper portions of the atmosphere were strongly ionized, meaning that hydrogen would become dissociated from atmospheric water vapor.[31] The parent star thus cannot be much more massive than the Sun, because more massive stars emit higher levels of ionizing radiation.

A spectral resolving power of order $\mathcal{R}_S \sim 10^{11}$, therefore, appears to be as high as might ever be needed for singling out systems harboring habitable planets through the reaction the planets induce on their parent star's orbital velocity. Since this limit is due to a gravitational effect, it holds identically for all electromagnetic wavelengths.

Habitable Exoplanets

As a powerful new tool in the search for all kinds of new planets and planetary systems, an entirely novel space mission named *Kepler* was launched in early 2009. It continuously stared at a wide swath of the sky, looking for variability and eclipses of all kinds, among others due to occultation of stars by planets, and vice versa. In the course of a few years this survey discovered more than 4000 new planets – as well as many of the lowest-mass stars – the *brown dwarfs*.[32]

The Search for Life Elsewhere

Considerably higher angular resolution than currently available could be useful for studies of the surfaces of exoplanets. The problem there is somewhat different from other angular resolving power requirements.

Because a Sun-like star is so luminous, and the reflected and thermally re-emitted light from a planet in an Earth-like orbit is just a tiny fraction of the starlight, a coronagraph or some other device for minimizing the star's strong glare is usually needed to directly detect the planet.

For unambiguous signs that a planet might be inhabited, its surface temperature might have to lie in the range of -20 to $+120$ degrees Celsius, the range of temperatures on Earth within which at least primitive life has been detected. At the extremes of this range water freezes or boils.

Searches for life elsewhere may also need to take into account the kinds of tools that might be required to detect living matter. Some of these are likely to involve spectroscopy, but whether these alone would suffice for an unambiguous recognition of the existence of life is not clear.

The Search for Intelligent Life

Partly because primitive life could be difficult to detect on a remote planet, the prospects for detecting technologically advanced civilizations have been pursued with considerable vigor. Only a small fraction of habitable planets are likely to be populated this way; but if most habitable planets were indeed inhabited, many more might at some epoch have developed means of communication that might be recognized. The *Search for Extraterrestrial Intelligence, SETI,* a systematic search for electromagnetic waves emanating from such planetary systems, originally introduced and spearheaded by Jill Tarter, has been active for several decades.

So far this has not yielded credible signs of intelligent life in the Universe. So the search continues. It involves sensitive detection techniques, over a wide range of potential electromagnetic frequencies, and the unscrambling and rejection of extraneous interference that might otherwise mask a coherent message. It also has begun to incorporate machine learning approaches, to discern whether messages that might escape a human observer's attention could be discerned through searches conducted to cover a wider range of recognition signs.

In a search for repeated millisecond-long bursts of radiation in the so-called Breakthrough Listen C-band (4–8 GHz) radio frequency domain, at the Green Bank Telescope in West Virginia, a team led by Yunfan Gerry Zhang at the University of California at Berkeley used machine learning techniques combined with *de-dispersion* of radiation to correct for the line-of-sight delays due to a concentration of electrons along the path to a fast radio burst source, FRB 121102, to detect bursts with far higher efficiency than had ever been previously attained. In the first half hour of operations these observations revealed 45 new pulses from the extragalactic source, an astonishingly high rate in a source whose pulses had previously been difficult to systematically study to characterize pulse trends.[33]

Such astronomical searches often share surveillance techniques that governmental and industrial efforts have developed by now, so that SETI efforts have been able to keep costs low.

6.16   The Spectral, Temporal, and Angular Resolution of Other Carriers of Information

Cosmic-Ray Particles

Highly energetic charged particles have primarily taught us that the Universe endows a small fraction of available messengers with enormous energies. This provides some of the clearest evidence that the Cosmos is wildly out of thermal equilibrium. Vast regions are extremely cold, at temperatures of just a few degrees Kelvin, while individual protons and more massive nuclei can attain energies of order 10^{20} eV, roughly equivalent to an out-of-equilibrium temperature of 10^{24} K.

In Chapter 3 we saw that Fermi acceleration triggered in explosive environments appears responsible for generating cosmic rays at high energies $\mathcal{E}$ moving at speeds approaching that of light. Galactic magnetic fields B can bend these into trajectories of spiral radius $R \sim \mathcal{E}/qB$, where q is the particle's electric charge.

At energies below roughly 10^{18} eV or 1.2×10^6 erg a proton's gyro-radius remains below 100 pc in a typical galaxy's magnetic field of 10^{-5} gauss. Even so, their trajectories become sufficiently complex that we cannot directly trace them back to their origins. At energies around 10^{18} eV some protons begin to spiral out of their parent galaxy's gaseous plane and its embedded magnetic field to escape into extragalactic space. At energies substantially higher than 10^{18} eV, most cosmic-ray protons escape their parent galaxy, often to be destroyed in crossing the vast distances between galaxies through collision with the ubiquitous microwave photons.

As messengers, cosmic-ray particles thus convey only limited information about where, when, and how they were formed. These limitations are endemic. Building instruments with higher angular, spectral, or temporal resolving powers will not shed greater light on their origins and subsequent history.

Gravitational Waves

In Chapter 4 we noted that, because the masses of neutron stars and stellar black holes are enormously higher than those of electrons and charged atomic nuclei, the spectral frequencies of the gravitational waves they radiate are many orders of magnitude lower than those of electromagnetic waves and their wavelengths λ are correspondingly longer. The duration $\Delta\tau$ of the wave trains explosively emitted in mergers of black holes or neutron stars, or in the collapse of massive stars preceding supernova explosions, also span just a few wavelengths, as Figure 4.3 shows, so that the ratio $c\Delta\tau/\lambda$ is extremely low and we are able to extract spectral information only at extremely low spectral resolving power.

Our ability to obtain high-angular-resolution information on the gravitationally radiating sources also is severely limited. Even for low-mass neutron-star binary mergers, the observed frequency f is no higher than $\sim$300 cycles per second, corresponding to wavelengths of order of $\lambda \sim 1000$ km. Terrestrial interferometric baselines D then cannot be much longer than a few wavelengths, limiting the angular resolving power to $\lambda/D \sim 3°$.

For the detection of the direction of arrival of a gravitational wave, both the phase and variable amplitude of a signal can be timed at several receiving stations. To obtain unambiguous locations of the source on the celestial sphere, at least four receiving stations distributed over the widest possible area are required. Increasing their mutual separation provides the highest angular resolution. Timing the arrival of a gravitational wave or even its undulations may then be the easiest part of determining the direction of arrival of the wave.

A considerably more difficult aspect of these observations is the establishment of sufficiently long and adequately stable baselines between the individual receiving stations to provide for high angular resolution. Plans are now going forward at the European Space Agency, with potential NASA collaboration, to place a gravitational wave observatory into an Earth-trailing Solar orbit, sometime in the mid-2030s.

With three receiving stations in the configuration of an equilateral triangle with sides spanning $D = 2.5$ million kilometers, this Laser Interferometer Space Antenna, LISA, is planned to sense gravitational waves in the 0.1 to 10^{-4} Hz frequency range, corresponding to wavelengths ranging from 3×10^6 to 3×10^9 km.[34] As this array floats in its circumsolar orbit with its 3×10^8 km diameter, it will gradually determine the direction of radiating sources – rather roughly at the longest wavelengths and with an accuracy of order of a degree, at least along directions tangential to its orbit, in its shortest wavelength ranges.

Gravitational waves can provide one particularly well-defined piece of information:

Binary merger events can yield an independent, fairly good estimate of the distance d_L at which a gravitational merger took place, as obtained through equation (4.6). This provides a check on the distances to remote cosmic sources, which otherwise largely depend on a complex sequence of calibrational steps involving triangulation, the periods and luminosities of pulsating stars, and the luminosities of supernovae of a given type, jointly constituting a *cosmological distance ladder*. Today, gravitational wave estimates of source distances are still crude compared to the more traditional distance ladder. If and when these estimates improve they may ultimately provide totally independent distance checks.

6.17   Sensitivity

Most astronomical observations obtained to date have provided information only on events occurring over limited time spans. Gradually evolving events have been reliably recorded for only a few centuries. Extremely rapid changes have often been difficult to follow as well, either because our instruments have been too sluggish or, more often, because narrowly concentrating on short time intervals prevented the collection of a sufficient number of photons to provide reliable signals.

Many instruments designed to follow rapid changes may thus appear to be sensitivity limited. Often, the yield of photons can be increased by constructing larger telescopes which then permits the detection of faint bursts of radiation and improved time resolution. Thus several currently available capabilities could be further extended, and are not intrinsically bounded.

Where affordable, extending the sensitivity or utility of a newly devised tool can rapidly be further honed, as testified by Figures 6.6, 6.7, and 6.8, demonstrating the speed at which instrumental capabilities evolved across broad astronomical fronts starting in the second half of the twentieth century.

Improvements on existing instrumental bounds may then be limited mainly by the cost of building ever-larger telescopes to gather more light to enable faster time resolution. Interferometers with longer baselines yielding improved angular resolution, or spectrometers with higher spectral resolution enabling the display of fainter spectral components, may likewise be limited by affordable cost.

Located at the peak of Cerro Pachón, a 2682-meter-high mountain in northern Chile, the Vera C. Rubin Observatory has a wide-field 8.4-meter-diameter reflecting telescope designed to survey the entire sky available at its site every few nights. In the course of a year, about 18,000 square degrees on the sky, roughly 45% of the celestial sphere, is to be covered, with about 825 distinct visits to each spot. Images will be obtained in six distinct wavelength bands ranging from 300 to 1080 *nanometers*, covering more than the entire visible wavelength band. The angular resolution of the images will be 0.7 arcseconds determined by atmospheric conditions, and will cover an exceptionally wide field of view extending 3.5 degrees in angular diameter across the sky.

Each exposure with this camera will be imaged onto a finer array of 3.2 billion pixels of charge-coupled devices, CCDs, each pixel covering a solid angle 0.2×0.2 arcseconds square on the celestial sphere.

In the course of routine mapping, each spot on the sky is to be imaged in two consecutive 15-second exposures to reject cosmic-ray hits on the pixel arrays. The telescope will then spend 5 seconds moving to its next adjacent field of view and stabilizing before starting its next exposure.

This sampling will permit the discovery of time variability in the optical regime on scales of 20 seconds, to several days, and to several years, in the course of the telescope's anticipated lifetime. Rapid computer-sifting of data is expected to produced of the order of a million alerts to detected changes each night. Each of these changes will be reported and available promptly within

60 seconds of detection. A log of prompt, day-to-day, and annual variability will then be compiled.

Ultimately, the performance of the Rubin Telescope at this mind-boggling clip will be limited by the ability to gather, process, store, and disseminate the deluge of obtained data. Available detector sensitivity does not appear to be a limiting factor to this enormously ambitious enterprise. This is not to say that improved detector sensitivity might not be important in other, less-developed wavelength regimes than the Rubin Telescope's optical domain, or for interferometry in narrow spectral bands. But it seems likely that astronomical advances may ultimately prove to be more dependent on other factors than improved detector sensitivity for studying the many complex processes astronomers often seek to better understand.

6.18   The Rate at Which Astronomical Information Is Lost

Impressive evidence showing the rate at which information about cosmic events is constantly being lost is provided by Figure 6.10, showing that only half the diffuse radiation reaching us from the remote Universe is contributed by visible and ultraviolet radiation. The other half lies in the infrared.

The diffuse radiation from these combined sources is called the *extragalactic background light, EBL*, defined as the integrated radiation over wavelengths in the range $10^{-5} \leq \lambda \leq 10^{-1}$ cm. The ultraviolet and visible components largely represent starlight and radiation from active galactic nuclei. The infrared component represents the portion of these emissions captured by interstellar dust and reradiated isotropically at lower frequencies, effectively erasing information on where the reradiated energy originated. Although the energy density of the EBL is 20 times lower than that of the cosmic microwave background radiation, it is a factor of ~100 higher than any other diffuse extragalactic radiation. If we keep in mind that the energy density of the EBL components generated during early epochs has, by now, undergone degradation through cosmic expansion, we can interpret its density to obtain a rough measure of all the energy ever radiated in star formation, accretion onto black holes, or any other dissipative processes. We encountered estimates of this general kind in the energy compilations of Fukugita and Peebles in Table 1.2.

Given the severity of the losses of information the EBL represents, it is worth going into some detail on work undertaken in recent years to at least recover insight on what these losses have entailed.

Efforts initially concentrated on observations of representative regions of the sky at low angular resolving powers, and correcting these by subtracting the radiation contributed by individual compact sources. Data provided by

Michael Hauser and colleagues on the Cosmic Background Explorer, COBE, and observations subsequently obtained by Hervé Dole's team on the Spitzer mission, gave a first set of relatively reliable direct observations.[35; 36]

At optical and ultraviolet wavelengths, Hubble Space Telescope observations reported by Rebecca A. Bernstein in 2007 were based on a summation of radiation from extragalactic sources fainter than 23rd magnitude corrected for Galactic foreground emission. These observations had to be supplemented with a more difficult subtraction for the diffuse foreground emission from the tenuous cloud of zodiacal dust grains orbiting within our own Solar System.[37]

More recent investigations compiled by Stephen K. Andrews, Simon P. Driver, and A. H. Wright, at the University of Western Australia, have depended on identifying myriad individual extragalactic sources, measuring their fluxes in some three dozen well-defined wavelength bands ranging from the far ultraviolet to the far infrared – at wavelengths ranging from about 1500 Å, or 1.5×10^{-5} cm, to ~ 0.05 cm – and integrating those findings into readily understood displays, such as Figure 6.10.[38; 39; 40] Toward the end of 2016, these separate lines of investigations tended to converge to mutually agreed values.

Observations compiled by Driver et al. (2016) ranged over a wide area of the celestial sphere and also cited data obtained by two spacecraft, Pioneer 10 and 11, as they gradually drifted out of the Solar System following launches, respectively, in 1972 and 1973. Included also were gamma-ray observations indicating the destruction of high-energy photons traversing extragalactic space to reach Earth.

The main result derived from their Figure 6.10 is that about half the extragalactic light, 52%, comprises mid- and far-infrared radiation and half, 48%, is contributed by ultraviolet, optical, and near-infrared photons. Half the information generated in starlight and in radiation emitted by AGNs has been absorbed and reradiated by dust, destroying half the information these sources of light could otherwise have transmitted. A slight compensating gain of information is that the absorbed radiation re-emitted in the infrared partially identifies the chemical composition of emitting dust grains, a mix of different size and chemical composition, both of which affect the integrated dust spectrum. Nevertheless the large-scale loss of information, through absorption of starlight by dust alone, lies close to 50%.

Reading the papers authored by Driver, Andrews, and Wright and their respective teams of co-authors shows the enormous amount of effort devoted to matching and assembling the data sets that went into the construction of Figure 6.10. Improving on these data are tasks that are less likely to be solved by constructing novel, more powerful tools.

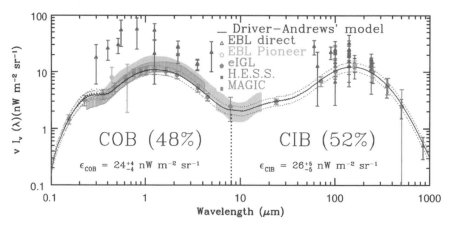

Fig. 6.10 The integrated galaxy light νI_ν displayed in units of nanowatts (10^{-9} watts) per square meter of telescope aperture and solid angle of a steradian. COB and CIB, respectively, refer to the optical and infrared cosmic background radiations pervading the Universe. The data compiled by Driver et al. (2016) are registered by data points and their deviations from a mean arrayed along the solid curve. These are compared to terrestrial and near-Earth measurements shown by triangles. Uncertainties are shown by the error bars attached to each set of points. The region marked with gray shading was derived from data generated by gamma-ray observatories. (From "Measurements of Extragalactic Background Light from the Far UV to the Far IR from Deep Ground- and Space-Based Galaxy Counts," Simon P. Driver, et al., *Astrophysical Journal* 827, 108, August 20, 2016, Fig. 6, p. 112, with permission of the authors and and the American Astronomical Society.)

To be sure, the angular resolution currently available with instruments in space at far-infrared wavelengths is ~36.3 arcseconds at ~500 μm, a factor roughly 40 times poorer than the angular resolution of ~0.9 arcseconds for ultraviolet observations. Considerably improved infrared data might thus provide decisive new information by matching far-infrared and ultraviolet data across smaller regions surrounding various classes of stars, distinct types of AGNs, as well as myriad other locales.

Nevertheless, the primary difficulties lie in the care that has to be taken in calibrating the effects of dust, whether it is through its absorption of light generated within the observed galaxies, the emission by the Solar System's zodiacal dust cloud or dust in interstellar space, or possibly even by radiation emitted or absorbed in extragalactic space. The matching of gamma-ray data to data obtained at ultraviolet and visual wavelengths, as Figure 6.10 does, is also complex.

These difficulties might eventually be overcome through assembly of larger data bases, though even this is not assured if those bases simply add more to the data already in hand without providing the information needed to arrive at closure on existing questions. Added observations are useful mainly if the number of questions they answer exceeds the number of new problems they raise!

We have a finite universe transmitting a finite amount of data, much of which already arrives polluted by passage through a complex dusty, partly ionized, gravitationally distorted, magnetically laced medium whose structure we cannot fully disentangle without further consideration.

6.19   Lost or Scrambled Information

As messengers cross the vast regions of interstellar or intergalactic space the information they convey can be altered in many ways. One particularly significant difficulty already sketched in Figure 5.8 is complex gravitational lensing. This is particularly insiduous because gravitational deflections affect all messengers equally. None can escape this fate or be used to calibrate it so we could purge its effects. But there are many other difficulties as well, illustrated by Figure 6.11 intended to provide a more comprehensive account of the difficulties different messengers need to overcome to arrive at Earth unscathed.

An observation of a radiating source in a remote region of the Universe, $L = n$, may not be correctly interpreted unless we take into account all the factors that may have affected its radiated emission on its journey to reach us. Figure 6.11 illustrates some of the common obstructions encountered by electromagnetic radiation from a source at some arbitrary location specified by its location on the sky, its right ascension and declination coordinates (α_i, δ_j).

Light following the indicated path first transits a layer labeled $L = (n - 1)$ without difficulty. Next, it may be slightly deflected by the gravitational attraction of a compact massive object such as a giant planet in some planetary system through which the light is passing. This deflection may not suffice to eject the radiation out of the field of view, but may be strong enough to slightly deflect its apparent position on the sky. Then, the radiation passes through a region in the corner of which a minor cloud of gas and dust may sap part of the star's emitted energy. From there on, nothing more may happen until the radiation also has to pass by a star in layer $L = 2$, whose radiation may not readily be separated from that of our primary source in layer $L = n$, unless the angular resolution of the observer's telescope suffices to distinguish the two sources.

Many of the hindrances listed in Figure 6.11 have long been understood and often can be calibrated away through judicious use of distinct messengers, some of which may easily overcome a particular hindrance. Gravitational lensing, in contrast, is a hindrance whose full impact has only recently been appreciated.

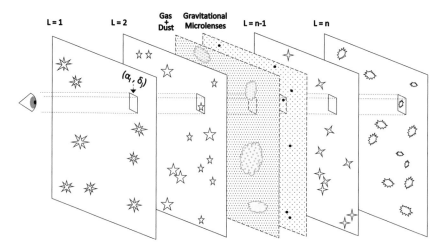

Fig. 6.11 An observer at left studies sources of radiation along a sight line toward an arbitrary coordinate position (α_i, δ_j) on the celestial sphere. On its trajectory toward the observer, radiation from a source in a remote region denoted by layer $L = n$ passes through several domains, which for all purposes may be considered to be layers populated by different types of stars of varied luminosities, concentrated in layers labeled $L = 1, L = 2, \ldots L = (n - 1)$. Interspersed layers may contain absorbing or scattering gas and dust, or massive objects gravitationally deflecting radiation. The observer's field of view around the sight line is indicated by the small squares carved out of each layer. Many of these squares may be empty. Others, like the square in layer $L = 2$, and in the more remote layers – respectively containing gas and dust or compact bodies gravitationally deflecting radiation through *microlensing* – may at least be partly populated. Some of the radiation from the source in layer $L = n$ may thus be absorbed or deflected along its trajectory, or difficult to separate from the radiation emanating from a luminous source in layer $L = 2$.

And so it is worth recalling the main difficulties it imposes, and thus identify just where increasingly powerful instruments may no longer help to advance observational astronomy.

6.20 Information Conveyed by Messengers Other than Photons

During the first few hundred thousand years of its existence, a fog of freely floating electrons permeated the Cosmos, absorbing, scattering, and scrambling all photons, thereby erasing most of the early history of the Universe, which otherwise might have revealed the foundations from which the Cosmos evolved.

Though this effectively sidelined all electromagnetic messengers, we remain confident that we may soon detect neutrinos at fluences sufficiently high for us to gain greater insight. And although we have no reason to be optimistic, potentially other primordial messengers might someday also reach us intact to suggest why the Universe today appears to primarily house matter, when all our theories suggest that matter and antimatter should have existed in equal abundances at the high temperatures prevailing at earliest epochs.

As Chapter 2 showed, two significant classes of messengers survived those primordial times to reach us intact:

The first were the light elements, hydrogen, helium, and lithium formed at the high temperatures prevailing during the first few minutes of cosmic history. Their relative isotopic abundances established during those earliest times, and subsequently maintained in recoverable ratios down to this day, were shown in Figure 2.5.

The second piece of evidence passed down over the eons consists of faint gravitational "ripples" in the otherwise smooth electromagnetic fog reaching us. The WMAP mission discerned these in minuscule temperature patterns these density fluctuations imprinted on the prevailing, otherwise largely featureless, microwave background radiation. Figure 2.7 displayed these.[41]

Even today the transmission of information across the Universe is distorted by interactions that need to be taken into account to faithfully portray ongoing cosmic events. Realms of freely floating interstellar electrons absorb long-wavelength radio waves that otherwise could reach us. As Figure 6.10 shows, about half of the visible and ultraviolet radiations emitted by stars and galaxies are quickly absorbed by ambient clouds of dust and gas, and reradiated at infrared wavelengths, erasing much of the information these emitting sources might otherwise have conveyed.

Interstellar magnetic fields deflect the trajectories of energetic electrons, protons, and charged atomic nuclei, so that we seldom can identify the sources that launched them. Streams of radiation are deflected and distorted, or even focused and potentially magnified on passage through gravitational gradients permeating the vicinity of massive galaxies or clusters of galaxies. Though the loss or distortion of information through such natural processes is beyond our control we can often take these losses into account and, where possible, correct for them on the basis of complementing data.

This jumble of transformed and at least partially compromised information then needs to be fitted into a panoramic display of the structure of the Universe and its evolution from earliest epochs – a vast jigsaw puzzle nature challenges us to reassemble.

6.21   Multi-Messenger Science

Messages conveyed by one class of messengers, or by messengers in one particular energy band, may succeed in reaching Earth practically unscathed by their journey through space, where other carriers of information potentially able to provide comparably useful data may arrive seriously impaired. This has led to the suggestion of an observing strategy in which each type of messenger is called on primarily for distinct observations at which it clearly excels.

Such a strategy sounds sensible, but differs from past assumptions by groups of specialists working with well-defined carriers of information such as neutrinos, or gravitational waves, or electromagnetic radiation, or in distinct wavelength bands, such as X-rays, infrared radiation, or radio waves. Those tacit assumptions have been that each discipline develops and perfects its own toolkits for both high- and low-resolution capabilities in all three of the primary phase space regimes defined by their spatial, spectral, and temporal dimensions, as presented in Figures 6.6, 6.7, and 6.8. A competing multi-messenger emphasis might have the advantage of avoiding duplication of tasks by mutually complementing messengers, concentrating instead on using each type of messenger primarily for purposes at which it excels. This might lead to a less costly approach to astronomy. It could, however, also rob us of warning signs, provided by mutual disagreements arising from distinct measurements carried out using different messengers, a strategy which often can hint that something in our understanding may be seriously amiss.

An example of the benefits of combined approaches was the discovery of the first directly certified binary neutron star merger to form a black hole. Gravitational radiation from such mergers exhibits wavelengths of order $\lambda = 1000$ km, only a factor of 10 lower than the maximum baselines separating gravitational wave detectors erected on our planet. This limits the location of individual sources on the celestial sphere to positional uncertainties of a few degrees.

For the sole neutron star merger detected to date, both through its emission of gravitational waves and X-rays, finding the outburst's location on the sky turned out to be simple. Neutron star mergers emit X-rays whose directions of arrival are readily determined with high accuracy. This enabled identification of the actual galaxy in which the merger had occurred. Perhaps this does not obviate a need for complex and presumably costly gravitational wave arrays with longer baselines spanning the Solar System, but it may at least assign them lower priority because, for purposes of detecting the merger of massive stars generating electromagnetic radiation, they could well be redundant.

A division and distribution of tasks, with each instrumental expertise focused on problems it is best and sometimes uniquely equipped to solve, could allow astrophysics to gain the most comprehensive insight at lowest cost. We know too little right now to recommend a particular emphasis. A careful analysis of alternative strategies for investing in the most economical mix of instrumentation and observatories to advance astronomy in the coming decades would certainly be worthwhile.

We will need to keep in mind, however, that multi-messenger methods will not overcome constraints imposed by gravitational lensing, to which we need to return one more time.

6.22   Gravitational Lensing

In Chapter 5 we found that passage past massive bodies cumulatively slows the transmission of radiation in passing close by such bodies. The Universe is filled with randomly distributed compact sources, large and small. At some level, any massive body gravitationally deflects radiation passing close by. These bodies can be asteroids, planets, stars, gas clouds, galaxies, or quasars – galactic nuclei harboring giant black holes – all more or less randomly distributed. Each of these bodies defines a gravitational terrain, sometimes guiding, sometimes disrupting the flow of messengers passing by and affecting the information they may convey.

No messengers can resist the gravitational deflections their trajectories suffer, because gravity constitutes the very fabric of space. Without gravitation we have no identifiable trajectories, no cosmic map, no organized Universe. Gravity channels and confines the trajectories all messengers are constrained to follow – no matter whether they are massive or massless.

Once the *gravitational deflection* limit is reached, building instruments with ever-higher angular resolving powers becomes ineffective. The Universe will be incapable of transmitting this information faithfully. Radiation reaching us from some regions will simply not have originated at the position from which the radiation falsely appears to be arriving.

As observational instruments become increasingly sensitive and we construct progressively larger telescopes capable of resolving sources at ever-greater distances, rays gravitationally deflected or delayed by ever-smaller, ever-less-massive astronomical bodies along the line of sight will lead to data becoming progressively more chaotic. Higher instrumental sensitivity, angular resolving power, spectral resolving power, or time resolution may then no longer help, because we will be facing too many unknown, uncontrollable sources of confusion.

6.23   What Will the Ultimate Cosmic Toolkit Entail?

Looking to the future, we can now begin to see how far we may need to improve existing observational capabilities to fruitfully tap whatever set of observations our unruly Universe may permit and make worthwhile. The assembly of this cosmic toolkit – this collection of instruments and observatories enabling us to reach this ultimate goal – may be described in terms of observational capabilities in hand already today.

Let us return to Figures 6.6, 6.7, and 6.8. They reveal two kinds of bounds. One set is defined by the Heisenberg uncertainty principle. Earlier in this chapter, we saw how this prevented clear distinction between successive pulses the Crab Nebula emits at nanosecond (10^{-9} s) spacings. Even the extraordinarily wide 2.5 GHz bandwidth of the original instrument recording these pulses in the 10.5 GHz radio band seemed unequal to the task.

In Figure 6.6 the Heisenberg uncertainty bound is shown by the sloping line at lower right. To obtain faster time resolution, we would need to observe the Crab Pulsar at higher radio or possible infrared or optical frequencies, as the slope of the dashed line dipping toward lower left indicates. The uncertainty principle will not permit observations below that line, which extends as far toward the lower left as we may wish.

Near the top of Figure 6.6 we encounter a complementing bound. Signals received near the top of the figure become progressively slower. And because most of the reliably established astronomical observations available today date back less than a century, we see that only a few observations at optical wavelengths have permitted us to directly witness gradual evolutionary trends over centuries. At the very top of the figure, we are bounded by processes so gradual they would take the age of the Universe to discern.

As also discussed earlier, detection of very fast pulses can similarly constrain spectral and angular resolving powers, as shown by the dashed lines marking limits for *Fast Coherent Pulses* at upper right in Figures 6.7 and 6.8.

Of course most astronomical signals are not expected to show such high pulse frequencies, and so we may well aim to build instruments with considerably higher spectral or angular resolving powers. These might well reveal new types of sources not yet discovered.

What specifically might we find? Extremely high spectral resolution is mainly useful in observations of sources exhibiting extremely low velocity spans. Very few interstellar gas clouds studied to detect their chemical constituents would require spectral resolving powers above, say, 10^{10} because their internal random velocities tend to be well above $\sim$1 meter per second. On the other hand, as we earlier saw, such high spectral resolving powers could reveal that some stars have habitable Earth-sized planets orbiting them.

So, we can see that we might usefully extend spectroscopic capabilities, at least in wavelength ranges running from the ultraviolet to the radio region, upward in Figure 6.7 to the level indicated by $\lambda/\Delta\lambda = 10^{10}$. And detection of highly coherent sources of radiation unknown to us today might require the use of even higher spectral resolving powers. Just as fast radio bursts initially revealed themselves only on rare occasions, sources of extreme spectral purity might also reveal themselves only rarely.

As discussed in Chapter 5, an angular resolution of 10 micro-arcseconds, currently available at a wavelength of 2 μm, with the GRAVITY interferometer at the European Southern Observatory in Chile, might succeed in singling out some of the more massive planets around stars within a 30 pc range, provided it were possible to couple such an interferometer to a high-performance coronagraph.

Similarly, singling out individual stars just beginning to form at redshifts $z \sim 10$ might require an angular resolving power of ~100 micro-arcseconds, to distinguish individual stars from their nearest neighbors. To be sure, the GRAVITY instrument operates in the 2 μm wavelength range, whereas the high redshifts of those earliest stars would make them more detectable at wavelengths of 10 μm, meaning that they would require observations from space, well beyond Earth's restricting atmosphere at these longer wavelengths.

In the radio frequency domain, as we also saw, the Event Horizon Telescope consisting of a network of powerful telescopes spanning our terrestrial globe has achieved an angular resolution of 25 micro-arcseconds, in partly resolving the region around a supermassive black hole in the galaxy M87, at wavelengths of 1.3 mm. Here also, going to higher angular resolving powers would require going into space, because the different telescopes forming the interferometer would need to span larger separations than Earth's diameter provides.

Perhaps extending currently available temporal, spectral, and angular resolving powers by factors of a hundred or even a thousand beyond those already available in astronomy today will eventually prove useful and worthwhile. Ambitious though this is, we should keep in mind that in the seven decades since the end of World War II, temporal resolving powers in astronomy have improved from nonexistence to ranges of order 10^6 to 10^{10}. Spectral resolving powers in many ranges again have frequently risen from nonexistence to 10^6 and sometimes higher, and angular resolving power has increased by factors of 10^8 or 10^9 in some ranges, as Figures 6.6, 6.7, and 6.8 so persuasively show.

This is why I believe that the Cosmic Toolkit will become available within just a century or two, and quite possibly sooner.

6.24   Epilogue

Let us now see how a useful upper limit to required angular, temporal, or spectral resolving powers may arise. This is likely to be influenced by an observer's priorities, but the examples shown may have an instructive generality.

Consider radiation reaching us from a remote source in the Galactic plane, say, at a distance of 10 kpc. To reach us it has to pass by myriad stars of roughly Solar mass $M_\odot$ distributed with a number density $n \sim 1$ per cubic parsec. For simplicity we take each of these stars to provide the sole gravitational forces encountered within a radial distance $R_{max} \sim 0.5$ pc $\sim 1.5 \times 10^{18}$ cm from the star, roughly corresponding to the radial distance from the star to the boundary of that cubic parsec.

Equation (5.4) tells us that the angular deflection θ the light experiences in passing a star on closest approach R obeys a simple product $R\theta = 4MG/c^2$. The value of θ remains constant over a cross-sectional area $2\pi R dR$ of a ring of radius R around the star and narrow width dR. The mean deflection $\langle \theta \rangle$ integrated over the range of potential distances from some closest passage R_{min} to a maximum R_{max} then is

$$\langle \theta \rangle = \frac{\int_{R_{min}}^{R_{max}} \theta [2\pi R] dR}{\int_{R_{min}}^{R_{max}} [2\pi R] dR} = \frac{8\pi GM}{c^2 [R_{max} + R_{min}]} \sim \frac{8\pi GM}{c^2 R_{max}}. \tag{6.1}$$

For $R_{max} \gg R_{min}$ this last approximation indicates that the mean deflection $\langle \theta \rangle$ is virtually independent of R_{min}, which we can now choose sufficiently large to exclude radiation never reaching the observer, through focusing at points X in Figure 6.12 lying between M and the observer O. Equation (5.4) shows that this generally holds for $R_{min} \geq 1.5 \times 10^{14}$ cm, a radius sufficiently large to prevent gravitational focusing of deflected radiation at distances less than a few kiloparsecs in the star's wake – well before ever reaching an observer at Earth. The ratio $(R_{min}/R_{max})^2$ indicates that only about one part in 10^8 of the incident radiation is lost through excessive deflection on closest approach to a star. For solar mass stars $\langle \theta \rangle \sim 0.5$ micro-arcseconds.

In transiting successive regions whose gravitational fields are dominated by individual stars, radiation is thus deflected laterally from side to side of the main trajectory in a largely random walk whose cumulative deflections grow as the square root of the number of individual deflections suffered. Across traversed distances of $\sim$10 kpc, past a succession of $n \sim 10^4$ individual stars, the accumulation of successive random deflections then gives rise to an angular deflection of order $n^{1/2} \langle \theta \rangle$, or $\sim$50 micro-arcseconds.

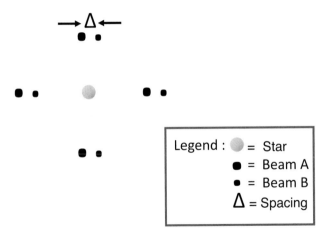

Fig. **6.12** Pairs of beams A and B deflected by a star. An observer sees approaching beams of radiation, A and B, arriving from beyond the disk of a foreground star. Originally they are separated by the horizontal displacement shown. If the two beams arrive from above or below the stellar disk they will respectively be gravitationally deflected downward or upward, but to lowest order the beam separation will not change. The beams arriving from beyond the left or right side of the disk will be deflected pairwise laterally toward the star, but the member of the pair passing closer to the star will be deflected more strongly, resulting in a shear systematically separating components A and B on each successive passage past a star.

The randomness arises because individual photons emitted from the remote source may follow slightly different paths, possibly interacting with a number of distinct stars before reaching the observer. This leads to a widened conical beam from which the radiation finally arrives at the observer, lowering the precision to which we can establish the actual position(s) on the sky from which the radiation emanated.

Angular precision differs from angular resolving power, which always refers to an ability to define the angle separating two observed sources. Let us consider beams of light reaching us from two neighboring sources A and B located in the same remote Galactic source as before. They are separated by a small distance $\Delta \ll R_{max}$, as shown in Figure 6.12. On average, the added angular separation between beams induced by each successive passage by a star now becomes

$$|\langle \theta(R)_A \rangle - \langle \theta(R)_B \rangle| = \frac{4\pi GM\Delta}{c^2 R_{max}^2} . \tag{6.2}$$

If we take $\Delta/R_{max} \sim 0.1$, corresponding to an angular separation of about 2 arcseconds between sources A and B at their original distance of 10 kpc, this implies a mean increase of angular separation of 0.025 micro-arcseconds per star encountered. This is substantially less than the random deflections individual beams or even pairs of beams jointly encounter. But the forces driving beam separation are systematic, rather than random, and add linearly, meaning that the angular separation between beams A and B will have increased by 250 micro-arcseconds after passing by 10^4 stars to reach us.

This is only an order of magnitude assessment because we don't know precisely how many stars any given pair of beams reaching us has encountered, nor what their individual masses were, nor their separations from nearest neighboring stars. Moreover, the choice of features separated by an angle $\Delta/R_{max} = 0.1$ may be deceptive. For an angular separation an order of magnitude lower, the gravitational shear per stellar encounter also diminishes correspondingly, so that the cumulative separation after 10^4 stellar encounters along the line of sight diminishes to less than the width of the random walk of the entire beam.

At this stage cumulative separations due to gravitational shear become progressively more difficult to distinguish from beams that potentially reach the observer along several distinct random walk paths. Quite generally, the angular resolution attainable cannot greatly exceed the angular precision of an observation limited by the coarse random walk uncertainties restricting the angular precision, which we just estimated as ~50 micro-arcseconds above. Instruments with angular resolving powers of order 10 micro-arcseconds should then suffice to derive just about all the astronomical information such an observation through the Galactic plane will normally yield.

The observations just described are relevant to studies of the massive black hole at the Galaxy's center, whose mass of $4 \times 10^6 M_\odot$ defines its diameter of 2.4×10^{12} cm. At a distance of 8 kpc, that diameter subtends an angle of 10^{-10} radians or 20 micro-arcseconds. This is somewhat lower than the best efforts just described, but perhaps still within reach of a carefully planned study.

Because gravitation deflects and channels all forms of energy equally, this shows that gravitational deflections will ultimately enforce a minimum angular noise level affecting all messengers, not merely electromagnetic radiation.

Here, I have acted as though the concept of angular resolving power is indistinguishably applicable to all observations. This needs more careful consideration: The spatial irregularities displayed by the cosmic microwave background radiation in Figure 5.12 indicate a coarse patchiness induced by massive extended foreground regions gravitationally lensing the background radiation on very large scales. Within individual patches, however, higher spatial resolution reliably exhibits much finer-scale structure. Applicable angular resolving power thus varies with the scale of regions observed.

Gravitational time delay also sets ultimate limits on temporal resolution. In the example of the dual beams just cited, beams A and B bypassing a star at radial separations of the order of 10^{17} cm are likely to suffer time delays differing by a few microseconds per stellar passage. In a random walk past 10^4 stars over a distance of 10 kpc, the cumulative time delays increasing as the square root of the number of stars encountered will rise to a level of order milliseconds.

Although I have treated the flow of radiation through the gravitational field of Galactic plane stars as largely a random walk, the recent observation of branched flow of light in a laboratory setting has demonstrated that electromagnetic waves propagating through a weak disordered potential, with correlation length larger than the light's wavelength, flow along narrow filaments that separate out like branches of a tree. Hence the name *branched flow*.[42] The authors of the paper vividly demonstrating this branching anticipate its occurring in curved general-relativistic fields, remarkably reminiscent of the undulating gravitational fields encountered along trajectories lying within the Galactic plane lined with myriad stars.

Equation (5.5) already provided some quantitative insight. At the heart of many of the limits involving time resolution again lie difficulties of judging gravitational time delay through deflections of arriving radiation reaching an observer simultaneously along different paths, though potentially radiated at distinct times by a single source, as depicted in Figure 5.8.

This same type of superposition of beams generated by a source at different epochs but fortuitously arriving at an observer simultaneously may also induce spectral smearing that cannot be reversed by increased instrumental resolving power.

In the absence of other factors, gravitational deflections ultimately will limit potential angular, spatial, and temporal resolving powers, even when the normal constraints of the Heisenberg uncertainty principle play no role.

Although we tend to think of gravitational deflection as dominated by truly massive deflectors, the amount of radiation blocked from reaching an observer from remote sources is likely to be influenced also by significantly smaller masses such as those of large or small planets, asteroids, or even cometary nuclei. Many of these smaller bodies emit little or no radiation, so that their gravitational deflection of radiation would be particularly difficult to predict.

Our knowledge of the abundance of these less massive objects is still fragmentary, even within the Solar System, but at some level of instrumental resolving power, information will inevitably be lost through random gravitational lensing even by some of the most common stars and minor planetary bodies already known to exist. This is where discovery space will shrink and ultimately vanish!

Notes

1 Dark Companions of Stars: Astrometric Commentary on the Lower End of the Main Sequence, P. van de Kamp, *Space Science Reviews* 43, 211–327, 1986, see pp. 317 ff

2 The Low-Level Radial Velocity Variability in Barnard's Star (= GJ 699): Secular Acceleration, Indications for Convective Redshift, and Planet Mass Limits, M. Kürster, et al., *Astronomy & Astrophysics* 403, 1077–87, 2003

3 A Candidate Super-Earth Planet Orbiting near the Snow Line of Barnard's Star, I. Ribas, et al., *Nature* 563, 365–68, 2018

4 A Planetary System around the Millisecond Pulsar PSR1257+12, A. Wolszczan & D. A. Frail, *Nature* 355, 145-47, 1992

5 Confirmation of Earth-Mass Planets Orbiting the Millisecond Pulsar PSR B1257+12, A. Wolszczan, *Science* 264, 538–42, 1994

6 Ibid., Wolszczan & Frail, 1992

7 Ibid., Wolszczan, 1994

8 A. Wolszczan, *email* to Martin Harwit, July 8, 2008

9 A Jupiter-Mass Companion to a Solar-Type Star, M. Mayor & D. Queloz, *Nature* 378, 355–59, 1995

10 Planetary Systems: Crossing the Jupiter Threshold, G. W. Marcy & R. P. Butler, *Bulletin of the American Astronomical Society* 27, 858, 1995

11 51 Pegasi, M. Mayor, et al., *IAU Circular* 6251, 1995

12 The Number of Class A Phenomena Characterizing the Universe, M. Harwit, *Quarterly Journal of the Astronomical Society* 16, 378–409, 1975

13 *Cosmic Discovery: The Search, Scope, and Heritage of Astronomy*, M. Harwit, Basic Books, New York, 1981; reissued by Cambridge University Press, 2019

14 A Bright Millisecond Radio Burst of Extragalactic Origin, D. R. Lorimer, et al., *Science* 318, 777–80, 2007

15 Fast Radio Burst Discovered in the Arecibo Pulsar ALFA Survey, L. G. Spitler, et al., *Astrophysical Journal* 790, 101 (9 pp.), 2014

16 The Host Galaxy of a Fast Radio Burst, E. F. Keane, et al., *Nature* 530, 453–56, 2016

17 *Planck* 2018 Results. VI. Cosmological Parameters, Planck Collaboration, *arXiv: 1807.06209v1 [astro-ph.co]* July 17, 2018

18 A Repeating Fast Radio Burst, L. G. Spitler, et al., *Nature* 531, 202–5, 2016

19 Radio Wave Propagation and the Provenance of Fast Radio Bursts, J. M. Cordes, et al., *arXiv:1605.05890v1*, 2016

20 A Magnetar-Powered X-Ray Transient as the Aftermath of a Binary Neutron-Star Merger, Y. Q. Xue, et al., *Nature* 568, 198–201, 2019

21 Extended Tail from SGR 1806-20 Bursts, E. Göğüş, et al., *Astrophysical Journal* 740, 55, 2011

22 Radio Emission Signatures in the Crab Pulsar, T. H. Hankins & J. A. Eilek, *Astrophysical Journal* 670, 693–701, 2007

23 Supergiant Pulses from Extragalactic Neutron Stars, J. M. Cordes & I. Wasserman, *Monthly Notices of the Royal Astronomical Society* 457, 232–70, 2016.

24 Theory of Communication, D. Gabor, *Journal of the Institution of Electrical Engineers* 93, 429–41, 1946

25 Ibid., Gabor, 1946

26 Nobel Lecture: Passion for Precision, T. W. Hänsch, *Reviews of Modern Physics* 78, 1297–309, October–December 2006

27 Resonant Electro-Optic Frequency Comb, A. Rueda, et al., *Nature* 568, 378–81, 2019

28 *Tools of Radio Astronomy*, K. Rohlfs & Thomas L. Wilson, A&A Library, Springer Verlag, 4th Revised and Enlarged Edition, pp. 192 and 201, 2003

29 Probing Dissipation Mechanisms in BL Lac Jets through X-Ray Polarimetry, F. Tavecchio, M. Landoni, L. Sironi, & P. Coppi, *Monthly Notices of the Royal Astronomical Society* 480, 2872–80, 2018

30 New Instrument Detects the Tug of Smaller Exoplanets, *Science* 358, 1231, 2017

31 Outgassing History and Escape of the Martian Atmosphere and Water Inventory, H. Lammer, et al., *Space Science Review* 174, 113–54, 2013

32 Denis Overbye article in the *International edition of the New York Times, INYT*, June 22, 2017

33 Fast Radio Burst 121102 Pulse Detection and Periodicity: A Machine Learning Approach, Y. G. Zhang, et al., *Astrophysical Journal* 866(2), 149 (18 pp.), 2018

34 http:sci.esa.int/lisa/61367-mission-summary/

35 The COBE Diffuse Infrared Background Experiment Search for the Cosmic Infrared Background. I. Limits and Detections, M. G. Hauser, *Astrophysical Journal* 508, 25–43, 1998

36 The Cosmic Infrared Background Resolved by Spitzer: Contributions of Mid-Infrared Galaxies to the Far-Infrared Background, H. Dole, et al., *Astronomy & Astrophysics* 451, 417–29, 2006

37 The Optical Extragalactic Background Light: Revisions and Further Comments, R. A. Bernstein, *Astrophysical Journal* 666, 663–73 , 2007

38 Measurements of Extragalactic Background Light from the Far UV to the Far IR from Deep Ground- and Space-Based Galaxy Counts, S. P. Driver, et al., *Astrophysical Journal* 827, 108, August 20, 2016

39 G10/COSMOS: 38 Band (Far-UV to Far-IR) Panchromatic Photometry Using LAMBDAR, S. K. Andrews, et al., *Monthly Notices of the Royal Astronomical Society* 464, 1569, October 3, 2016

40 Galaxy and Mass Assembly (GAMA): Accurate Panchromatic Photometry from Optical Priors Using LAMBDAR, A. H. Wright, et al., *Monthly Notices of the Royal Astronomical Society* 460, 765, 2016

41 Three-Year Wilkinson Microwave Anisotropy Probe (WMAP) Observations: Temperature Analysis, G. Hinshaw, et al., *Astrophysical Journal Supplement Series* 170, 288–334, June 2007; see Fig. 21, p. 323

42 Observation of Branched Flow of Light, A. Patsyk, U. Sivan, M. Segev, & M. A. Bandres, *Nature* 583, 60–65, July 2, 2020

Part IV

The Pace of Progress

7

An Era of Surveys

7.1 New Communication Channels, New Phenomena, and Extended Bounds

Gravitational waves first detected directly by the Advanced LIGO experiment in late 2015 illustrated two generic traits:

They revealed a new carrier of information available to astrophysics, and showed that novel instruments could be engineered to detect such theoretically predicted messengers. Separately, they also confirmed observationally that stellar-mass black holes could merge in a cataclysmic collapse general relativity had predicted.

Recognizing these separate aspects of a single discovery is useful. It contrasts the development of a new tool to search for messengers never detected before to the tool's ability to identify entirely novel cosmic phenomena. Both efforts are valuable, but each pursues a distinct priority.

Frequently, the introduction of a novel tool soon reveals the major phenomena it is particularly adept at detecting. Older, more established tools in contrast often are better suited to analyzing existing phenomena, classifying their properties and defining their variants. In part this is in line with a related characteristic first documented extensively by Peter Galison, historian of science at Harvard.

7.2 Golden Events

In his book *Image and Logic*, Galison distinguishes two traditions in high-energy physics. Each appears to also play a key role in astronomy.[1]

First, he notes the power of a single event – a *golden event* so totally decoupled from any conceivable extraneous disturbances that the insights it provides are incontrovertibly convincing. In astronomy such powerful events are rare but singularly persuasive, enabling theory, speculation, computer simulations, and overarching world views to all fall into place.

Among the golden events of high-energy physics, Galison lists Carl D. Anderson's 1932 photograph of the trail left by a cosmic-ray positron. It convincingly showed the existence of antimatter in nature – a new class of matter never detected before, but neatly fitting into Paul Dirac's reformulation of quantum theory. Through its discovery of a new cosmic constituent and a previously unanticipated potential messenger, it became a golden event also for astrophysics.

The first direct detection of gravitational waves, just mentioned, similarly was such a golden event for astronomy, in clearly identifying also the existence of massive binary stellar black holes. For physicists this was a golden event as well, confirming in exquisite detail general relativity's predictions of the structure these waves should exhibit.

In Chapter 4 we encountered the discovery of a "bizarre" source SS 433, whose spectrum changed relativistically from day to day. For Bruce Margon and his colleagues it departed from anything ever reported by astronomers before. Surely this was a golden event! The subsequent decades-long search for what this observation actually meant, and its conclusion that it showed the existence of quasar-like objects on stellar-mass scales, far better fits into a quite different way of advancing science which Gallison named the *logic tradition.*

7.3   The Logic Tradition

We may consider this more complex tradition in terms of Figure 6.11 encountered in Chapter 6. This raised a question we should now revisit: Given the complex layers of activity through which messengers reporting on remote cosmic events need to pass to reach an observer on Earth, how can we be certain of anything we observe? Or will we need to resign ourselves to the sobering thought that any astronomical measurement we undertake will inevitably entail considerable uncertainty?

Some of the large surveys undertaken today provide insight on how astronomers deal with such difficulties. The question can be mathematically rephrased in terms of the many different sight lines, each defining a different declination and right ascension on the sky, labeled (α_i, δ_j) in Figure 6.11. Radiation reaching us along such a sight line may have passed through a dust cloud, gaseous regions of different chemical compositions and densities, or been gravitationally lensed long before reaching us.

Can we then gain sufficient information from all the distinct $\ell \times m$ combinations of sight lines with combinations of subscripts $i = 1, \ldots, \ell$ and $j = 1, \ldots, m$, so that, using the laws of physics, we would have enough information to derive a self-consistent physical model for an event occurring in the remote volume $(\alpha_i \delta_j, n)$ in layer n of Figure 6.11?

In the most general case, the answer is "No." A completely self-consistent physical model would need to conform to $\ell \times m$ mutually independent equations summing all the signals from n different locations along each sight line. We thus would have $n \times m \times \ell$ variables with only $m \times \ell$ equations plus a handful of laws of physics providing additional information.

One reason we are able to overcome such odds in astronomy is that the occupancy of the Universe is sparse. Most of space is empty. A few sight lines often are occupied by just a single dominating astronomical body that can then be studied in isolation, where the laws of physics can help us understand the nature and behavior of that body.

We tend to learn from these clearly isolated events, essentially devoid of peripheral influences. Their isolation leaves no doubt about their inherent nature. Once a number of these golden events are known and the physical processes they illustrate are understood, we often can infer how these events might change in more complex environments.

Verifying those inferences may then require surveys, which individually may not teach us much until an entire trove of survey data is sifted in search of a fraction of distinct conditions under which the observed systems tend to evolve identically. Step by step this can lead to insight on increasingly complex systems. This is the essence of Galison's *logic tradition*. If a lack of golden events puts us in a bind, the logic tradition can still provide persuasive insight. Galison applied the concepts of *golden events* and *logic tradition* to portray how high-energy physicists deal with complex questions. The two traditions apply in astronomy as well.

Earlier, we noted how the accidental discovery of two virtually identical, well-isolated images of the quasar 0957 + 561 had led astronomers Walsh, Carswell, and Weymann to conclude, in 1979, that they were dealing with gravitationally lensed images theoretically predicted by Sjur Refsdal 15 years earlier.[2; 3] The existence and importance of gravitational lensing was therefore quickly accepted by the astronomical community persuaded by this *golden event*.

In contrast, statistical attempts to recover the true shapes of myriad remote galaxies, whose lateral dimensions appear to have been distorted through gravitational lensing by foreground galaxies, have proven to be far more difficult. Here, the logic tradition comes into play as we attempt to find earmarks of gravitational lens distortions by comparing the shapes of galaxies that have been gravitationally lensed to the shapes of galaxies that clearly have not.

The logic tradition, however, also has its limits. Inevitably, as David Weinberg remarked in Chapter 5, a proper accounting of microlensing tends to call for ever-larger surveys within the *Hubble radius* to ensure nothing has been overlooked. Eventually, when we have surveyed all the galaxies within this sphere, we reach the *cosmic variance limit* at which we have exhausted all the data the Universe provides.[4]

Whether a discovery should be considered due to a golden event or in the logic tradition is never completely clear. Most astronomical findings have elements of both. But it is useful to understand how recognition of fundamental advances can come about through two such distinct approaches, and how these may be perceived by different astronomers, some of whom may be more readily convinced by one of these forms of evidence rather than the other.

7.4   A Logic of Extracting Faint Commonalities

Galison notes that, in contrast to golden events, particle physicists more often extract statistical information from successions of hundreds of billions of particle collisions at high-energy colliders, each producing an uninformative-looking spray of hadrons and leptons. Only when many billions of these events are compiled and examined do common traits emerge. If the first hundred billion collisions don't resolve a question, a further hundred billion events may. This is how Galison's *logic tradition* tends to advance high-energy physics.

Astronomy is only beginning to confront such tasks. We do not have the luxury of repeating analyses on successive sets of hundreds of billions of galaxies. A hundred billion is the total number of galaxies in the portion of the Universe we survey. When we run out of galaxies, we encounter nature's firm limits. If the shared properties revealed by logic then appear universal, such findings can become persuasive by default.

Galison cites the 1953 demonstration by Frederick Reines and Clyde L. Cowan for the existence of neutrinos as falling into this tradition. They based "their experimental argument for the free neutrino entirely on statistical measurements from a sophisticated array of counters wired in coincidence and anticoincidence around a large vat of liquid scintillator."

An astronomical counterpart in the logic tradition may be the conclusion that, if they exist, highly massive dark matter objects cannot constitute a significant fraction of the gravitational attraction the Galaxy exerts. As we saw in Chapter 5, when the MACHO team had photometrically observed 8.5 million stars in the Large Magellanic Cloud and found only eight candidate microlensing events that stellar-mass bodies of dark matter conceivably might have produced, they concluded that invisible clumps with masses comparable to those of stars could not account for the gravitational forces attributed to dark matter.

In the logic tradition it is also significant that the MACHO consortium had needed to "learn how to perform event selection from the data; it was not possible to develop meaningful selection criteria independent of the data." This may constitute the ultimate description of logical conclusions based on and derived from data examined without prejudice or reliance on preconceived theory.

The difference between golden events and logical derivation may best be shown by the two main thrusts introducing neutrino astronomy in the 1980s: The totally independent detection by the Kamiokande and IMB groups of the powerful burst of neutrinos and antineutrinos from supernova SN 1987A was clearly a golden event. In contrast, the logic tradition was definitely involved in the laborious extraction by Raymond Davis Jr. of the argon isotope ^{37}Ar, collected atom by atom, year after year, in his 390,000 litre vat of carbon tetrachloride, almost a mile underground. He rejected all types of distractions to finally assemble enough data to claim detection of neutrinos from the radioactive decay of unstable boron nuclei deep in the Sun. Initially it was not clear why the number of ^{37}Ar atoms Davis collected was substantially lower than theory had predicted. Ultimately this did not matter; once neutrinos were understood to have rest-mass, and that electron neutrinos arriving from the distance of the Sun could oscillate to states that no longer interacted with chlorine ^{37}Cl, the detections Davis had documented began to neatly fit theoretical predictions.

7.5   An Era of Surveys

The rapid technological advances brought about by World War II and the Cold War era profoundly affected the pace of astronomical advances. Initially, first discoveries often were random and remained isolated events. Not until the 1980s were more systematic surveys widely undertaken in virtually all electromagnetic energy ranges. Detection of cosmic neutrinos, analyses of the most energetic cosmic-ray particles, and direct measurement of gravitational waves provided further means of probing the Universe.

Three factors made such surveys not only possible but affordable.

The first was the introduction of highly sensitive, low-noise, solid-state detectors. These tiny devices, by then often arranged in large two-dimensional arrays, provided images clearer than those those of photographic plates, at vastly faster rates and over a far wider wavelength range.

The second was the industrial community's simultaneous and technologically related advance in the development of computing power, speed, and data storage devices, all available at incredibly affordable prices.

The third was the reliable launch of satellites conveying complex payloads into space. Once aloft, a telescope could operate virtually around the clock. Day or night no longer mattered. Gamma-ray, X-ray, or far-infrared radiations unable to penetrate Earth's atmosphere could be readily observed.

Space also offered an unanticipated added bonus – a degree of instrumental stability that had never before been attained. Observations could be taken one day and then repeated months or years later to yield an essentially identical result. If a difference was indeed found, it probably reflected a real change in an observed source. The space environment was immune to Earth's atmospheric turbulence, weather, seasons, volcanic activities, warfare, traffic noises and vibrations, or the urge of scientists to tinker with instruments to keep them continually in prime condition.

How and where had these advances so rapidly emerged?

All were part and parcel of two urgent demands – the need of the military's Cold War ballistic missile guidance and early warning defense systems, and an ever-increasing public thirst for improved entertainment and telecommunication services. Add to this a mounting emphasis on public health improvement and the invention of non-invasive medical diagnostic imaging, and it is easy to see how and why military and industrial investments largely paid for the advanced sensors, electronics, and data processing systems astronomers required for increasingly ambitious undertakings. All were delivered at breathtakingly low cost to a curiosity-driven astronomical enterprise.

In hindsight, these low costs were understandable. Most of the required electronic components merely consisted of tiny specks of doped silicon. The cost of the raw materials was hardly worth discussing. The main cost in successfully engineering these novel devices had gone into discovering the means for inexpensively purifying silicon or germanium, finding the optimum levels of chemically doping these basic materials, and determining the best ways of transmitting electric currents through the devices noiselessly. With these manufacturing advances – often made possible by advances in quantum physics and chemistry – prices dropped, and astronomers could take advantage of novel devices at minimal expense.

For astronomy a main problem became the development of sensitive apparatus built to survive the accelerations and vibrations of a rocket launch and the subsequent need to operate reliably on arrival in space. Longevity in space was also a concern. Instruments had to be designed and developed to operate faultlessly for many years, indirectly powered solely by sunlight, with little or no need for expendable propellants or coolants.

Spectral and timing capabilities enabling investigations previously beyond reach now became possible. Large surveys conducted at wavelengths that had earlier been inaccessible could now flourish. Observations orders of magnitude more complex than any undertaken before could now be undertaken, their results electronically processed, archived, and shared via the internet with colleagues at any number of astronomical institutes.

Extensive surveys in all these wavelength ranges would not only lead to new discoveries, but would also show how rare or how common some recently discovered phenomena might be, and whether, or how well, prevailing theory could account for them.

Some of these surveys, no doubt, would be far costlier than any contemplated before and would involve larger consortia of astronomers than had ever been previously assembled. This had to be expected: The cost of launching equipment into space continued to remain high, and made sense only if large segments of the astronomical community expressed their interest by contributing their efforts. The total funding allotted to astronomy was certainly limited, and an expensive survey would inevitably require curtailing a large number of smaller investigations, even if the overall amounts spent on astronomy might gradually expand and permit larger numbers of scientists, often young post-docs, to enter the field.

The high cost of supporting the work of this expanding community required a new emphasis on increased efficiency. Most ground-based national observatories, and certainly all space observatories, began restricting themselves to limited sets of standard observing modes, upgraded only on a regulated schedule. The rigorous attention to conformity made anomalous observational results much easier to check and verify. Minute deviations from anticipated results now could be readily checked and their origin determined. Operating in this revised mode, observatories launched into space, as well as their more traditional ground-based counterparts, began to yield a legacy of further discoveries.

Starting with nearby occurring events and working our way out to the furthest reaches of the Cosmos, we may now sketch a succession of major discoveries.

7.6   Debris and Pre-Planetary Accretion Disks (1984)

The *Infrared Astronomical Satellite (IRAS)* launched into space by a consortium of US, Dutch, and British scientists in 1983 provided an unbiased survey of nearly the entire sky. It charted the celestial sphere in four broad spectral wavelength bands at 12, 25, 60, and 100 micrometers, μm. One of its major discoveries was the existence of extended clouds of debris around stars born within the past hundred million years.

By the start of the mission, Fred Gillett of the Kitt Peak National Observatory and H. H. (George) Aumann of the Jet Propulsion Laboratory had assumed responsibility for guiding the orbit-by-orbit calibration of the all-sky survey in the four wavelength bands. One of the stars they had selected to serve as a primary standard against which radiation received from other sources would be calibrated was Vega, α-Lyrae, a star displaying a spectrum roughly matching that of a blackbody at a temperature of 10,000 K.

Soon, something appeared amiss either with the instrumentation, the data reduction software, or the anticipated spectrum of Vega: At far-infrared wavelengths, the star appeared far more luminous than expected – 7 times more luminous at 60 μm and 16 times more luminous at 100 μm than a blackbody at the originally anticipated stellar temperature of 10,000 K. The region around the star from which this excess radiation arrived was also far more extended than the stellar surface. Led by Aumann, the IRAS team reported:[5; 6]

> The source of the 60 μm emission has a diameter of about 20″. This is the first detection of a large infrared excess from a main-sequence star without significant mass loss. The most likely origin of the excess is thermal radiation from solid particles more than a millimeter in radius located approximately 85 AU from α Lyrae and heated by the star to an equilibrium temperature of 85 K. These results provide the first direct evidence outside of the solar system for the growth of large particles from the residual of the prenatal cloud of gas and dust.

A decade later, a follow-on survey conducted with the Infrared Space Observatory (ISO), launched by the European Space Agency, further elucidated the nature of circumstellar disks. Observations undertaken by the Dutch astrophysicist Harm J. Habing and his colleagues showed that stars which could be identified as older than about 400 million years had shed their disks, while stars younger than this still retained them.[7] In ISO observations conducted at wavelengths ranging from 90 to 200 μm, Derek Ward-Thompson of Cardiff University in Wales and Philippe André working at Saclay in France also found some of the first indications of how the onset of star formation may be observed in cool interstellar clouds at temperatures less than 20 K. At this low temperature, nothing could prevent these clouds from collapsing gravitationally.[8]

Radio observations of the central portions of a number of interstellar clouds had shown them to be rich in carbon monoxide, CO, and ammonia, NH_3.

Ward-Thompson and André called these regions *pre-stellar cores*. A lack of emission at shorter infrared wavelengths indicated that they did "not contain embedded, accreting protostars or young stellar objects, but rather [were] in the pre-collapse phase of evolution." Infrared and radio astronomical surveys such as these were beginning to lead to theories of how stars and planetary systems begin to form.

The history of these insights on how planetary systems form, shortly after a star is born, indicates how carefully key observations need to be followed up to obtain credible evidence on how nature actually works. Here the logic tradition prevailed in piecing together a credible sequence of events.

7.7   The Emergence of Brown Dwarfs (1995)

A question that had been lingering for several decades concerned the smallest, least massive stars that we might imagine, and how the difference between such stars and ordinary planets might best be defined. By the late twentieth century, we knew that a star must be sufficiently massive to gravitationally induce high temperatures and pressures deep in its interior to trigger nuclear reactions releasing energy that made the star shine. Stars like the Sun radiate energy by converting hydrogen into helium.[9]

As early as 1963, Shiv S. Kumar, at NASA's Institute for Space Studies in New York, and Chushiro Hayashi and Takenori Nakano, at Kyoto University in Japan, were wondering whether stars of extremely low mass might find other ways of generating energy.[10; 11; 12] Specifically, it seemed that such a class of objects, while not sufficiently massive to convert hydrogen into helium, could be sufficiently massive to at least convert the relatively rare hydrogen isotope deuterium, or the light elements lithium and beryllium, into helium. Because these light atomic species are rare, their energy yield might be meager; but Kumar's calculations suggested that these small bodies could continue to shine for about a billion years.

By the early 1990s agreement had emerged that bodies more than 10 times as massive as Jupiter, and 1 to 8% the mass of the Sun, should be able to radiate energy through these nuclear reactions.[13] Their predicted luminosity was quite low, partly because they had to be small – comparable to the size of Jupiter – in order to generate a sufficiently high central temperature. The small amount of energy generated would suffice only to heat such a star's small surface area to low temperatures, at which they would glow feebly; the color they displayed would then appear brown. Hence the name "brown dwarf." Once such a star had converted all its light-element nuclei into helium, it would gradually cool but be unable to contract further because the pressure at the brown dwarf's center would not suffice to further compress the densely packed cloud of *degenerate electrons* squeezed out of atoms at the star's center.

The first convincingly identified brown dwarf was Gliese 229b, a faint companion to the star Gliese 229 in a listing compiled by the Heidelberg astronomer Wilhelm Gliese.[14] This star, first identified as a brown dwarf by Benjamin R. (Rebecca) Oppenheimer, a graduate student of Shrinivas Kulkarni at Caltech, and their co-discoverers Keith Matthews and Tadashi Nakajima, lies at a distance of only 5.7 pc or 18.6 light years from the Solar System.[15] Its surface temperature is only about 1000 K, but significantly higher than that of Solar System planets, and its spectrum exhibited methane absorption, never seen in more massive stars because at higher temperatures carbon forms carbon monoxide molecules, CO, whereas below 1000 K CO tends to combine with hydrogen to form water, H_2O, and methane, CH_4.

Over the next five years, this discovery was followed by the unveiling of another 11 brown dwarfs, thanks to two ground-based, large-area near-infrared sky surveys that came on line in 1996 and 1997 – the European *Deep Near Infrared Survey of the Southern Sky, DENIS*, and the US *Two Micron All Sky Survey, 2MASS*. By the year 2005, 450 brown dwarfs had been catalogued and separated into two new classes by their respective surface temperatures – L stars ranging from 2500 K down to 1300–1500 K, and T stars, from 1300–1500 K to about 700 K.[16]

The careful interplay of theory and observations, to identify the potential existence of a class of stars that could be faint but nevertheless self-sustaining for long periods, without depending on the conversion of abundant hydrogen or helium into heavier elements, made this a discovery in the logic tradition.

7.8   Galaxy Evolution and Mergers (1984)

An initial indication on how galaxies first formed when the Universe was still young was glimpsed through an audacious observation arranged for the Christmas period, December 18 to 30, 1995, when many US astronomers were quite happy to stay home with their families. Over those days the Hubble Space Telescope was steadily gathering data on a small dark northern patch of sky in continuous view of the telescope as it orbited Earth every 90 minutes. The region was devoid of bright Milky Way foreground stars and thus provided a clear view of the distant Universe.[17]

The result of these observations revolutionized our ways of thinking about the growth of galaxies: At great distances and thus at early times in the evolution of the Universe, the *Hubble Deep Field*, as the patch of sky displayed in Figure 7.1

Fig. 7.1 A Portion of the Hubble Deep Field. A major, unanticipated finding of the Hubble Space Telescope was that a portion of the sky, deliberately selected for its apparent lack of galaxies or stars, when stared at for sufficiently long periods, revealed a large number of small, distant, blue galaxies, quite different from the larger, redder galaxies observed at low redshift nearby. Over the eons, these small galaxies evidently merged to form the larger galaxies we see locally today. In such mergers the gas in the colliding galaxies gives rise to new generations of massive stars, producing neutron-rich heavy chemical elements they eventually eject into their environs. Evolution along these lines appeared consistent also with the finding that, in some of the earliest-formed stars still found in our Galaxy and in the intergalactic medium at high redshifts, these heavy chemical elements are detectable only in minute concentrations, whereas more recently formed stars contain these elements in abundance. (This work was originally published in the *Astronomical Journal* 112, 1335–89, plates: 1734–52, 1996. Credit: NASA/JPL/STScI Hubble Deep Field Team.)

came to be known, showed a preponderance of small blue galaxies – blue presumably because massive young stars were lighting them up. Nearer to us, meaning in more recent evolutionary eras, the Universe exhibited fewer but markedly larger galaxies. Evidently the galaxies we see around us today had not been born the way they appear today. Over the eons, the small galaxies formed earlier had been continually merging and evolving.

What distinguished this survey was not so much that it had no specific preconceived aims. Rather, it was that the Hubble deep field survey conducted at visible wavelengths could be compared to other deep surveys that astronomers had recently begun to compile in previously inaccessible wavelength domains. Often these were all-sky surveys to search for whatever one might find.

One of these was the IRAS mission discussed above, which had catalogued some 250,000 celestial sources, the vast majority of which had never before been detected in the infrared. A most striking finding was the discovery of galaxies emitting up to 80 times more energy at far-infrared wavelengths than in the optical domain. Many of these galaxies also were a hundred to a thousand times more luminous than our own Milky Way. Some of these luminous sources appeared to be colliding galaxies, apparently merging. Others exhibited active nuclei reminiscent of *quasars – quasistellar objects*.[18]

At first, these mergers and their enormous luminosities were puzzling, indicating the existence of highly energetic processes that had not been anticipated in theories of galaxy evolution. Soon several evolutionary trends began to emerge. The ultraluminous mid- and far-infrared galaxies were forming unbelievable numbers of massive stars, at rates and in ways we still do not properly understand. The mergers also were transporting matter toward the center of mass of the merged configuration, with the result that a supermassive black hole could form, giving rise to additional energetic processes – quasar-like relativistic outflows, the generation of cosmic rays, and the emission of X-rays.

A dramatically changed view of galaxy formation came to the fore. We began to see that both in the earliest-formed small galaxies, and in galaxies that had undergone later mergers, massive stars were far more common than in galaxies forming stars today in our Galaxy's immediate cosmic neighborhood. These massive stars are the main sources of heavy elements formed deep in their interiors. At the end of their lives, they eject heavy elements into their surroundings through supernova explosions or, as we saw in Chapter 4, through neutron star mergers.

The growth of galaxies, the formation of stars, and a steady enrichment of heavier chemical elements in the Universe at early times, now appear to have been intertwined consequences of mergers originating in the populated centers of galaxy clusters. Rapid star formation is also triggered by extragalactic gases falling into massive galaxies.

The discovery by IRAS that merging galaxies give rise to massive star formation, and the Hubble Space Telescope's striking discovery that galaxies are born small, and only then evolve into the large galaxies observed in today's Universe, came as a surprise. Both appear to have been discoveries in the golden events tradition. The realization that large, deep, blind surveys can yield pay-offs through a single new overview, rather than solely through hints and conclusions steadily emerging from accumulating large data sets, is worth noting here. The next section points to similar conclusions that emerged from another wide-ranging blind survey.

7.9   Galaxy Clusters, Filaments, Sheets, and Voids (1989)

Over a number of years preceding 1986, two distinct groups had begun charting the two-dimensional distribution of galaxies across the sky, and determining the galaxies' individual redshifts as a measure of their distance. This yielded a three-dimensional distribution of galaxies in the nearby Universe.

Martha Haynes at Cornell University and Riccardo Giovanelli at the Arecibo Observatory had by then mapped the distribution of 2743 galaxies in a portion of the sky containing a large aggregate of galaxies called the Pisces–Perseus supercluster. For a distance indicator they made use of the observed spectral redshift of their galaxies' neutral hydrogen 21-cm radio emission.[19]

Four months earlier, Valérie de Lapparent, Margaret Geller, and John Huchra of the Harvard–Smithsonian Center for Astrophysics had published a similar paper based on redshifts for 1100 galaxies observed in the hydrogen H-α 6562 Å spectral features radiated in similar galaxies, by gaseous regions ionized by massive young stars.[20]

Between them, the two teams thus were dealing with both the emission from cold neutral regions and hot ionized regions in galaxies, respectively observed in the radio and optical domains.

A recessional speed, for example, of $v = 5000$ km s^{-1} in either of these sets of observed spectra corresponded to a distance of $5000/H$ Mpc. If the value of the Hubble constant H, the rate at which the Universe is expanding, is taken to be $H \sim 70$ km s^{-1} Mpc^{-1}, the distance at which galaxies recede at 5000 km s^{-1} lies around $\sim$71 Mpc.

By 1989 the number of galaxies mapped by Geller and Huchra and shown in Figure 7.2 was about twice the number that had earlier been available to Haynes and Giovanelli.[21] The essential finding revealed by both surveys was that galaxies are not distributed randomly in space. Rather, they are aligned along narrow extended filaments, or spread across thin sheets separating and possibly enclosing voids.

As Geller and Huchra reported:

> The extent of the largest features is limited only by the size of the survey. Voids with a density typically 20 percent of the mean and with diameters of 5000 km s^{-1} are present in every survey large enough to contain them. Many galaxies lie in thin sheet-like structures. The

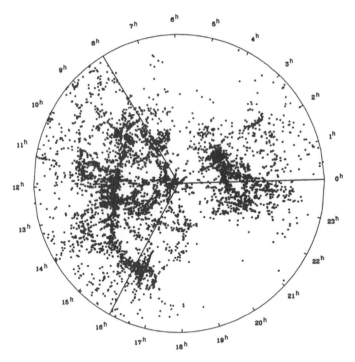

Fig. 7.2 A planar 360° view of 6112 galaxies in the declination range δ on the sky, $20° \leq \delta \leq 40°$. In this plot redshift z ranging up to velocities $cz \leq 15,000$ km s^{-1} increases radially outward. Geller and Huchra called the narrow concentrated band of galaxies with recession velocities of roughly 10,000 km s^{-1}, i.e., at a radial distance about half-way out to the figure's rim between 8^h and 17^h, the *Great Wall*. Note also the voids encircled by lesser walls of galaxies. The large blank spaces in the figure correspond to regions obscured by foreground Galactic clouds. (From "Mapping the Universe," Margaret J. Geller & John P. Huchra, *Science* 246(4932), 897–903, November 17, 1989, DOI:10.1126/science.246.4932.897.)

largest sheet detected so far is the Great Wall with [minimum extent of 85 Mpc by 240 Mpc if the Hubble constant is $H = 70$ km s^{-1}]. The frequent occurrence of these structures is one of several serious challenges to our current understanding of the origin and evolution of the large-scale distribution of matter in the Universe.

With the same choice of Hubble constant, voids with diameters ~ 70 Mpc also were ubiquitous.

The large walls of galaxies encircling voids were features that had not been previously mapped because astronomers had lacked sensitive instrumentation to rapidly survey large areas of the sky. Today, the number of mapped galaxies, whether isolated or identified with clusters and voids, has increased

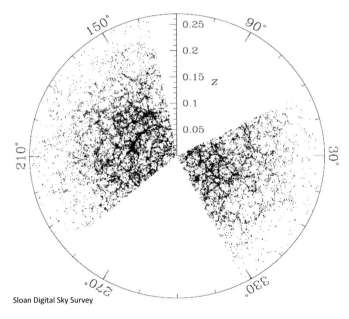

Sloan Digital Sky Survey

Fig. 7.3 A frothy galaxy distribution observed by the Sloan Digital Sky Survey stretching out to higher redshifts than the early efforts of Geller, Giovanelli, Haynes, and Huchra had surveyed. (Courtesy of the Sloan 2D Quasar Survey, providing images on a Creative Commons Attribution license (CC-BY). Funding for the Sloan Digital Sky Survey has been provided by the Alfred P. Sloan Foundation, the US Department of Energy Office of Science, and the Participating Institutions. SDSS acknowledges support and resources from the Center for High-Performance Computing at the University of Utah. The SDSS website is www.sdss.org.)

dramatically into the multimillions, often separately mapped at a number of particularly informative wavelengths.

The processes governing the distribution of galaxies, clusters, and voids across the Universe shown in Figure 7.3 are still not adequately understood. The data in hand may well have to await insights that ultimately will only emerge from the logic tradition. Computer simulations might eventually help, but whether or not these can be shown to be unique and correspond to cosmic realities may then require further proof.

7.10   Cosmic Re-ionization: The Gunn–Peterson Effect (2001)

An early model of the Universe, in which extragalactic gases might have remained neutral once star formation in the Universe had begun, received a rude dismissal in 1965 when an observation of the highly redshifted spectrum of the quasar 3C9 by Maarten Schmidt showed no trace of the ultraviolet absorption lines that neutral hydrogen atoms in the intergalactic medium should have displayed. The spectrum solely displayed a redshifted 1216 Å Lyman-α emission line and an ionized carbon line [C IV] at identical redshifts of $z = 2.01$. Schmidt attributed both to the quasar itself.[22]

Two young astronomers, the 27-year-old James E. Gunn and the 24-year-old Bruce Peterson, first remarked on this absence of neutral extragalactic hydrogen. At the time, both were at Caltech where Maarten Schmidt had just announced his findings. Gunn and Peterson suggested that this absence of atomic hydrogen in the intergalactic medium most likely meant that the medium was highly ionized and was kept ionized by its high temperature, which made recombination of protons and electrons rare, preventing the formation of neutral hydrogen.

They predicted that the spectra of highly redshifted quasars should all be found devoid of ultraviolet emission sufficiently energetic to ionize hydrogen. All such energetic photons would have been consumed in dissociating any neutral hydrogen atoms along our line of sight to the quasar.[23] It took another 35 years until telescopes could reach far enough back in time to verify this prediction, which came to be known as the *Gunn–Peterson effect*.[24]

Figure 7.4 displays the spectra of three quasars at redshifts $z \sim 6$, where their ionizing radiation is redshifted into the visible regime, to wavelengths around 6000 Å. The absence of quasar radiation shortward of the Lyman-α wavelength, in all three of these spectra, tells us that, although the quasar could emit energy in those wavelength ranges, the cosmic expansion would progressively redshift this radiation into the Lyman-α wavelength band, where extragalactic hydrogen atoms would readily absorb it.

These observational spectra of quasars at redshifts $z \sim 6$, when the Universe was only a billion years old, required establishment of a massive, systematic survey that ultimately produced an inventory of more than 300 million images and about a million spectra of galaxies or quasars. The Sloan Digital Sky Survey, a decades-long venture spearheaded by Jim Gunn himself, provided the dedicated means to sift through this vast accumulation of data, ferret out highly redshifted candidate quasars, obtain their spectra, and recognize what they were telling us.

Rarely is it possible for one individual to make a significant prediction early in life, and then construct the necessary massive apparatus and lead a dedicated team to see the prediction verified more than 35 years later.

This definitely was a result obtained in the logic tradition. Setting up a survey that could produce visible spectra of a million galaxies or quasars, finding highly redshifted quasars in this population and noting the strong absorption of radiation at wavelength shorter than the quasars' hydrogen Lyman-α emission, indicated that the quasars were embedded in intergalactic gaseous hydrogen, which their ultraviolet emission was systematically ionizing.

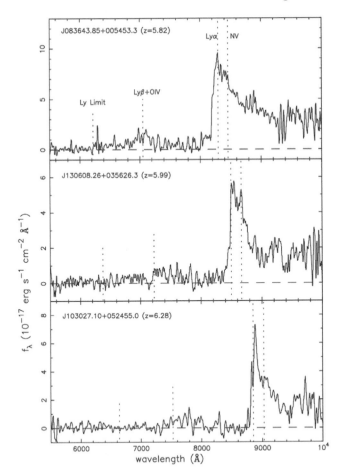

Fig. 7.4 The Gunn–Peterson effect observed in the emission from three quasars at redshifts $z \sim 6$. Radiation at wavelengths shorter than the Lyman-α emission from the quasars is strongly absorbed and appears totally absent short of hydrogen's ionization wavelength, the Lyman limit. Dashed vertical lines show the positions of potential emission lines at the redshifts of the quasars. The wavelengths indicated refer to our local rest frame. (From "A Survey of $z > 5.8$ Quasars in the Sloan Digital Sky Survey. I. Discovery of Three New Quasars and the Spatial Density of Luminous Quasars at $z \sim 6$," Xiaohui Fan, et al., *Astronomical Journal* 122, 2833–49, 2001 with permission of the authors and the American Astronomical Society.)

7.11 The Intergalactic Medium (1980)

In 1971, Roger Lynds of the Kitt Peak National Observatory in Arizona obtained an optical spectrum of the quasar 4C 05.34. Unexpectedly, it showed five discrete series of absorption lines due to Lyman-α transitions of atomic hydrogen and associated spectral lines, mostly of ionized carbon, nitrogen, and oxygen, at essentially identical redshifts.[25] All of the redshifts were considerably lower than that of the quasar.

Nine years later, a group led by Wallace Sargent observed six other quasars, and obtained similar results. They concluded:

> [The] single Lyman-α absorption lines and the metal line systems arise from physically distinct populations of clouds. Both types of systems are produced by cosmologically distributed intervening material not associated with QSOs. The metal line systems probably arise in galaxy halos, and the Lyman-α clouds must be an intergalactic population which is not associated with galaxies. If so, the clouds may be used to probe the intergalactic medium.[26] [a]

Since then, these clouds have been intensively studied. Some of the abundant Lyman-α absorbing clouds, referred to as the *Lyman-α forest*, indeed appear to be intergalactic features.[27] Others are translucent portions of galaxies along the line of sight to a quasar. As Sargent and his colleagues pointed out, these regions must be appreciably shielded from the X-ray emission from another form of extremely hot intergalactic material that pervades many large clusters of galaxies. X-ray spectra characterizing these gases tend to be rich in emission features of ionized iron, silicon, sulfur, and nickel, with a notable absence of other heavy elements such as calcium or argon.

The recognition of a gaseous intergalactic medium came gradually, in the logic tradition. No strikingly novel technological advances enabled the finding, though gradual improvements in available instrumentation did play a role, as might be expected.

7.12   Dark Matter Haloes (1979)

In 1933, Fritz Zwicky noticed that galaxies he had observed in clusters similar to the one shown in Figure 5.2 were exhibiting such high line-of-sight velocities that they would escape the gravitational attraction of the cluster unless the cluster mass was substantially higher than the apparent masses of the individual galaxies. He concluded, "If this were confirmed, the surprising result would emerge that dark matter is present in far greater density than luminous matter."[28]

---

[a] In astronomical jargon the heavy chemical elements are often referred to as *metals*, even though the designation includes constituents such as carbon, oxygen, and sulfur, among other heavy elements not exhibiting the high electric conductivities that physicists normally attribute to metals such as copper, iron, or lead.

Three decades later, in 1964, Zwicky re-examined this problem in joint observations with Milton Humason of the Mount Wilson Observatories. In the intervening years the Hubble constant had been lowered from an earlier false value of $\sim$500 km s^{-1} Mpc^{-1}, to $\sim$70 km s^{-1} Mpc^{-1}, quite close to today's best estimates. While this required Zwicky's earlier ratio of dark to luminous matter, $R*$, to be lowered by a factor of about 7, the surfeit remained significant and demanded explanation: Zwicky and Humason referred to this surfeit as *missing mass.*[29]

The following year, Vera Rubin and Kent Ford at the Carnegie Institution in Washington, DC, began finding that many, if not most, individual galaxies contained considerably more mass than expected from the combined masses of stars and interstellar matter. This puzzling anomaly became known as the *dark matter problem.*

We still do not know whether dark matter exists or whether something could be amiss with our understanding of gravitation. If dark matter does exist, we still do not know whether it comprises any messengers – carriers of information on cosmic events – in the same class as electromagnetic radiation, neutrinos, gravitational waves, or cosmic rays, all of which can reach Earth from far out in the Universe to convey information on the sources that produced them. But we cannot discount the possibility that, some day, we will be able to observe dark matter directly and discern its properties with the help of appropriate instruments. Accordingly, it is worth noting what we do and don't understand about dark matter today.

The optical spectrophotographic techniques available in the 1950s were slow; each galaxy was a separate challenge. But by the time Vera Rubin joined the Carnegie Institution she and Ford found ways to quickly move forward. She recalled,[30]

> It was Kent's image tube spectrograph that made the project succeed. When I arrived ... in 1965 he was just completing the instrument, and we estimated gains of 50 in speed over conventional spectra, but with a loss of resolution. For equal resolution the gain was about 10, still impressive. That ... made it all work, with integrations of 1–6 hours each. Exposures 10 times longer would not work ... spectrographs were not stable enough ...

Earlier than most other US astronomers, Vera Rubin also realized the value of mapping not only the velocities of visible stars but also the motions of interstellar gases emitting the 21-cm radio-astronomical radiation of atomic hydrogen. By 1979, working with radio astronomer Morton Roberts, Rubin and Ford were able to establish an unexpectedly high mass, of order $10^{12}M_\odot$, for the spiral galaxy NGC1961, confirmed by an equally surprising gaseous atomic hydrogen component exceeding $10^{11} M_\odot$. At the time, they established it as "the most massive galaxy known."[31]

The following year, Rubin, Ford, and Nobert Thonnard reported on the internal motions of 21 spiral galaxies. There was no question but that the velocities at which ionized gas clouds in these galaxies rotated about the centers of their galaxies remained constant at least out to distances at which their visible radiation could still be detected. They wrote, "The conclusion is inescapable that non-luminous matter exists beyond the optical galaxy."[32]

From these and other papers Rubin and Ford were publishing, it was becoming clear that galaxies were far more massive than inferred from their visible stars and gaseous components. Some other form of matter, *dark matter*, appeared to be present.

Whether or not Zwicky and Humason's *missing mass* of galaxy clusters is identical to the *non-luminous matter* of Rubin, Ford, and Thonnard is still unclear. We do not know whether more than one type of dark matter might exist. Searches for a number of different kinds of dark matter, with names such as *axions*, *WIMPS*, and *MACHOS* among others, are currently being pursued.

This is a topic that has been evolving since the early 1930s and still is not satisfactorily understood. The changing observational approaches have largely been in the logic tradition. Gradually evolving technological improvements have played a role but key astronomical questions remain unanswered.

7.13   Large-Scale Cosmic Flows (1996, 2008)

By 1990, radio observations had abundantly shown that the microwave radiation background originally discovered in 1965 had a temperature of about 3 degrees Kelvin. But we did not know whether this ambient radiation bath had a temperature that was precisely 3 K. If this radiation permeating the Universe was in thermal equilibrium, it should exhibit a *blackbody spectrum*, the spectrum emitted by a body that completely absorbs radiation at all wavelengths, and reradiates thermal energy into its surroundings with 100% emission efficiency.

The *Cosmic Background Explorer* satellite, *COBE*, launched late in 1989, was designed to measure the temperature of the microwave background with great precision by observing the amplitude and spectral energy distribution of the radiation emitted by the Cosmos. It succeeded spectacularly in this, determining the temperature to be 2.725 K. As Figure 2.6 shows, even the first published results already made clear that the spectrum differed from that of blackbody by less than 1% of peak emission at all wavelengths.[33] But COBE also showed that, rather than being entirely isotropic, the microwave background radiation peaked in one region of the sky – now called its *pole* – and was faintest at the antipole. Together, these two regions exhibited the maximum and minimum of a generally smooth dipole surface brightness distribution. The temperature excess at the peak was 3.353±0.024 mK, and its direction on the sky, expressed in Galactic coordinates, was $(\ell, b) = (263°.85 \pm 0°.1, 48°.25 \pm 0°.04)$ or, in equatorial coordinates of epoch 2000, $(\alpha, \delta) = (11^h 12^m.2 \pm 0^m.8, -7°.06 \pm 0°.16)$.[34]

If the microwave background radiation was intrinsically isotropic, this meant that the Solar System was moving with a velocity of about 369 km/s toward the direction of the peak background signal. But the motion of the Solar System is superposed on its motion relative to our Milky Way galaxy, which in turn can be shown to be moving toward the center of mass of the *Local Group* of galaxies largely dominated by the Milky Way and the Andromeda Galaxy, M31. The Local Group velocity relative to the microwave background, in turn, is of the order of 630 km/s.

The Local Group itself appears to be falling toward a large mass aggregate, initially named the *Great Attractor* and attributed to a distant mass concentration of order $5 \times 10^{16} M_\odot$ with which the galaxy cluster *Abell 3627*, whose mass was estimated to be $5 \times 10^{15} M_\odot$, may be associated. If so, the mass of the Great Attractor would roughly equal that of $\sim 10^5$ Milky Way galaxies.[35]

The origin of this particular mass distribution is difficult to define because it lies in a direction where the dust in the Milky Way prevents optical studies of the region, and X-ray as well as radio observations conducted to date raise a new set of questions that will require further investigation before they can be sorted out.[36; 37]

If our local motion through the Universe is not unique, then such velocities should be observable elsewhere as well, and the lumpiness, i.e., the density inhomogeneities in the Universe, as well as the local gravitational fields the inhomogeneities produce, should be traceable. Information of the velocity structure on smaller scales than COBE was able to provide has now become available through the Wilkinson Microwave Anisotropy Probe, WMAP, and its observations of the kinetic Sunyaev–Zel'dovich effect.[38]

The X-ray emission from many clusters of galaxies indicates the presence of intensely hot gases permeating the spaces between their galaxies. In the early 1970s the Soviet astrophysicists Rashid Alievich Sunyaev and Yakov Borisovich Zel'dovich predicted that these gases would be found to distort the spectrum of the microwave background radiation observed at the clusters' locations. The electrons in the fully ionized hot gases would scatter the background radiation. Electrons moving toward us in the hot gases should backscatter the impinging background radiation toward us, slightly shifting its radiation spectrum toward higher frequencies. Similarly, electrons moving away from us should backscatter radiation arriving from beyond the cluster and thus reduce the amount of the background radiation reaching us from beyond the cluster. This prediction has been borne out. The combined shift in the background spectrum due to these two scattering tendencies is called the *Sunyaev–Zel'dovich (S-Z) effect*.

Superimposed on this S-Z effect, however, is a second but similar scattering of the microwave background, due to any systematic motion of the entire galaxy cluster along the line of sight. This is the *kinetic Sunyaev–Zel'dovich effect*. Wherever the two effects can be separated it becomes possible to discern the line-of-sight motion of a cluster relative to the Hubble flow, i.e., relative to the general expansion of the Universe.

The two effects provided the means for charting the line-of-sight radial motions of clusters of galaxies relative to the cosmic microwave radiation marking the frame of rest in the Universe. Though clearly predicted, the actual observations were difficult and, following a logic tradition, required considerable care.[39; 40; 41]

7.14   The Cosmic Microwave Background Spatial Fluctuations (2001)

A decade following the observations conducted by the Cosmic Background Explorer, COBE, maps of the cosmic microwave background radiation began to emerge at higher spatial resolution and with far higher precision.

A balloon experiment launched from the McMurdo Station in Antarctica in 1998 provided considerably more detail than COBE, exhibiting ubiquitous small spatial fluctuations in the background radiation in the limited swath of the sky the observations covered.[42]

The *Wilkinson Microwave Anisotropy Probe, WMAP*, launched into space that same year, 2001, provided far more detailed maps of the microwave background radiation's spatial distribution, spectral variations, and polarization changes across a much larger portion of the entire sky.[43; 44; 45] This showed that the fluctuations were consistent with and provided reasonably well-defined values for a set of six fundamental cosmological parameters consisting of (i) the total mass–energy density of the Universe consistent with a flat (zero curvature) universe; (ii) the mass–energy density attributable to a cosmological constant, Λ; (iii) the density attributable to a complementing non-relativistic mass distribution; (iv) the density of the baryonic component of this non-relativistic mass; (v) the optical depth τ_c due to Thomson scattering by electrons once the Universe was re-ionized by radiation from massive stars and/or quasars that had begun to form; and (vi) a parameter designated n_s of order unity, consistent with an early cosmic inflation model that was nearly scale invariant.

These conclusions of the WMAP mission were re-analyzed in greater depth by its successor, the *Planck mission*, which further refined the values of these six parameters, and confirmed their consistency with an ever-widening network of cosmological observations. But the Planck mission was also coming up against hints that random features inherent to the structure of the Universe itself were beginning to limit how much more we might reasonably expect to learn from ever-more incisive cosmological surveys.[46; 47]

The initial discovery of the microwave background radiation took many astronomers by surprise, even though the radiation had been predicted by Ralph Alpher and Robert Herman 17 years earlier. Gradual improvements in technology, and in particular the launch of the Cosmic Microwave Background Explorer, which detected the background's pure thermal spectrum, made its thermal equilibrium crystal clear, establishing this cosmological finding as an indubitable golden event. The theoretically predicted background fluctuations, in contrast, then had to be sought, again through the logic tradition. The minuscule scale of the fluctuations, the reliable sensitivities needed to establish them, the polarization capabilities and the short wavelength end at which they were successively implemented by the WMAP and Planck collaborations were essential technological advances for these missions to succeed.

7.15   Dark Energy (1998)

For the past century, astronomers have been at pains to study distant portions of the Universe – initially because we wished to learn more about the nature of galaxies. Later, as the expansion of the Universe became apparent, studies of the distant Universe became tantamount to learning about its ancient past. But to understand how far away a given galaxy or cluster of galaxies might be, some standard indicator of distance needed to be found.

In 1912, Henrietta Leavitt had examined the periodicities of pulsations in stars resembling the variable star Delta Cephei in the Small Magellanic Cloud. Since these stars all were colocated in this small galaxy, they could be taken to lie at virtually identical distances from Earth. Leavitt noticed that the magnitudes, and thus the luminosities, of these stars were correlated with their pulsation periods, a finding that revolutionized astronomy!

Cepheid variables are luminous and can be readily discerned at great distances. Their apparent magnitudes in other galaxies can be compared to their apparent magnitudes in the Milky Way, where their distances are reliably determined by other means. Care must, however, be taken because Cepheids in the disks of galaxies called Type I Cepheids are about 1.5 magnitudes brighter than globular cluster Cepheids, called Type II Cepheids.

As Earth orbits the Sun the distance to nearby stars can be determined from their trigonometric parallax – their slight recurrent annual shift on the sky, relative to the vast majority of more remote stars. The distances of nearby Cepheid variables may thus be used to calibrate the distances of corresponding

Cepheid variables in remote galaxies, yielding a direct relation between a galaxy's redshift and its distance.

But even the high luminosity of Cepheid variables permits their observation only out to a limited distance. Supernova explosions enable measurement of distances to even more remote galaxies. Among supernovae, a type designated as SN Ia share virtually identical luminosities. Supernovae of Type Ia originate in close stellar binaries, in which the more massive star evolves more rapidly, becoming a red giant – a phase in which its high luminosity radiatively blows away its outer layers, eventually leaving nothing but a stellar core of spent fuel, consisting mainly of oxygen and carbon.

At this stage the star collapses to become a white dwarf with a radius comparable to that of Earth. For a while it evolves no further, but as its companion star subsequently also becomes a red giant and begins to shed its outer layers, in turn, the white dwarf gravitationally captures this material, gradually gaining mass. When its mass eventually reaches a tipping point, called the Chandrasekhar limit, where its internal structure no longer can resist its own gravitational pull, it undergoes a terminal collapse, followed by a giant explosion. Since this tipping point comes at precisely the same mass for each white dwarf, the consequent explosion is always nearly identical, leading to supernova explosions that, for practical purposes, exhibit close to identical luminosities.

The apparent brightness of Type Ia supernova explosions thus reliably indicates how distant a supernova's host galaxy lies. Spectroscopy of the galaxy's starlight then provides its redshift, telling us the velocity of expansion of the Universe at the distance of the explosion. By observing explosions at different distances, we can determine just how rapidly the Universe is expanding at each distance.

In the mid-1990s two consortia of astronomers took somewhat different approaches to this search. The Supernova Cosmology Project Collaboration, led by Saul Perlmutter of the E. O. Lawrence Berkeley National Laboratory in California, submitted a letter to *Nature* in October 1997 in which they reported the discovery of a supernova at a redshift $z = 0.83$, corresponding to a look back in time to an era when the Universe was roughly half its current age.[48] In their introduction they pointed out:

> The ultimate fate of the Universe, infinite expansion or a big crunch, can be determined by using redshifts and distances of very distant supernovae to monitor changes in the expansion rate. We can now find large numbers of these distant supernovae, and measure their redshifts

and apparent brightnesses; moreover, recent studies of nearby type Ia supernovae have shown how to determine their intrinsic bolometric luminosities – and therefore with their apparent brightness obtain their distances. The >50 distant supernovae discovered so far provide a record of changes in the expansion rate over the past several billion years. However, it is necessary to extend this expansion history still farther away (hence back further in time) in order to begin to distinguish causes of the expansion-rate changes – such as the slowing caused by the gravitational attraction of the Universe's mass density, and the possibly counteracting effect of the cosmological constant.

Only weeks later, on December 30, 1997, the High-z Supernova Search Team, spearheaded by the Australian Brian P. Schmidt, submitted a paper to the *Astrophysical Journal* in which they reported observations obtained using two high-throughput filters to obtain accurate two-color light curves with B (blue, roughly 4400 Å) and V (visual, roughly 5500 Å) filters for large numbers of supernovae, and especially one that provided a distance to a galaxy at redshift $z = 0.479$. The use of two filters permitted them to account for any dust absorption within a supernova's host galaxy – an important feature that permitted distinction between a low apparent brightness due to extreme distance and one due to absorption within a host galaxy at appreciably nearer distances.[49] The question of whether the observed expansion rates were due to cosmic curvature, to a cosmological constant, or some other form of dark energy was addressed a few months later in a paper inspired by Adam G. Riess, in which the High-z Supernova Search Team summarized their observational evidence from supernovae of Type Ia for an accelerating universe and a cosmological constant.[50] It took the astronomical community several more years to accept the evidence for a cosmological constant – a mysterious form of cosmic energy density whose origins remain unknown even two decades later. But over time and as further observations accumulated, the reality of this surprising finding gained increasing support.

As the observed number of these supernovae has accumulated, the luminosities of the most highly redshifted among them appears to be systematically lower than those at low redshift. The most likely explanation for this is that the lower luminosity supernovae must be more distant than an extrapolation of today's cosmic expansion rate suggests.

To see why this is so, we may consider observations in a universe that initially was not expanding rapidly, but then switched over to accelerated expansion. If we were to observe two galaxies, one distant, the other nearby, light from the nearby galaxy might have been emitted only after the acceleration had already set in and would thus have been more highly redshifted throughout its journey toward us. In contrast, light from the more distant galaxy could initially have experienced a lesser redshift as it traveled a considerable distance toward us preceding accelerated expansion. Its redshift might then not be much higher than that of the nearby galaxy, despite its greater distance. Its low redshift would make us believe it was nearer than it actually is, and it would appear appreciably fainter than its redshift would indicate if we were to assume the Universe had been expanding at a steady rate all along.

Insisting that the physical processes giving rise to Type Ia supernovae require them all to be equally luminous then leaves no recourse except to conclude that the expansion of the Universe cannot have been steady and must have accelerated, at least over recent eons.

The question we still cannot answer, today, is whether the cosmological constant is a characteristic feature of the geometry of space as Einstein's original form of general relativity had suggested, or whether it signifies the universal existence of some novel form of matter or energy – generally referred to as *dark energy*, potentially related to the *dark matter* that we don't understand either.

The discovery of dark energy through the efforts of the Supernova Cosmology Project Collaboration and the High-z Supernova Search Team followed the logic tradition of gradually establishing the reality of a previously unrecognized new factor. The primary technological means involved were the increased availability of sizeable telescopes and the rapid advances in the performance of pixel arrays partially expressed in Moore's law, which for a significant number of years correctly predicted a doubling of instrumental capabilities on time scales of 18 months, or a factor of one thousand every 15 years. Both permitted the rapid scan of large portions of the sky in searches for increasingly remote Type Ia supernovae explosions.

7.16   A Summary of Discoveries Since 1979

We may now summarize the list of major observational discoveries discussed in this book, so far. Those shown in Table 7.1 succeeded a list I had compiled earlier, in *Cosmic Discovery* published in 1980. Taken together, they provide a historical sequence we will be studying next, in Chapter 8.

TABLE 7.1

Factors Contributing to Major Astronomical Discoveries Since 1979

Discovery	Year	Tradition	Novel Instrument	Section
Stellar-Mass Black Holes	1979	Golden Event/Logic	No	4.3
Dark Matter Haloes	1979	Logic	No	7.12
Gravitational Lenses	1979	Golden Event	No	5.6
Intergalactic Medium	1980	Logic	No	7.11
Gravitational Radiation	1982	Golden Event	Yes	4.4
Pre-stellar Disk	1984	Logic	No	7.6
Galaxy Evolution & Mergers	1984	Golden Event	Yes	7.8
Neutrinos from Supernovae	1987	Golden Event	Yes	3.11
Large-Scale Cosmic Structures	1989	Logic	No	7.9
Extrasolar Planets	1992–95	Logic	No	6.1
Solar Neutrinos & Oscillations	1994	Logic	Yes	3.12
Brown Dwarfs	1995	Logic	No	7.7
Hot Jupiters	1995	Golden Event	Yes	6.1
Evolving Chemical Abundances	1996	Logic	No	2.24
Highest-Energy Cosmic Rays	1997	Logic	Yes	3.7
Dark Energy	1998	Logic	Yes	7.15
Cosmic Re-ionization	2001	Logic	Yes	7.10
Supermassive Black Holes	2004	Logic	No	4.8
Fast Radio Bursts	2007	Golden Event	No	6.7
Large-Scale Cosmic Flows	2008	Logic	No	7.13
Background Fluctuations	2013	Logic	Yes	7.14
Black Hole Gravitational Mergers	2016	Golden Event	Yes	4.5

Notes

1 *Image and Logic: A Material Culture of Microphysics*, P. Galison, The University of Chicago Press, 1997, pp. 22–23.

2 0957+561 A, B: Twin Quasistellar Objects or Gravitational Lens? D. Walsh, R. F. Carswell, & R. J. Weymann, *Nature* 279, 381–84, 1979

3 The Gravitational Lens Effect, S. Refsdal, *Monthly Notices of the Royal Astronomical Society* 128, 295–306, 1964

4 Observational Probes of Cosmic Acceleration, D. H. Weinberg, et al., *Physics Reports* 530, 87–255, 2013

5 Fred Gillett's Role in the Discovery of Planetary Disks: A Commemorative History, F. J. Low & H. H. Aumann, in "Debris Disks and the Formation of Planets: A Symposium in Memory of Fred Gillett," *ASP Conference Series* 324, editors L. Caroff, L. J. Moon, D. Backman, & E. Praton, 2004, pp. 3–8

6 Discovery of a Shell around Alpha Lyrae, H. H. Aumann, et al., *Astrophysical Journal* 278, L23–27, 1984

7 Disappearance of Stellar Debris Disks around Main-Sequence Stars after 400 Million Years, H. J. Habing, et al., *Nature* 401, 456–58, 1999

8 An ISOPHOT Survey of Pre-Stellar Cores, D. Ward-Thompson & P. André, in *ISO Surveys of a Dusty Universe*, editors D. Lemke, M. Stickel, & K. Wilke, Springer Lecture Notes in Physics 548, 309–16, 2000

9 The Formation of Deuterons by Proton Combination, H. A. Bethe & C. L. Critchfield, *Physical Review* 54, 248–54, 1938

10 The Structure of Stars of Very Low Mass, S. S. Kumar, *Astrophysical Journal* 137, 1121–25, 1963

11 The Helmholtz-Kelvin Time Scale for Stars of Very Low Mass, S. S. Kumar, *Astrophysical Journal* 137, 1126–28, 1963

12 Evolution of Stars of Small Masses in the Pre-Main-Sequence Stages, C. Hayashi & T. Nakano, *Progress of Theoretical Physics* 30, No. 4, 460–74, 1963

13 The Search for Brown Dwarfs, D. H. Stevenson, *Annual Reviews of Astronomy & Astrophysics* 29, 163–93, 1991

14 Catalogue of Nearby Stars, W. Gliese, *Veröffentlichungen des Astronomischen Rechen-Instituts Heidelberg* Nr. 22, Verlag G. Braun, Karlsruhe, 1969

15 Infrared Spectrum of the Cool Brown Dwarf Gl 229B, B. R. Oppenheimer, S. R. Kulkarni, K. Matthews, & T. Nakajima, *Science* 270, 1478–79, 1995

16 New Spectral Types L and T, J. Davy Kirkpatrick, *Annual Reviews of Astronomy & Astrophysics* 43, 195–245, 2005

17 The Hubble Deep Field: Observations, Data Reduction, and Galaxy Photometry, R. E. Williams, et al., *Astronomical Journal* 112, 1335–89, image plates: 1734–52, 1996

18 The Remarkable Infrared Galaxy Arp 220 = IC 4553, B. T. Soifer, et al., *Astrophysical Journal* 283, L1–4, 1984

19 The Connection between Pisces-Perseus and the Local Supercluster, M. P. Haynes & R. Giovanelli, *Astrophysical Journal* 306, L55–59, 1986

20 A Slice of the Universe, V. de Lapparent, M. J. Geller, & J. P. Huchra, *Astrophysical Journal* 302, L1–5, 1986

21 Mapping the Universe, M. J. Geller & J. P. Huchra, *Science* 246, 897–903, 1989

22 Large Redshifts of Five Quasi-Stellar Sources, M. Schmidt, *Astrophysical Journal* 141, 1295–300, 1965

23 On the Density of Neutral Hydrogen in Intergalactic Space, J. E. Gunn & B. A. Peterson, *Astrophysical Journal* 142, 1633–36, 1965

24 A Survey of $z > 5.8$ Quasars in the Sloan Digital Sky Survey. I. Discovery of Three New Quasars and the Spatial Density of Luminous Quasars at $z \sim 6$, Xiaohui Fan, et al., *Astronomical Journal* 122, 2833–49, 2001

25 The Absorption-Line Spectrum of 4C 05.34, R. Lynds, *Astrophysical Journal* 164, L73–78, 1971

26 The Distribution of Lyman-Alpha Absorption Lines in the Spectra of Six QSOs: Evidence for an Intergalactic Origin, W. L. W. Sargent, P. J. Young, A. Boksenberg, & D. Tytler, *Astrophysical Journal Supplement Series* 42, 41–81, 1980

27 Heavy-Element Enrichment in Low-Density Regions of the Intergalactic Medium, L. L. Cowie & A. Songaila, *Nature* 394, 44–46, 1998

28 Die Rotverschiebung von extragalaktischen Nebeln, F. Zwicky, *Helvetica Physica Acta* 6, 110–27, 1933, see p. 126

29 Spectra and Other Characteristics of Interconnected Galaxies and of Galaxies in Groups and in Clusters III, F. Zwicky & M. L. Humason, *Astrophysical Journal* 139, 269–83, and plates 11, 63, and 72, 1964, see also p. 283

30 Vera Rubin to Martin Harwit, *email* dated June 9, 2008

31 Extended Rotation Curves of High-Luminosity Spiral Galaxies V. NGC 1961, the Most Massive Spiral Known, V. C. Rubin, W. Kent Ford, Jr., & Morton S. Roberts, *Astronomical Journal* 230, 35–39, 1979

32 Rotational Properties of 21 Sc Galaxies with a Large Range of Luminosities and Radii, from NGC 4605 (R = 4 kpc) to UGC 2885 (R = 122 kpc), V. C. Rubin, W. Kent Ford, Jr., & N. Thonnard, *Astrophysical Journal* 238, 471, 1980

33 A Preliminary Measurement of the Cosmic Microwave Background Spectrum by the Cosmic Background Explorer (COBE) Satellite, J. C. Mather, et al., *Astrophysical Journal* 354, L37–40, 1990

34 Four-Year COBE DMR Cosmic Microwave Background Observations: Maps and Basic Results, C. L. Bennett, et al., *Astrophysical Journal* 464, L1–4, 1996

35 A Nearby Massive Galaxy Cluster behind the Milky Way, R. C. Kraan-Korteweg, et al., *Nature* 379, 519, 1996

36 A Systematic X-Ray Search for Clusters of Galaxies behind the Milky Way. II. The Second CIZA Subsample, D. D. Kocevski, H. Ebeling, C. R. Mullis, & R. B. Tully, *Astrophysical Journal* 662, 224–35, 2007

37 MC^2: Galaxy Imaging and Redshift Analysis of the Merging Cluster CIZA J2232.8+5301, W. A. Dawson, et al., *Astrophysical Journal* 805,143 (18 pp.), 2015

38 The Observation of Relic Radiation as a Test of the Nature of X-Ray Radiation from the Clusters of Galaxies, R. A. Sunyaev & Ya. B. Zel'dovich, *Comments on Astrophysics and Space Science* 4, 173, 1972

39 The Velocity of Clusters of Galaxies Relative to the Microwave Background. The Possibility of Its Measurement, R. A. Sunyaev & Ya. B. Zel'dovich, *Monthly Notices of the Royal Astronomical Society* 190, 413–20, 1980

40 A Measurement of Large-Scale Peculiar Velocities of Clusters of Galaxies: Results and Cosmological Implications, A. Kashlinsky, et al., *Astrophysical Journal* 686, L49–52, 2008

41 Measurement of the Electron-Pressure Profile of Galaxy Clusters in 3 Year Wilkinson Microwave Anisotropy Probe (WMAP) Data, F. Atrio-Barandela, et al., *Astrophysical Journal* 675, L57–60, 2008

42 Cosmological Parameters from the First Results of Boomerang, A. E. Lange, et al., *Physical Review D* 63, 042001, January 19, 2001

43 First-Year Wilkinson Microwave Anisotropy Probe (WMAP) Observations: Preliminary Maps and Basic Results, C. L. Bennett, et al., *Astrophysical Journal Supplement Series* 148, 1–27, 2003

44 Nine-Year Wilkinson Microwave Anisotropy Probe (WMAP) Observations: Final Maps and Results, C. L. Bennett, et al., *Astrophysical Journal Supplement Series* 208, 20 (54 pp.), 2013

45 Nine-Year Wilkinson Microwave Anisotropy Probe (WMAP) Observations: Cosmological Parameter Results, G. Hinshaw, et al., *Astrophysical Journal Supplement Series* 208, 19 (25 pp.), 2013

46 Planck 2015 Results XIII. Cosmological Parameters, Planck Collaboration, P. A. R. Ade et al., *Astronomy & Astrophysics* 594, A13, 63 pp., 2016

47 Planck 2015 Results: XV. Gravitational Lensing, The Planck Collaboration, *Astronomy & Astrophysics* 594, A15, 28 pp., 2016

48 Discovery of a Supernova Explosion at Half the Age of the Universe, S. Perlmutter, et al., *Nature* 391, 51–54, 1998

49 The High-z Supernova Search: Measuring Cosmic Deceleration and Global Curvature of the Universe Using Type Ia Supernovae, B. P. Schmidt, et al., *Astrophysical Journal* 507, 46–63, 1998

50 Observational Evidence from Supernovae for an Accelerating Universe and a Cosmological Constant, A. G. Riess, et al., *Astronomical Journal* 116, 1009–38, 1998

8

The Accumulation of Discoverable Phenomena

8.1   Discoveries over Time

More than four decades ago, in 1975, I wrote a paper on the possibility of statistically assessing the number of major cosmic phenomena characterizing the Universe. Published in the *Quarterly Journal of the Royal Astronomical Society*, it also outlined the efforts it might take for the astronomical community to discover all the phenomena not yet in hand.[1] This and two related articles[2,3] led to the book *Cosmic Discovery: The Search, Scope, and Heritage of Astronomy*, in 1981, presenting the case in greater depth.[4] The underlying methodology on which these estimates were based was not novel. In a seminal paper of 1943, titled "The Relation Between the Number of Species and the Number of Individuals in a Random Sample of an Animal Population," the statisticians R. A. Fisher, A. S. Corbet, and C. B. Williams had presented the basic outlines of the approach.[5] Ecologists have been making use of this and related statistical estimates ever since.

Cosmic Discovery listed a set of criteria that defined what we might mean by a "major cosmic phenomenon." It provided a list of 43 then-known astronomical features that met these criteria, and estimated that a total of some 130 such phenomena might ultimately be discovered if each exhibited one or more observable traits that jointly distinguished it from all other discoveries of comparable significance.

In returning to this topic now I will recall how these estimates were derived at the time, and how the past 40 years have broadened our understanding. Over those years, colleagues have urged me to revisit the subject, perhaps in part because a universe which for all we know could well be infinite might at first sight be expected to also exhibit an infinite variety of major phenomena. How could one then claim that the number of existing phenomena might actually be finite?

Reopening this question now allows us to introduce a new perspective. If the number of major phenomena characterizing the Universe were finite because the number of different types of messengers able to reach us with their messages reliably intact is finite, this would directly affect our strategies for optimally conducting observational astronomy: Observational discovery of new cosmic phenomena no longer might depend on sampling ever-larger portions of the Universe. Instead, it could shift emphasis to sampling ever-larger portions of the phase space of observations and providing ever-more powerful tools for detecting and analyzing information that did reach us unscathed.

In order to see how this newer point of view had arisen, it may be simplest to directly quote the prevailing mindset in the mid-1970s, at least as it appeared to me at the time.

8.2   How Do We Define and Distinguish Major Phenomena?

The 1975 *Quarterly Journal* article had noted that astronomers usually invent entirely novel names for observational discoveries that resemble no astronomical phenomena ever detected before. Preferably, the assigned names cryptically describe the discovery's unique features without attempting to guess the peculiar mechanism at work – which often does not become clear until years later.

The article pointed out that each of the major phenomena for which astronomers had invented new names appeared to differ from all astronomical phenomena already known by at least a factor of order 1000 in some observed attribute.[6]

I wondered where this factor of 1000, rather than, say 100 or 10,000, might come from. Was it some inherent human trait we all share? In support of this possibility, the article recalled:

> Most of us have a rough intuitive grasp of what we mean by a
> *phenomenon*, and we also have a similar intuitive feel for ranking their
> importance. This trait seems to be present since childhood and, for
> example, forms the basis on which the children's game 'Twenty
> Questions' becomes possible. If this game is to be played at all, the
> contestants must agree to limit the choice of objects to some set
> containing perhaps 100,000 members. Some 16 'Yes-No' questions
> would be needed to single out one particular member in this set

of $\sim 2^{16}$; and four more questions are tossed in to take care of inefficiency. While a child could perceive tiny individual specks of dust that would number far in excess of 100,000, in practice such detectable details are considered insufficiently important for membership in the set. Such minutiae would make a game of 'Twenty Questions' unworkable, although they would be quite appropriate in a new game, 'Fifty Questions,' which could include some 10^9 trivial features of insignificant interest.

In astronomy the situation is similar. We are able to detect far more sources than we can discuss in detail, and we need to recognize those that seem distinct and therefore of particular interest in some sense or other.

8.3   Appearance and Reality

One way in which the number of major cosmic phenomena exhibited by the Universe might be determined, at least somewhat objectively, is to restrict ourselves to the premise that cosmic features – objects, events – that appear different actually are different. And cosmic features that appear strikingly distinct identify distinct cosmic phenomena.

Such a designation may appear superficial because no attempt is made to understand the nature or astrophysical implication of any phenomenon. Nor do we ask about interrelations among them. A second difficulty is that differing phenomena specified in this way do not necessarily represent events produced by vastly different physical processes. According to our premise, phenomena are classified simply by noting their gross formal appearance.

Nevertheless, appearance and reality do often go hand in hand in nature.

I have therefore provisorily assumed that any two cosmic sources transmitting radically different-appearing signals also represent radically different phenomena. This should not disturb us: Pulsars were given a name and considered a new phenomenon well before we had any consensus on what kind of entity might be blinking out there. We had never before seen anything in the sky that pulsed so regularly, emitting a sharp radio pulse on a time scale of seconds or fractions of seconds. This was a new phenomenon!

Quasars were recognized initially because they had a high surface brightness in radio waves and a stellar appearance at optical wavelengths. These traits were unmatched by any other sources in the sky. Gamma-ray bursts were discovered because nothing else we had ever seen gave off a single sharp pulse of gamma rays for an interval of a few seconds with no repeat pulse to follow for at least months thereafter, perhaps forever.

For similar reasons I chose to distinguish two sets of cosmic events as representing different *major phenomena* if their appearance differed by a factor of 1000 or more in at least one observational trait. The factor of 1000 was somewhat arbitrary but brought the definition into rather good agreement with lists of cosmic phenomena on which most astronomers might agree. A scale factor of 1000 distinguishes the two main types of star clusters. Galactic clusters generally contain somewhere between a hundred and a thousand stars. Globular clusters contain a hundred thousand to a million stars. The distinction is clear.

A factor of 1000 permits differentiation between planetary nebulae that emit almost all their light in two or three narrow spectral lines, and globular clusters emitting a spectrum mainly exhibiting a continuum. It permits the identification of cosmic masers through their point-like radio appearance, their high spectral purity, and their high polarization. X-ray stars are recognized because of their high ratio of X-ray emission to visible light, and so on.

A factor of 1000 also permits distinction between novae, stars that suddenly flare in brightness by a factor of 10,000, from supernovae, explosions that are intrinsically another factor of ten-to-a-hundred thousand more luminous. But detailed differences between different classes of supernovae are not noted on this scale of coarseness, nor are all the minor differences between all kinds of peculiar variable stars emphasized. And this is what we would wish to see in a system that classifies distinct phenomena. It should emphasize clearly important factors or traits, and neglect small variations among subspecies.

8.4   Rediscovery

In addition to providing new discoveries, novel astronomical techniques have also provided another, almost equally important datum, the rediscovery of phenomena already recognized from earlier observations. Frequently even this rediscovery occurs unexpectedly. Detection of strong radio emission from Jupiter by Bernard Burke and Kenneth Franklin in 1955 represented the first, unexpected measure of radio emission from a planet. Previously, planets had been considered to be undetectable with available radio telescopes because their theoretically predicted flux was too low to permit detection.

We say that a phenomenon is recognized by two completely independent means if it could be discovered equally well with separate instruments that differ by at least a factor of 1000 in one of their observing capabilities, such as their response to the energies of distinct information-conveying messengers, or the angular, spectral, or temporal resolving power required to identify the messenger. Spiral galaxies, for example, can be recognized equally well through radio observations and through optical studies at wavelengths a million times shorter.

This independent recognition of phenomena through widely differing channels of information provides a statistical key to the total number of observational discoveries we might ultimately hope to make.

At our present stage of development in astronomy, this is a satisfactorily close estimate which should continually improve as we learn more about the phenomena we uncover. We can see at least that we are not dealing with a list of phenomena numbering in the thousands or the millions, an estimate that could have been quite conceivable without the approach developed here.

8.5   The Number of Major Observable Cosmic Phenomena

With new observational tools for detecting radio, infrared, X-ray, and gamma radiation made available by the military in the wake of World War II and the Cold War that followed, US astronomers rapidly discovered a large number of unanticipated major astronomical phenomena. This adoption and adaption of novel techniques required specialists not only to overcome peculiar technical problems, but to systematically strive for improved sensitivity and reliability.

Each group of these specialists initially searched the celestial sphere for astronomical sources sufficiently powerful to be readily detected and further investigated. As a result, each group at first was studying its own small collection of sources until, sometime around the mid-1970s, it became clear that some of these independently discovered X-ray, radio, and infrared sources were powerful emitters in two, or even all three of these wavelength ranges and occasionally shared other traits as well.

The discovery and independent rediscovery of individual phenomena in entirely distinct wavelength regimes now led to a realization that the number of major new phenomena the Universe would ultimately reveal might be finite rather than infinite. Though the size of the Universe might be undefined and potentially infinite, the number of different types of bodies or processes we could ultimately identify in the parts of the Universe we can survey appeared to be finite.

This point of view was based on statistical sampling theories that Ronald Aylmer Fisher had developed in Britain in the 1940s. Later, we will return to Fisher's work, and its implications for the scope of potential astronomical discoveries.[7] But the thrust of his work can be easily grasped by anyone familiar with another childhood activity, as the article of 1975 had also suggested.[8]

Many of us have children who collect baseball or football cards. At the beginning of each season, when the collection is still small, all the cards differ from each other. But as the collection grows, an increasingly large fraction of the cards become duplicates. Assume that all the athletes' faces are equally represented and that the cards are obtained in some statistically random order. Then the very first duplicate card obtained tells us an important characteristic of the set – namely that it is finite. As the collection of cards grows and the number of duplicates increases, the relationship between the number of single cards in our collection and the number of duplicates can yield an increasingly accurate estimate of the total number of [distinct] cards in a complete set. Poisson statistics apply to this type of sampling and the calculation is straightforward …

… [O]ur search for cosmic phenomena through different channels of information is similar to collecting cards. Initially, as our technical expertise grows, we discover an increasing number of new phenomena. But, as the number of discoveries grows a survey carried out through a completely new channel of communication will uncover an increasing number of 'duplicates' – phenomena already known from previous discovery through an established channel.

Figure 8.1 is a fanciful abstraction of a space designed to show how different cosmic phenomena might similarly be arrayed in a five-dimensional phase space that directly matches the physical traits most clearly identifying each phenomenon by the instrumental capabilities best able to detect it.

The most striking implication of the mutually independent astronomical rediscoveries, such as those sketched in the figure – effectively equivalent to finding duplicates of previously collected baseball cards – was that the number of cosmic phenomena appeared finite and could be estimated through application of binomial statistics.

This approach appears to have been largely unbiased. We had originally come to recognize astronomical discoveries and unanticipated rediscoveries by means of tools earlier developed for the military – hardly a characteristic that would prejudge the astronomical findings that might ensue.

A list of major astronomical phenomena known in 2019 is provided in Figure 8.2 so that readers might obtain an impression of the measure of distinction that qualifies a new finding as a novel *astronomical phenomenon*.

An important caveat should however be noted. The list of currently known phenomena that Figure 8.2 exhibits admittedly is subjective. Most astronomers making up their own lists of known phenomena would quite likely arrive at a somewhat different set. But equally likely their lists and mine would probably

Entire volume of phase space

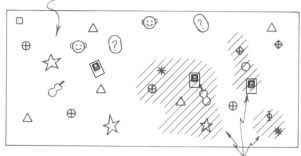

Subvolumes observed to date

Independently observed phenomena **and**

Fig. 8.1 Distribution of Observed Phenomena in Phase Space. The phase space of observations is a multidimensional space, each point of which corresponds to a distinct observing capability. The rectangular border of the figure represents a region of this space containing all conceivable astronomical observations that can be carried out in our Universe. Although clearly not five-dimensional, the sketch attempts to summarize all that might be learned through use of the more realistic depictions in Figures 6.6, 6.7, and 6.8, which show projections of such a space onto the two-dimensional page of a book. Distinct cosmic phenomena are represented by a variety of familiar symbols, some of which appear in several portions of the space. Violins appear both in a region to which we already have instrumental access, indicated by its shading, and another in which we currently lack observational capabilities. Some phenomena, such as the one represented by the Greek letter Φ appear only once. Some, like the question marks or the little faces – which might represent life in the Universe – appear only beyond current instrumental reach. These phenomena remain to be discovered. Others still have been observed in two widely separated shaded regions. These are exemplified by the asterisks and baseball cards identified at bottom right, and represent phenomena already observed in two distinct ways.

agree and overlap on something like two-thirds of the sources Figure 8.2 displays. And if those other astronomers then applied the same statistical approach to their own lists, they should arrive at reasonably comparable estimates of the total number of cosmic phenomena the Universe comprises.

For the phenomena listed in Figure 8.2 in hand, the present chapter will provide an estimate of the total number of cosmic phenomena that may ultimately be observationally identified. Four decades ago, *Cosmic Discovery* already obtained such an estimate; but with the additional information available by the year 2019, a re-computed estimate of that number has become available and, as we will see, appears in reasonable agreement with those earlier predictions.

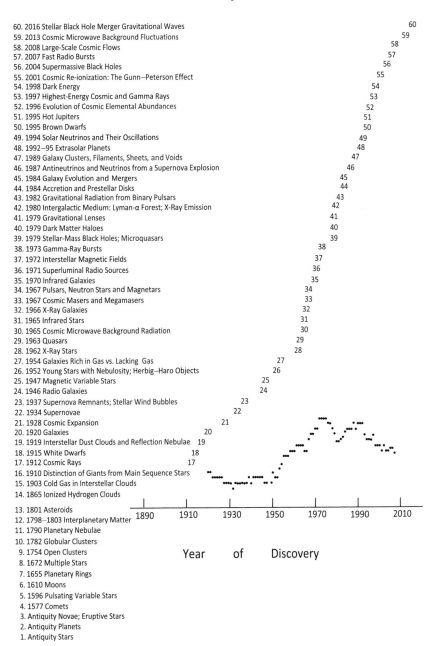

60. 2016 Stellar Black Hole Merger Gravitational Waves
59. 2013 Cosmic Microwave Background Fluctuations
58. 2008 Large-Scale Cosmic Flows
57. 2007 Fast Radio Bursts
56. 2004 Supermassive Black Holes
55. 2001 Cosmic Re-ionization: The Gunn–Peterson Effect
54. 1998 Dark Energy
53. 1997 Highest-Energy Cosmic and Gamma Rays
52. 1996 Evolution of Cosmic Elemental Abundances
51. 1995 Hot Jupiters
50. 1995 Brown Dwarfs
49. 1994 Solar Neutrinos and Their Oscillations
48. 1992–95 Extrasolar Planets
47. 1989 Galaxy Clusters, Filaments, Sheets, and Voids
46. 1987 Antineutrinos and Neutrinos from a Supernova Explosion
45. 1984 Galaxy Evolution and Mergers
44. 1984 Accretion and Prestellar Disks
43. 1982 Gravitational Radiation from Binary Pulsars
42. 1980 Intergalactic Medium: Lyman-α Forest; X-Ray Emission
41. 1979 Gravitational Lenses
40. 1979 Dark Matter Haloes
39. 1979 Stellar-Mass Black Holes; Microquasars
38. 1973 Gamma-Ray Bursts
37. 1972 Interstellar Magnetic Fields
36. 1971 Superluminal Radio Sources
35. 1970 Infrared Galaxies
34. 1967 Pulsars, Neutron Stars and Magnetars
33. 1967 Cosmic Masers and Megamasers
32. 1966 X-Ray Galaxies
31. 1965 Infrared Stars
30. 1965 Cosmic Microwave Background Radiation
29. 1963 Quasars
28. 1962 X-Ray Stars
27. 1954 Galaxies Rich in Gas vs. Lacking Gas
26. 1952 Young Stars with Nebulosity; Herbig–Haro Objects
25. 1947 Magnetic Variable Stars
24. 1946 Radio Galaxies
23. 1937 Supernova Remnants; Stellar Wind Bubbles
22. 1934 Supernovae
21. 1928 Cosmic Expansion
20. 1920 Galaxies
19. 1919 Interstellar Dust Clouds and Reflection Nebulae
18. 1915 White Dwarfs
17. 1912 Cosmic Rays
16. 1910 Distinction of Giants from Main Sequence Stars
15. 1903 Cold Gas in Interstellar Clouds
14. 1865 Ionized Hydrogen Clouds
13. 1801 Asteroids
12. 1798–1803 Interplanetary Matter
11. 1790 Planetary Nebulae
10. 1782 Globular Clusters
 9. 1754 Open Clusters
 8. 1672 Multiple Stars
 7. 1655 Planetary Rings
 6. 1610 Moons
 5. 1596 Pulsating Variable Stars
 4. 1577 Comets
 3. Antiquity Novae; Eruptive Stars
 2. Antiquity Planets
 1. Antiquity Stars

Fig. 8.2 A Record of Cosmic Discoveries. The 60 major discoveries listed at left are ordered by their year of discovery. The steadily rising curve shows the accumulated discoveries by year. The assembly of dots at bottom displays a 25-year running average of the number of major discoveries per quarter century centered on the years indicated on the abscissa. This ranges from a minimum of four during the early 1930s to a maximum of 18 in the 1970s to 1990s when radio, X-ray, and infrared astronomy were all hitting their strides, and neutrino astronomy was announcing its first successes.

8.6   Practical Consequences

One way in which these ideas had a major influence on space exploration is that two successive directors of NASA's Astrophysics Division of the early and mid-1980s, Franklin D. Martin and Charles J. Pellerin, recognized that observations across the entire electromagnetic range would ultimately have to be undertaken if different phenomena that all emitted, say, strong X-ray emission, were to be differentiated from one another through their emission in the optical, gamma-ray, or infrared wavelength bands. This would only be possible if observations could be undertaken in all these wavelength regimes, preferably simultaneously but at least at roughly coeval epochs.[9] Pellerin and the astronomical advisory committee he asked me to chair then used this argument to persuade the US Congress to agree to the need for a family of four observatories, soon named the *Great Observatories*, to be launched into space in close succession – an agreement reached even before the first of these observatories had been launched.[10] All four of these observatories were successfully launched into space, each respectively named for a leading astronomer or cosmic-ray physicist: the *Compton Gamma Ray Observatory (CGRO)*, the *Hubble Space Telescope (HST)*, the *Chandra X-Ray Observatory (CXRO)*, and the *Spitzer Infrared Telescope Facility (SIRTF)*. The total cost of launching all four observatories was approximately $8B (8 billion dollars at launch measured in late-twentieth-century dollars).

Observations obtained by means of the Great Observatories have transformed the way we view our Universe today. They also reinforced the need for attempting observations in previously inaccessible portions of *discovery space*. Though this is hardly ever the sole rationale for proposing a new space mission, it is where unanticipated new phenomena are most likely to be discovered, and therefore particularly excites curiosity.

One purpose of the present book has been to determine how many new major phenomena have been discovered since the earlier list recognized around 1980, and to seek insight on how these additional discoveries came about.

To this end, it may be worth re-examining some of the conclusions *Cosmic Discovery* had reached regarding the means by which discoveries made before 1980 had been obtained; the types of instruments that had enabled the discoveries; the length of time these instruments had existed before their use had led to a prized discovery; and the technical background and expertise that had led the discoverer or discoverers of a major novel phenomenon to succeed, where others, less-well equipped, had previously failed.

Cosmic Discovery pointed to some of the distinguishing features of these phenomena: (i) the importance assigned to them by their widely accepted, clearly defined name – often a descriptive but noncommittal name, such as "quasar" or "pulsar," conjured up because nobody knew what the phenomenon represented; (ii) by the major conferences and conference proceedings devoted to them; (iii) by the books and review articles written about them; in short, because a large number of astronomers needed to understand these phenomena better in order to remain active in research.

Many of the phenomena listed in *Cosmic Discovery* published in 1981 can also be found included in the inventory of cosmic energy sources compiled by Masataka Fukugita of the University of Tokyo and P. J. E. (Jim) Peebles at Princeton University, in 2004, as shown in Table 1.2. To simplify their arguments, Fukugita and Peebles assumed that the *Hubble constant*, the rate at which the Universe expands, though still somewhat uncertain even today, was approximately $H_0 = 70$ km s^{-1} Mpc^{-1}. To justify the timeliness of their efforts they explained:

> There is now a substantial observational basis for estimates of the cosmic mean densities of all the known and more significant forms of matter and energy in the present-day universe. The compilation of the energy inventory offers an overview of the integrated effects of the energy transfers involved in all the physical processes of cosmic evolution operating on scales ranging from the Hubble length to black holes and atomic nuclei. The compilation also offers a way to assess how well we understand the physics of cosmic evolution, by the degree of consistency among related entries.

Table 1.2 provides an extract of some of the 40 mass–energy sources Fukugita and Peebles examined. Their table's original entries have in part been updated by others in the past dozen years, but the thrust of their compilation has remained intact.[11; 12; 13; 14]

Fukugita and Peebles concluded that *dark energy* constituting about 72% of the total cosmic mass–energy reservoir largely controls the expansion of the Universe. *Dark matter* amounting to roughly another 23% of the total energy accounts for most of the observed kinetic energy within and among galaxies. The remaining roughly 5% cumulatively account for the roughly 40 distinct energetic processes occurring in various types of galaxies, black holes, stars, planets, or interstellar and intergalactic clouds of gas and dust populating the Cosmos.[a]

---

[a] The term *energy*, ϵ, as used here, includes any form of *mass*, m, whose equivalent energy is the product of mass and the speed of light squared: $\epsilon = mc^2$.

Even though so much cosmic energy was thus cached in largely unexplored reservoirs, Fugukita and Peebles drew the distinction between forms of energy consequential in shaping the landscape of the Universe in individual locales, as contrasted to forms of energy acting globally through gravitational interaction, though otherwise remaining largely inert.

8.7   Why Might the Number of Major Phenomena Appear Finite?

Four decades ago, there were no clear reasons why the total number of cosmic phenomena we might ultimately come to recognize should be finite. By now, several factors suggest why.

First, as Chapter 3 showed, the energies of all the messengers we now recognize – whether they be photons, cosmic rays, neutrinos, gravitons, or distinct isotopes of atoms – appear to be bounded. Almost equally important, the accuracies with which we are able to determine these messengers' directions and instants of arrival, or the degree to which their spectral characteristics can be determined, are largely bounded as well, as Chapters 5 and 6 indicated. And finally, the Universe has systematically erased a significant fraction of the information it might ideally have transmitted, including much of the history of primeval phenomena.

As a result of these various bounds and erasures, it should perhaps not surprise us that the number of phenomena we can actually observe today should similarly be bounded!

This argument of necessity disregards the vastly higher amounts of cosmic energy stored in the form of *dark energy*, and the somewhat smaller but still impressive energy stored in *dark matter*.

Dark energy drives the cosmic expansion. Dark matter primarily influences the rotational velocity of galaxies and assures the stability of galaxy clusters. To date, neither form of mass–energy has been found to lead to phenomena other than those attributed solely to its associated gravitational field.

Both appear to lack *Gibbs free energy*, the part of a system's energy that can be converted into useful work. In cosmology, one can think of free energy as the energy available for shaping complex cosmic events. The lack of free energy – the inertness – of dark energy and dark matter appears to be most clearly expressed in their dearth of interaction with atomic or subatomic matter through any known processes aside from their gravitational influence. Nevertheless, we may ultimately come to realize that a whole complex of processes exist mediated by dark matter or dark energy messengers we do not yet know how to access, let alone decipher. This is one of the major unresolved questions of our times.

8.8   The Number of Distinct Cosmic Phenomena

By the year 2019, rather larger numbers of cosmic phenomena were recognized than had been apparent in 1981 when *Cosmic Discovery* first published its list. I cite these more recent discoveries in the chronologically ordered listing of Figure 8.2.

Some phenomena are readily detected only in a single wavelength band. Long-duration gamma-ray bursts, for example, are recognized solely through their seconds-long burst of gamma radiation, although once this burst has been detected other varieties of telescopes can be swiftly turned in the direction from which the burst arrived to follow the subsequent evolution of the source at optical, X-ray, or radio frequencies.

Other phenomena, like the large-scale cosmic expansion, although first detected at visible wavelengths, in the 1920s, could subsequently have been independently discovered in the radio domain, where the atomic lines of hydrogen observed at their characteristic 21 cm wavelengths have by now clearly shown the same effect.

Similarly, extrasolar planets could have been, and actually were, independently discovered at radio as well as optical wavelengths. As we saw earlier, the presence of planets orbiting a pulsar was first detected at radio frequencies, where the orbital period of the planets was clearly discerned in the phasing of the pulsar's sequence of pulses. Ever since, however, a much larger fraction of these planetary systems has been located through partial occultations of stars by their orbiting planets.

8.9   The Statistics of Discoveries

Discoveries of new phenomena have often been unexpected and surprising. Many emerged through use of new instruments introduced only weeks or months earlier. It seemed as though each powerful new telescope was randomly looking at the sky through a new set of eyes offering an entirely new perspective.

A question that naturally arises is "Can we estimate, at least roughly, how many more such discoveries we should expect?"

The mid-twentieth century statistician and population geneticist Ronald Aylmer Fisher (Figure 8.3) and his colleagues had noted that a reasonable estimate of the number of distinct species of butterflies, birds, plants, or trees in a particular habitat could be determined by randomly searching through just a limited portion of that realm, and then listing the number of individuals of each species encountered in the search.[15] Once this census was complete, Fisher's membership of distinct species for each of which only a single representative

Fig. 8.3 Portrait of Ronald Aylmer Fisher (1890–1962). Fisher, shown here in a photograph dated around 1932, revolutionized twentieth-century genetics through his statistical approaches. His methods, however, transcended genetics and became appreciated as keys to other scientific efforts. In the present chapter, they have been used to resolve implications of the rates at which astronomical discoveries have emerged. Fisher was interested in estimating the number of distinct plant or animal species in a forest, based on samples collected in limited parts of the forest. The same statistical sampling applies to the number of distinct cosmic phenomena based on sampling limited portions of the phase space of observations. (This image was originally published in "The Genetical Theory of Natural Selection," by A. W. F. Edwards in the journal *Genetics* 154(4) 1419–26, April 1, 2000. It is reproduced here with the permission of the Genetics Society of America.)

had been encountered might then be labeled A. The number of species for which two representatives had been encountered might be labeled B. The number of distinct species for which three members had been encountered in this random search would be labeled C, and so forth.

I assumed that the habitat a corresponding astronomical search might be sampling was the phase space of observations. Novel instruments permitted astronomers to more or less randomly search through this habitat to see how many distinct phenomena had been discovered only by means of a single observational technique, but by none other, such as the merger of two stellar-mass black holes, which has to date been discovered solely through the detection of the emitted gravitational waves, to the exclusion of all other signals. Stellar black holes thus would be listed as a phenomenon contributing to the number of singly encountered species of phenomena, A.

Correspondingly, the discovery of exoplanets came about through the implementation of high-resolution timing studies at radio wavelengths and, soon thereafter, independently through high spectral resolution at optical frequencies. We might therefore include exoplanets in our compilation of twice-encountered species, B. More recently, however, we have also discovered exoplanets through occultation observations. In hindsight, such observations could have been possible even earlier, before either of the other techniques had succeeded in discovering exoplanets. But historically, the Kepler mission, which by now has discovered by far the largest number of known exoplanets, was not a randomly selected technical approach. The mission was launched primarily because by then we knew that exoplanets exist and wanted to take a more detailed census. So, should exoplanets be listed members of group B, for which they certainly qualify, or group C for discoveries we now realize could have been made by means of three entirely distinct instrumental techniques?

Either approach – the historical sequence of discovery, or the almost equiprobable potential sequence – would be valid, but we would need to be clear on which of these two methods our census covers, because the conclusions of these respective approaches will differ.

Often, several distinct methods might exist by means of which the discovery of a particular astronomical phenomenon could be independently implemented in the absence of any others. Thus, galaxies rich in gas content could be independently discovered in at least three distinct ways: e.g., through the detection of visible/infrared spectral lines emitted by gas clouds heated by massive luminous stars; radio emission from electrons accelerated to relativistic speeds through explosion of massive stars; and far-infrared continuum radiation from heated dust invariably embedded in interstellar gas clouds. In contrast, it isn't clear whether other phenomena, such as the recently discovered fast radio bursts, FRB, often luminous for no more than a few milliseconds at a time, will ever be discoverable by some other, totally independent means.

For all we know, every major cosmic phenomenon might someday be discovered through a multitude, m, of distinct observational means – though, once a novel phenomenon is discovered through one instrumental approach, the astronomical community tends to further pursue the finding to determine whether more can be learned through any other available means. This makes it more difficult to decide, after the fact, whether a rediscovery was truly random or based on prior knowledge.

8.10   Coding of Instrumental Capabilities

Messengers reaching us from afar may be electromagnetic waves, neutrinos, cosmic rays, gravitational waves, or occasional dust grains or radioactive atoms drifting into the Solar System from its surroundings. Tools for detecting and identifying the properties of individual messengers can be labeled with monograms specifying their defining properties – the type of messenger they are able to detect, the range of messenger energies to which they are tuned, the ranges of angular, spectral, and temporal resolving powers they can define, and their ability to recognize different types of polarization. Finally, the need for a separate instrument capable of providing a measure of the distance from which a messenger arrived is often needed in order to clearly distinguish phenomena that might appear similar until one realizes that their radiation arrives from vastly different distances and reflects source *luminosities* differing by many orders of magnitude.

In the listings I will use, letters enclosed in square brackets identify the instruments required for studying different types of messengers or distinct electromagnetic wavelength ranges by means of which individual phenomena are recognized: [C] Cosmic-ray particles; [Ch] Cherenkov radiation; [D] Direct impact on Earth; [G] Gamma rays; [GW] Gravitational waves; [I] Infrared radiation; [N] Neutrinos/Antineutrinos; [R] Radio; [T] Transition radiation; [V] Visual/near-infrared radiation; [X] X-rays.

Table 8.1 enables us to sort out distinct messenger properties by identifying the instrumental traits required to detect them. Note that each band listed in the first column on the left sorts out traits that differ by factors of a thousand in ascending or descending order in the different columns to its right. A messenger arriving in the energy range listed for band 3 typically differs by a factor of order one thousand from one that lies immediately above it in band 2 or below it in band 4. And similar differences by factors of a thousand exist for the various rows for each of the instrumental resolving powers listed in the next four columns of the table, as well as for the final column dedicated to source luminosity, the rate at which the source emits energy.

The table permits us to generate six-character monograms such as [4115U1] or [1215U1] we will find in Tables 8.2 and 8.3 as well as in Table 8.4. This pair of monograms characterizes the observational capabilities that historically led to the discovery of planets, first by their reflection of visible light, and later

TABLE 8.1

Parameters of Instrumental Capabilities Required to Sort Cosmic Phenomena

Band	Messenger Energy (eV)	Angular Resolution (log radians)	Spectral Resolving Power $\log(\lambda/\Delta\lambda)$	Time Resolution $\log t$	Polari -zation Type	Source Luminosity (erg/s)
1	$\leq 10^{-7}$	isotropy $\to 10^{-3}$	photometry $\to 10^3$	$\leq 10^{-2}$ s	L	$\leq 10^{29}$
2	$10^{-7} \to 10^{-4}$	$10^{-3} \to 10^{-6}$	$10^3 \to 10^6$	$10^{-2} \to 10$ s	C	$10^{29} \to 10^{32}$
3	$10^{-4} \to 10^{-1}$	$10^{-6} \to 10^{-9}$	$10^6 \to 10^9$	10 s $\to$ 3 hr	E	$10^{32} \to 10^{35}$
4	$10^{-1} \to 10^2$	$\leq 10^{-9}$	$\geq 10^9$	3 hr $\to$ 0.3 yr	U	$10^{35} \to 10^{38}$
5	$10^2 \to 10^5$			0.3 $\to$ 300 yr		$10^{38} \to 10^{41}$
6	$10^5 \to 10^8$			300 $\to$ 3×10^5 yr		$10^{41} \to 10^{44}$
7	$10^8 \to 10^{11}$			$3 \times 10^5 \to$ 3×10^8 yr		$10^{44} \to 10^{47}$
8	$10^{11} \to 10^{14}$			$> 10^{8.5}$ yr or Unresolved		$10^{47} \to 10^{50}$
9	$10^{14} \to 10^{17}$					$10^{50} \to 10^{53}$
!	$\geq 10^{17}$					$\geq 10^{53}$

through their emission of radio waves. The first of these two falls in the photon energy range of band 4; the modest angular resolution required, band 1; spectral resolution, band 1; and time resolution, band 5, originally needed to detect Solar System planetary motions on time scales of 0.3 to 300 years; the unpolarized light they emit, U; and their low radiated power, less than 10^{29} erg/s, band 1. The rediscovery of planets in the radio band required sensitivity to far-lower-energy radio photons, band 1, and rather improved angular resolving power corresponding to band 2.

The specific tools that enabled the discovery of a new phenomenon can be thought of as unique filters that permitted us to detect an individual phenomenon most clearly.

As Table 8.1 shows, the properties of such a filter can be summed up in terms of the individual energies of the conveying messengers, the angular, spectral, and temporal resolving powers, or polarization that readily enable these messengers to identify a phenomenon of interest, as well as a range of typical source luminosities characterizing the phenomenon, displayed in ergs per second, in the final column on the right.

The ranges of messenger traits that a complete toolkit needs to recognize are thus determined solely by the range of messengers that reach us intact to reliably register the traits of observable cosmic phenomena. Constructing ever-more powerful instruments for detecting Galactic or extragalactic radio waves with wavelengths exceeding a few kilometers thus may be unnecessary knowing, as we now do, that interstellar gases strongly absorb these radiations, preventing their reaching us. Other messengers face similar restrictions, recognition of which can prevent us from undertaking futile investigations. Table 8.1 attempts to take these factors into account.

Where a Table 8.1 entry straddles the dividing lines between the ranges shown, I record its entry in italics, thus [4214U5], for Cepheid variables, to indicate that both their time variations and their luminosities straddle a neighboring region, but the value assigned is the best indicator to use.

Monograms presented in square brackets refer to individual filters contributing to the identification of a listed phenomenon. Monograms straddled by asterisks, such as *[4115U1]* in Table 8.2, contribute to the identification of several distinct phenomena, in this case of stars, comets, moons, planetary rings, and asteroids.

8.11   A List of Cosmic Phenomena Recognized by 2019

The list of 60 major cosmic phenomena compiled, below, by date of discovery includes 20 discovered in the past 40 years. Another 40 among those listed now were already known when *Cosmic Discovery* was published in 1981. Some phenomena that appeared distinct at earlier times now are known to constitute a single phenomenon observed from different perspectives. I have combined these in Table 8.2, even though they were once considered distinct.

In a complementing list presented immediately below, Table 8.3, the designation (1), (2), or (3) appearing in parentheses to the right of each phenomenon cited shows the number of mutually independent tools, or equivalently the number of independent types of messengers, by means of which each named phenomenon had been discovered or rediscovered by 2019. This list no longer comprises literature references included in the more detailed entries found in Table 8.2. Instead, it highlights the instrumental means and, correspondingly, each type of messenger, that enabled the listed discovery.

Summing the entries in parentheses we find that the number of singly recognized phenomena designated (1) is $A = 38$. The numbers of doubly and triply recognized phenomena, respectively designated (2) and (3), are $B = 18$ and $C = 4$.

Clearly, some judgement on how reliably each phenomenon is established by existing observations determines whether it should be considered as recognized only by the entire evidence in hand, or whether the wealth of data obtained can be divided into separate portions, each of which independently verifies the phenomenon's existence.

TABLE 8.2

Cosmic Phenomena by Date of Discovery

Cosmic Phenomena and Discovery Dates	Coded Designations (see Table 8.1)
1. Antiquity **Stars**	[V] Fixed position on the sky *[421-U3]*
2. Antiquity **Planets**	[V] Orbital motions *[4115U1]* [R] Radio emission [1215U1][16]
3. Antiquity **Novae/ Eruptive Variables**	[V] [4113U4] [V] Rapid rise time; slow decline over months [4115U] [V] Flare Stars [4213U2]
4. 1577 **Comets**	[V, X] More remote than the Moon.[17] [V] Orbital motions *[4115U1]* [V, X] Unrelated brightness variations *[4114U1]*, [5115U1][18]
5. 1596 **Pulsating Variable Stars**	[V] Cepheids: luminosity variations [4214U5][19] [V] Spectral shift *[4224U4]* [V] RR Lyrae stars: luminosity variations *[4214U4]*
6. 1610 **Moons**	[V] Solar orbits *[4115U1]*[20] [V] Planetary orbits [4214U1] [V] Size [421-U1]
7. 1655 **Planetary Rings**	[V] Solar orbits *[4115U1]*[21] [V] Changing perspective [4215U1]
8. 1672 **Multiple Stars**	[V] Eclipses [4114 U3][22] [V] Doppler shift periods [4125U3] [V] Astrometric orbits [4215U3]
9. 1754 **Open Clusters**	[V] Irregular grouping of stars [421-U5][23] [V] Magnitudes of individual stars [421-U4]
10. 1782 **Globular Clusters**	[V] Spherical grouping of stars [411-U5][24] [V] Magnitudes of typical stars *[421-U3]* [V] Periods and magnitudes of individual variable stars *[4214U3]*[25]
11. 1790 **Planetary Nebulae**	[V, R] [4226U3][26] [V] Expansion age [4226U3] [R] Radio continuum emission [221-U2] [R, I] Radio line emission [221-U2], [322-U3][27, 28]
12. 1798–1803 **Interplanetary Matter**	[V, I] Zodiacal glow: visual [411-L1][29] infrared [311-U1] [D] Meteorites: direct collection [V] Meteors: visual observations [4112U1] [R] Radar reflections off infalling masses [2112U-]
13. 1801 **Asteroids**	[V] Orbital variations *[4115U1]*[30] [V] Brightness variations *[4114U1]*

TABLE 8.2

(Continued)

Cosmic Phenomena and Discovery Dates	Coded Designations (see Table 8.1)
14. 1865 **Ionized Hydrogen Clouds**	[V] Visual spectral lines [422-U5][31] [I] Infrared continuum [321-U5], Infrared emission lines *[322-U4]*[32] [R] Radio recombination lines [222-U2] [R] Radio thermal continuum [221-U3]
15. 1903 **Cold Interstellar Gas Clouds**	[V] Interstellar absorption lines *[412-U-]*[33] [V] Interstellar absorption continuum [41-6U-] [R] Radio emission and absorption lines [222-U3][34, 35] [I] Infrared emission continuum [311-U3] [I] Infrared Emission Lines *[322-U4]*[36]
16. 1910 **Distinction of Subgiants and Red Giants from Main Sequence Stars**	[V] Visually resolved broadband variables;[37] and mass loss and nuclear evolution rates [4217U4]
17. 1912 **Cosmic Rays**	Direct Impact of low-energy particles; [Ch] Cherenkov radiation from less abundant high-energy components [-11-U-][38] [Ch] Highly energetic ions, leptons, and gamma rays, all distinguishable from one another, but generally able to induce Cherenkov radiation in the atmosphere and, at lower energies, with elemental composition and energy determined by Cherenkov, scintillation and [T] [-11-U-][39] transition radiation detectors
18. 1915 **White Dwarfs**	[V] Motion across the sky [4315U2][40] [V] In the binary Sirius A/B, the white dwarf Sirius B is roughly 400 times fainter than Sirius A at visual wavelengths, but it [X] outshines Sirius A in X-ray emission [5215U2], continuum
19. 1919 **Interstellar Reflection Nebulae**	[V] Dust clouds with an optical reflection continuum [411-L3][41] [V] Spectral lines [412-L3]
20. 1920 **Galaxies Containing Gas**[42, 43]	[V] Doppler shift [422-U6] [V] Rotation rate [4227U6] [I] Infrared continuum emission *[321-U6]* [R] Radio continuum emission [211-U4] [R] Radio Doppler shift [212-U3] [R] Radio rotation frequency [2227U3][44]
21. 1928 **Cosmic Expansion**[45]	[V] *[412-U-]*[46] [R] [212-U-][47]

TABLE 8.2

(Continued)

Cosmic Phenomena and Discovery Dates	Coded Designations (see Table 8.1)
22. 1934 **Supernovae**[48]	[V] Power [4214U6] [V] Spectrum [4224U6]
23. 1937 **Supernova Remnants**	[V] Crab Nebula continuum expansion at a rate of 0.2 arcsec per year [4215L4] [V] Filamentary structure [4225U4] [R] Polarized radio continuum [2115L3] [X] X-ray continuum [511-U4][49]
24. 1946 **Radio Galaxies**	[V, R] Extended sources [411-U6], [211-U6] [V] Doppler appearance *[422-U6]* [R] Polarized radio continuum [211-L6][50] [R] Compact sources: radio scintillation [2113L5][51, 52]
25. 1947 **Magnetic Variable Stars**	[V] [4224C4][53]
26. 1952 **Young Stars with Nebulosity; Herbig Haro Objects**[54]	[V] Stellar variability [4214U3] [V] Nebular variability [4115L2] [I] Far-infrared line emission [3224U2] [I] Far-infrared continuum [3214U2][55] [R] Radio variability [2214U1][56] [X, R] Flaring X-rays sometimes accompanied by radio flares and vice versa [5213U3][57]
27. 1954 **Galaxies Lacking Gas**	[V] *[422-U6]*
28. 1962 **X-ray Stars**[58]	[V, X] Crab pulsar [5111U4][59] [X] X-ray binaries with variable continuum [6114U4] [V] Similar variations in the visible domain [4114U4] [X] X-ray bursts [6112U4],[60] [5213U6][61]
29. 1963 **Quasars**	[V] Redshift measurements [4214U7] [R] [2214U6] [X, V] Subsequent optical observations by Maarten Schmidt, using the 200-inch Mount Palomar telescope, showed the spectrum of 3C273 to also be highly redshifted. [5214U7],[62] [4214U7][63] [I] Quasar integrated infrared luminosities [3225U7] [I] Kinetic energy of the molecular outflows [3225U6][64]
30. 1965 **Cosmic Microwave Background Radiation**	[R] [3118U-][65, 66]

TABLE 8.2

(Continued)

Cosmic Phenomena and Discovery Dates	Coded Designations (see Table 8.1)
31. 1965 **Infrared Stars, Circumstellar Dust Clouds**	[V] Slow Variable Continuum [4317U4][67] [I] Debris disks [3217U3].[68] The maximum age for disks is of order 400 Myr[69]
32. 1966 **X-ray Galaxies**	[V, X] Visual appearance [4114U6], [5214U6] [V] Visual appearance of clusters of galaxies [411-U7] [X] X-ray appearance of clusters of galaxies [5114U7][70]
33. 1967 **Cosmic Masers and Megamasers**	[R] Circularly, linearly or elliptically polarized emission [2224E2][71]
34. 1967 **Pulsars, Neutron Stars, and Magnetars**	[R] Pulsar radio pulse rate [2212L2] [R] Millisecond pulsars [2211L2] [X] Magnetars [5212-3][72, 73]
35. 1970 **Infrared Galaxies**	[V] [421-U6] [I] Infrared dust continuum *[321-U6]*[74, 75]
36. 1971 **Superluminal Radio Sources**	[R] Source separation accuracy $\sim 10^{-4}$ arcsec. Spectral resolution due to a rubidium clock. Measurements based on 110 sec integration times [2443L6][76, 77]
37. 1972 **Interstellar Magnetic Fields**	[R] Zeeman effect: [212-C-] [R] Faraday rotation on pulsars with 100 μs pulse arrival times [2111L-][78, 79]
38. 1973 **Gamma-Ray Bursts**	[G] [6112U9][80, 81]
39. 1979 **Stellar-Mass Black Holes, Microquasars**	[R] [2314U1] [X] [6214U5] [I] *[4214U4]* [V] Kinetic energy of the bulk motion of the expanding superluminal source is $\sim 3 \times 10^{46}$ erg [4214U4][82, 83]
40. 1979 **Dark Matter Haloes**	[V] *[422-U6]* [R] 21 cm radiation at distance of 82 Mpc and 0.1 Jy [212-U4][84]
41. 1979 **Gravitational Lenses**	[V, R] Identical adjacent time-shifted images [4225U6],[85, 86] [2225U6][87]
42. 1980 **Intergalactic Medium: Lyman-α Forest**	[V, X] X-ray Emission *[422-U-]*,[88] [511-U6] [X] Indications that the Iron abundance in nearby galaxy clusters is about 30% the Solar abundance *[511-U6/7]*[89]
43. 1982 **Gravitational Radiation from Binary Pulsars**	[R] Instantaneous time resolution [2321U-] [R] Long-term changes [2325U-][90]

TABLE 8.2

(Continued)

Cosmic Phenomena and Discovery Dates	Coded Designations (see Table 8.1)
44. 1984 **Accretion and Prestellar Disks**	[I] [3217U3][91, 92]
45. 1984 **Galaxy Evolution & Mergers**	[I] *[321-U7]*[93] [R] [221-U5][94] [V] [431-U6][95, 96]
46. 1987 **Antineutrinos and Neutrinos from a Supernova Explosion**	[N] [6112-9][97, 98]
47. 1986–89 **Galaxy Clusters, Filaments, Sheets &Voids**	[V, R] General distribution [411-U7], [211-U6] [R, V] Redshifts [222-U4],[99] *[422-U7]*[100]
48. 1992–95 **Extrasolar Planets**	[R] Pulsar time resolution [2321U2] [R, V] Planetary orbital period [2324U-],[101] *[4233U3]*[102] [V] Kepler planetary transit observations, mainly integrated on Solar type stars and read out at half hour intervals [4213U3]
49. 1994 **Solar Neutrinos and their Oscillations**	[N] [611-U3][103, 104]
50. 1995 **Brown Dwarfs**	[I] Proper motion [4315U1] [I] Spectrum [4325U1][105]
51. 1995 **Hot Jupiters**	[V] *[4233U3]*[106] [V] Kepler planetary transit observations, read out at half hour intervals [4213U3]
52. 1996 **Evolution of Cosmic Elemental Abundances**	[V] [422-U3],[107] *[422-U-]*[108]
53. 1997 **Highest-Energy Cosmic & Gamma Rays**	[Ch] 10^{22} eV [!111U-][109]
54. 1998 **Dark Energy**	[V] Type Ia supernova luminosity distance [4114U6] [V] Spectral redshift [4124U6][110]
55. 2001 **Cosmic Re-ionization – The Gunn-Peterson Effect**	[V] [4218U6][111]
56. 2004 **Supermassive Black Holes**	[V] Orbital position and period of encircling stars [4326U-] [V] Orbital period and position uncertainty [4324U-][112, 113]
57. 2007 **Fast Radio Bursts**	[R] [2221E6][114, 115, 116]

TABLE 8.2

(Continued)

Cosmic Phenomena and Discovery Dates	Coded Designations (see Table 8.1)
58. 2008 **Large-Scale Cosmic Flows**[117]	[V] Galaxy X-ray Cluster visual data *[422-U7]* [X] *[511-U6/7]*[118] [I] *[3118E-]*[119]
59. 2013 **Cosmic Microwave Background Fluctuations**	[I] *[3118E-]*[120, 121]
60. 2016 **Black Hole Mergers & their Gravitational Waves**	[GW] [GW1112-!][122]

The list restricts itself to observationally obtained evidence. It deliberately excludes discovery through exploration – a mode of astronomical research to date available solely to Solar System investigations. It thus lists moons and asteroids, respectively, as recognized reliably in merely a single way, even though spacecraft landing and possibly returning with surface samples have by now yielded an extraordinarily richer lode of evidence, at present out of reach for all other astronomical investigations. Leaving exploration aside thus provides us with a more uniform set of data sampling all parts of astronomy – Solar System, stellar, exoplanetary system, Galactic, extragalactic, and cosmological. Inclusion of knowledge gained through exploration would have sampled Solar System phenomena differently from all others.

A complete list of all the distinct monograms characterizing the instruments and messengers involved in discovering the entire set of major phenomena recognized to date is provided in Table 8.4.

Each monogram listed is accompanied in its adjacent column by the phenomenon or set of phenomena it helped to discover, numbered 1 through 60, as in Figure 8.2 and Table 8.2 above.

A similar compilation I provided four decades ago in *Cosmic Discovery,* published in 1981, listed 43 phenomena. That I now list 17 more indicates considerable progress. But the difference is actually even greater. A number of phenomena included in the *Cosmic Discovery* list have now been dropped. More sensitive instruments revealed the earlier entries "unidentified radio sources" to be readily identified; and "gamma-ray background radiation" and "X-ray background radiation" now are both known to merely reflect myriad individual sources – unlike the cosmic microwave background, which is a true spatial continuum, even if permeated by faint ripples.

Some added clarification may also help: I have kept the former designation "infrared star" even though some of these are now known to be highly evolved

TABLE 8.3

Cosmic Phenomena with Number of Independent Messengers

Cosmic Phenomena	Number of Independent Messengers	Coded Designations (see Table 8.1)
1. Stars	1	*[421-U3]*
2. Planets	2	*[4115U1]*, [1215U1][16]
3. Novae/Eruptive Variables	1	[4113U4], [4115U4], [4213U2]
4. Comets	2	*[4115U1]*, *[4114U1]*, [5115U1][18]
5. Pulsating Variable Stars	1	[4214U5], *[4224U4]*, *[4214U4]*
6. Moons	1	*[4115U1]*, [4214U1], [421-U1]
7. Planetary Rings	1	*[4115U1]*, [4215U1]
8. Multiple Stars	1	[4114 U3], [4125U3], [4215U3]
9. Open Clusters	1	[421-U5], [421-U4]
10. Globular Clusters	1	[411-U5], *[421-U3]*, *[4214U3]*
11. Planetary Nebulae	2	[4226U3], [4226U3], [221-U2], [221-U1], [322-U3]
12. Interplanetary Matter	3	[411-L1], [311-U1], [4112U1], [2112U-], [D]
13. Asteroids	1	*[4115U1]*, *[4114U1]*
14. Ionized Hydrogen Clouds	3	[422-U5], [321-U5], *[322-U4]*, [222-U2], [221-U3]
15. Cold Interstellar Gas Clouds	2	*[412-U-]*, [41-6U-], [222-U3], [311-U3], *[322-U4]*
16. Distinction of Subgiants and Red Giants from Main Sequence Stars	1	[4217U4]
17. Cosmic Rays	1	[Ch] [-11-U-]
18. White Dwarfs	1	[4315U2], [5215U2]
19. Interstellar Reflection Nebulae	1	[411-L3], [412-L3]
20. Galaxies Containing Gas	2	[422-U6], [4227U6], *[321-U6]*, [211-U4], [212-U3], [2227U3]
21. Cosmic Expansion	2	*[412-U-]*, [212-U-]
22. Supernovae	1	[4214U6], [4224U6]
23. Supernova Remnants	2	[4215L4], [4225U4], [2115L3], [511-U4]
24. Radio Galaxies	1	[211-U6], [422-U6], [211-L6], [411-U6], [2113L5]
25. Magnetic Variable Stars	1	[4224C4]
26. Young Stars with Nebulosity; Herbig Haro Objects	2	[4214U3], [4115L2], [3224U2], [3214U2], [2214U1], [5213U3]

TABLE 8.3

(Continued)

Cosmic Phenomena	Number of Independent Messengers	Coded Designations (see Table 8.1)
27. Galaxies Lacking Gas	1	*[422-U6]*
28. X-ray Stars[58]	1	[5111U4], [6114U4], [4114U4], [6112U4], [5213U6]
29. Quasars	3	[2214U6], [4214U7], [5214U7], [3225U7], [3225U6]
30. Cosmic Microwave Background Radiation	1	[3118U-]
31. Infrared Stars, Circumstellar Dust Clouds	1	[4317U4], [3217U3]
32. X-ray Galaxies	1	[4114U6], [5214U6], [411-U7], [5114U7]
33. Cosmic Masers and Megamasers	1	[2224E2]
34. Pulsars, Neutron Stars and Magnetars	2	[2212L2], [2211L2], [5212-3]
35. Infrared Galaxies	1	[421-U6], *[321-U6]*
36. Superluminal Radio Sources	1	[2443L6]
37. Interstellar Magnetic Fields	2	[212-C-], [2111L-]
38. Gamma-Ray Bursts	1	[6112U9]
39. Stellar-Mass Black Holes, Microquasars	1	[2314U1], [6214U5], *[4214U4]*, [4214U4]
40. Dark Matter Haloes	2	*[422-U6]*, [212-U4]
41. Gravitational Lenses	2	[4225U6], [2225U6]
42. Intergalactic Medium: Lyman-α Forest	2	*[422-U-]*, [511-U6], *[511-U6/7]*
43. Gravitational Radiation from Binary Pulsars	1	[2321U-], [2325U-]
44. Accretion and Prestellar Disks	1	[3217U3]
45. Galaxy Evolution & Mergers	2	*[321-U7]*, [221-U5], [431-U6]
46. Antineutrinos and Neutrinos from a Supernova Explosion	1	[6112-9]
47. Galaxy Clusters, Filaments, Sheets &Voids	2	[411-U7], [211-U6], [222-U4], *[422-U7]*
48. Extrasolar Planets	3	[2321U2], [2324U-], *[4233U3]*, *[4213U3]*
49. Solar Neutrinos and their Oscillations	1	[611-U3]

TABLE 8.3

(Continued)

Cosmic Phenomena	Number of Independent Messengers	Coded Designations (see Table 8.1)
50. Brown Dwarfs	1	[4315U1], [4325U1]
51. Hot Jupiters	2	*[4233U3]*, *[4213U3]*
52. Evolution of Cosmic Elemental Abundances	2	[422-U3], *[422-U-]*
53. Highest-Energy Cosmic & Gamma Rays	1	[!111U-]
54. Dark Energy	1	[4114U6], [4124U6]
55. Cosmic Re-ionization – The Gunn-Peterson Effect	1	[4218U6]
56. Supermassive Black Holes	1	[4326U-], [4324U-]
57. Fast Radio Bursts	1	[2221E6]
58. Large-Scale Cosmic Flows	2	*[422-U7]*, *[511-U6/7]*, *[3118E-]*
59. Cosmic Microwave Background Fluctuations	1	*[3118E-]*
60. Black Hole Mergers & their Gravitational Waves	1	[GW1112-!]

red giants; adding this newer entry indicates the need for distinctions beyond those between giants and main sequence stars (entry 16 of the current table). Similarly, I have kept "X-ray stars" even though these divide into a number of different phenomena, such as pulsars, neutron stars, magnetars, and stellar black holes, while more sensitive instrumentation now shows that many young stars also emit X-rays copiously. On the present list "flare stars" are no longer listed separately; they are subsumed among "novae; eruptive stars." I have kept the designations "Infrared galaxies," "Radio galaxies," and "X-ray galaxies," even though many now appear merged with "Quasars." On the other hand, some infrared galaxies arise through galaxy mergers, and thus have a clearly distinct origin.

Three phenomena discovered but not yet clearly recognized in 1979 are only now included; their full significance emerged some years later. Entries numbered 39 and higher list phenomena established since 1979; they constitute the 22 discoveries listed in italics, none of which were included on the list of phenomena published in *Cosmic Discovery* in 1981.

TABLE 8.4

Filter Monograms Tracing the Characteristics of Major Phenomena

Monogram	Phenomena	Monogram	Phenomena	Monogram	Phenomena
1215U1	2	3225U7	29	4224C4	25
2111L-	37	322-U3	11	4224U4	5
2112U-	12	322-U4	14,15	4224U6	22
2113L5	24	4112U1	12	4225U4	23
2115L3	23	4113U4	3	4225U6	41
211-L6	24	4114U1	4,13	4226U3	11
211-U4	20	4114U3	8	4226U3	11
211-U6	24	4114U4	28	4227U6	20
211-U6	47	4114U6	32	422-U3	52
212-C-	37	4114U6	54	422-U5	14
212-U3	20	4115L2	26	422-U6	20,24,27,40
212-U4	40	4115U4	3	422-U7	47,58
212-U-	21	4115U1	2,4,6,7,13	422-U-	42,52
2211L2	34	411-L1	12	4233U3	48,51
2212L2	34	411-L3	19	4315U1	50
2214U1	26	411-U5	10	4315U2	18
2214U6	29	411-U6	24	4317U4	31
221-U2	11	411-U7	47	431-U6	45
221-U3	14	411-U7	32	4324U-	56
221-U5	45	4124U6	54	4325U1	50
2221E6	57	4125U3	8	4326U-	56
2224E2	33	412-L3	19	5111U4	28
2225U6	41	412-U-	15,21	5114U7	32
222-U1	11	41-6U-	15	5115U1	4
222-U2	14	4213U2	3	511-U4	23
222-U3	15	4214U1	6	511-U6	42
222-U4	47	4214U3	26	511-U6/7	42,58
2227U3	20	4214U3	10	5212-3	34
2314U1	39	4214U4	5,39	5213U3	26
2321U2	48	4214U4	39	5213U6	28
2321U-	43	4214U5	5	5214U6	32
2324U-	48	4214U6	22	5214U7	29
2325U-	43	4214U7	29	5215U2	18
2443L6	36	4215L4	23	6112U4	28
3118E-	58,59	4215U1	7	6112U9	38
3118U-	30	4215U3	8	6214U5	28
311-U1	12	4217U4	16	6114U4	39
311-U3	15	4218U6	55	[Ch]:-11-U-	17
3217U3	44,31	421-U1	6	[Ch]:!111U-	53
321-U5	14	421-U3	1,10	[GW]:1112-!	60
321-U6	20,35	421-U4	9	[N]:6112-9	46
321-U7	35,45	421-U5	9	[N]:611-U3	49
3214U2	26	421-U6	35	[T]	17
3224U2	26	4213U3	48,51	[D]	12
3225U6	29				

8.12   The Bernoulli Binomial Distribution

In the early post-World War II years, discovery of a novel phenomenon by independent instrumental means, such as James Stanley Hey's discovery of the radio galaxy Cygnus A, gave reasonable assurance that each such discovery had occurred purely by chance.[123] In the three decades before I wrote *Cosmic Discovery*, in 1981, astronomy had inherited a sizeable influx of discarded military techniques that could be fruitfully incorporated into the nascent fields of radio, infrared, X-ray, or gamma-ray astronomy. Each of the novel techniques had to be tested out.

For a number of years, the contributions of each of these novel techniques progressed independently, sampling the phase space of instrumentation more or less randomly. Random sampling of cosmic phenomena was a reasonable assumption also because most physicists and engineers applying their new instruments to astronomical detection were primarily interested in determining whether their instruments had any applications at all that might prove useful to the field. They had no particular astrophysical agenda and often were unaware of the few theoretical predictions that did exist.

Discovery of new phenomena under these circumstances was similar to randomly sampling the previously unexplored portions of the phase space of observational capabilities – in short, sampling discovery space, and coming up with the numbers, A, B, or C, of phenomena discovered solely by a single means, by two technically distinct means, or in three different ways. Numbers of higher rediscovery rates, labeled E, F, ..., might eventually also be included. But, at the time, three independent discoveries of a single phenomenon was the maximum that had been encountered. The total number of distinct phenomena known was then $N = A + B + C$.

We may now examine what the values $A = 38$, $B = 18$, and $C = 4$ provided by the table of monograms may teach us, if we make the single assumption that multiplicity m, the number of different ways that any phenomenon can be independently established in our Universe, is always the same. In that case, the somewhat haphazard ways in which tools originally designed to meet industrial or military needs became available to astronomers suggest that a *binomial distribution* might provide a useful first approximation to the relative values A, B, and C encountered. The binomial distribution, proposed by the seventeenth-century Swiss mathematician Jacob Bernoulli, might then be expected to predict the approximate total number of distinct cosmic phenomena we might ultimately discover.

By no means will an arbitrarily selected combination of values A, B, C necessarily conform to Bernoulli's distribution. Whether or not it does becomes a first test the data need to undergo.

Two potentially simplifying assumptions also must be included to pass this test. The first is that m, the *multiplicity* of ways a phenomenon is randomly distributed in the phase space of observations, is consistent with having one and the same value for all discoveries to date, is positive and, as the present listings of phenomena indicate, lies in the range $m \geq 3$. The second is that the fraction of the discovery space, r, to date, lies in the range $0 < r \leq 1$.

Let V be the entire volume of the accessible phase space of cosmic observations and v be a subvolume of this space to which we currently have access. For the moment, let us assume that a search through the entire volume would lead us to find the phenomenon displayed in m distinct locales of V, meaning that it is detectable by m distinct instrumental means. Based on this information, we now ask what the probabilities are of finding q independent sightings of this phenomenon in subvolume v. This again is equivalent to asking for the probability that q different instrumental techniques would lead to positive detections of the phenomenon within this subvolume.

If we call the ratio of these volumes $r \equiv v/V$, the probability P_q of finding the phenomenon of interest, by virtue of any number q of distinct instrumental means, is

$$P_q = \frac{m!}{q!\,[m-q]!} r^q [1-r]^{(m-q)} , \tag{8.1}$$

where $m! \equiv m[m-1][m-2]\ldots[2][1]$ is the factorial product, and $0! \equiv 1$.

The probability distribution P_q is called *Bernoulli's binomial distribution*. The sum of all the probabilities P_q is unity, meaning that some astronomical phenomenon is observable through at least one of the means by which the Universe transmits information. The mean value of q is the product of the total population, or multiplicity m of the phenomenon in V, and the fraction r of the volume observed:

$$\sum_{q=0}^{m} P_q = 1 \quad ; \quad \langle q \rangle = \sum_{q=0}^{m} q P_q = mr. \tag{8.2}$$

The total number of distinct, ultimately discoverable cosmic phenomena, n, can be derived from the distributions defined by expressions (8.1) and (8.2), respectively, for values of $q = 1, 2, 3,$ and 4, as:

$$A = nmr(1-r)^{m-1}, \tag{8.3}$$

$$B = \frac{nm(m-1)}{2} r^2 (1-r)^{m-2}, \tag{8.4}$$

$$C = \frac{nm(m-1)(m-2)}{6} r^3 (1-r)^{m-3}, \tag{8.5}$$

and

$$D = \frac{nm(m-1)(m-2)(m-3)}{24}r^4(1-r)^{m-4}. \tag{8.6}$$

Earlier we found that current observations yield estimates of $A = 38$, $B = 18$, $C = 4$, and $D = 0$. We can enter the first three of these estimates in equations (8.3), (8.4), and (8.5) to derive the quantities m, r, and n, and then verify that equation (8.6) is compatible with a value for D close to zero.

We first combine our values for A, B, and C to derive an expression for the multiplicity m:

$$\frac{3AC}{2B^2} = \left(\frac{m-2}{m-1}\right), \quad \text{or} \quad m = 1 - \left(\frac{1}{3AC/2B^2 - 1}\right). \tag{8.7}$$

This yields $m = 4.375$. The ratio of expressions (8.3) and (8.4) yields

$$r = \frac{2B}{A(m-1) + 2B} = 0.2192. \tag{8.8}$$

This is reassuring, because the emerging product $mr = 0.959$ is so close to unity. If discovery space is finite with each phenomenon randomly confined to m distinct regions in this space, and if a fraction r of these regions has been investigated to date, the probability of having discovered any given phenomenon by now should be approaching $mr \sim 1$ if we are nearing discovery of all available phenomena by at least some single instrumental means.

The final piece of information we wish to derive is the number of phenomena n we may ultimately hope to discover in the Universe. Equations (8.3) and (8.4) provide this as well:

$$n = \left(\frac{A^2}{2B}\right)\left(\frac{m-1}{m}\right)\frac{1}{(1-r)^m}. \tag{8.9}$$

As our observations of the Universe become increasingly trustworthy, n should approach a constant value. At first, it may seem a bit magical that we should be able to estimate the number of cosmic phenomena we may ultimately hope to discover. But because we found the product mr to be close to unity, the assumption that all phenomena exhibit an identical multiplicity suggests that we should by now have discovered all phenomena roughly once, although because their distribution in the phase space of observations is random, some of them will have escaped discovery to date, just as a compensating number $(B + 2C) = 26$ has been discovered twice or three times through this random process. So n should be larger than $A+B+C = 60$, but only by a number of order

$(B + 2C)$. Substituting the values for $A, B, m,$ and r in equation (8.9) leads to an estimate $n = 91.34$, which roughly meets these criteria, in that $A + 2B + 3C = 86$.

When data for all four entries A, B, C, and D are available, we can form the ratio

$$\frac{AD}{BC} = \frac{m-3}{2m-2} \quad \text{and} \quad D = \frac{(m-3)BC}{(2m-2)A}, \tag{8.10}$$

which yields $D \sim 0.386$ for $m \sim 4.375$. This is closer to $D = 0$ than to $D = 1$, and thus in coarse agreement with the estimate $D \sim 0$ the observations in hand yield. In summary, the values $A = 38$, $B = 18$, $C = 4$, $D = 0$, available in 2019, lead to values for the multiplicity $m = 4.375$; the ratio of the phase space accessible, $r = 0.2193$; and the estimated number of phenomena characterizing the Universe, $n = 91.34$. As a consistency check, we may reinsert the derived values for r, m, and n back into equations (8.3), (8.4), and (8.5), to obtain $A = 38.0025$, $B = 18.0140$, $C = 4.0060$, and for fidelity purposes we also calculate $D = 0.3668$, which would have been expected to be zero. But, satisfactorily, $A + B + C = 60.0225$, though $A + B + C + D = 60.3893$.

Had we initially insisted that $D = 0$ precisely, equation (8.10) could have been satisfied only by $m = 3$, which would have been incompatible with equation (8.7) obtained from the far more reliably established values of A, B, and C.

8.13   Compatibility of Data with a Bernoulli Distribution

Given our assumption that all phenomena are equally distributed across mutually exclusive realms of the phase space of observations, how can we verify that discovery and rediscovery rates of cosmic phenomena are governed by the statistics of a Bernoulli distribution? One sign of this is that the range of data in hand should be compatible with physically meaningful values of the multiplicity m appearing in equations (8.3) to (8.10), and that the derived values of n, m, and r mutually agree when different formulae for them are employed.

By physically meaningful, we may mean, for example, that m must exclude values $m \leq 3$. Otherwise no major phenomena could be recognized in at least four mutually independent ways – an unlikely circumstance, given that data by now in hand indicate the existence of several triply recognized phenomena, meaning that $m = 3$ already emerges from at least those findings.

A useful illustration of this is data sets that plausibly could have emerged from existing data, but did not. One such set would include a single quadruply recognized phenomenon. This appears not to be a particularly unlikely possibility, given that the distribution emerging from the data right now are well fitted by values $A = 38$, $B = 18$, and $C = 4$. We ask, what would have happened if one of the triply recognized phenomena actually had been mistakenly called a quadruply recognized phenomenon, so that we actually were dealing, instead, with a data set $A = 38, B = 18, C = 3, D = 1$? This would leave the currently recognized phenomena at 60, but two distinct values of m would emerge from the Bernoulli distribution, depending on whether we used equation (8.7) or (8.10). Both turn out to be unphysical. Equation (8.10) leads to the negative value, $m = -3.91$; equation (8.7) yields $m = 3.118$, incompatible with the fourfold recognition of at least the one phenomenon that is fit by $D = 1$.

A similar perturbation of our present data set of 60 distinct phenomena, but leading to $A = 37, B = 18, C = 4, D = 1$, still leads to unphysical values of m that strongly differ. It is not until we lower A and increase C, to a hypothetical data set $A = 35, B = 18, C = 5, D = 1$, that roughly compatible, physically meaningful values of m emerge from the two distinct equations we have been testing. Just as m may be derived in two different ways, by these means, r may also be derived in two different ways from the Bernoulli relations, leading to a total of four r values, and to two distinct estimates of n, respectively $n = 141$, and $n = 104$, which only coarsely agree with each other.

The lesson indicated by this example is that the emergence of quadruply recognized phenomena in data sets compatible with a Bernoulli distribution requires a considerably flatter distribution $A : B : C : D$ than our current data set in Section 8.11 presents. Distributions such as those examined in the present section may exist, but they would be incompatible with a Bernoulli distribution.

8.14   Two Stability Tests

Given the restrictions just encountered by quadruply recognized phenomena in fitting into a Bernoulli distribution, we need to check whether the data compiled in Section 8.11 face similar instabilities when individual entries A, B, or C are perturbed with $A + B + C$ kept constant at a value 60, and $D = 0$. Figures 8.4 and 8.5 demonstrate that perturbing any of the values A or B or C by one or two units, in the environs of our best fit to the data, $A = 38, B = 18$, $C = 4$, show the number of phenomena n, their multiplicity m, and the ratio of phase space r observed to date, to be remarkably constant, though gradually and systematically drifting away from the best central value the observational data indicate.

A further test can be run to discern how a division of the data into two sets of phenomena, one exhibiting higher, the other lower multiplicities would affect a prediction of the total expected number of phenomena, n. Many ways exist for dividing our main data into two complementing sets. Not all are likely to be as stable as the very first division of the data tried, but it may be a sufficient indicator. It is the almost equal split of phenomena into two sets, respectively with $N_1 = 31$ in the first set with $A_1 = 18, B_1 = 10, C_1 = 3$; and $N_2 = 29$ in the second set, with $A_2 = 20, B_2 = 8, C_2 = 1$.

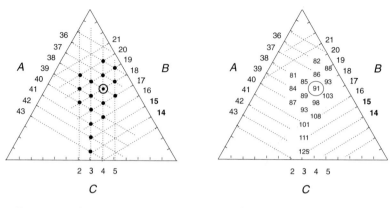

The Predicted Total Number of Cosmic Phenomena, *n*, Derived from Their Adherence to Bernoulli Binomial Statistics

Plausible A-B-C Coordinate Ranges for the 60 Phenomena Investigated

Bernoulli Statistics Estimates for *n*, the Total Number of Cosmic Phenomena

Fig. 8.4 The predicted number of cosmic phenomena, *n*, based on data for $N = A + B + C = 60$ major phenomena. $A = 38$ of these are recognized through only one set of observations, $B = 18$ by two independent means, and $C = 4$, by virtue of three mutually independent observations. The best values of *A*, *B*, and *C* the data exhibit are highlighted by the encircled coordinate point on the left. Surrounding points in this figure maintain this total though permitting, *A*, *B*, and *C* to vary individually. Plausible uncertainties in their values may then make the total number of predicted phenomena, *n*, more uncertain. The diagram on the right shows the *n* values obtained correspondingly derived for these respective coordinate locations, all of which could fit Bernoulli's binomial distribution for $N = 60$. The derived number *n* obtained in this way clusters around the data's best prediction of $n \sim 91$, across a range from $n = 85$ to 103, in the central position's immediate surroundings, and from 81 to 108, one step further removed. Systematic trends of *n* along coordinates *A*, *B*, and *C* also are clearly evident.

When we calculate the multiplicities in these respective sets within a Bernoulli distribution, we find that $m_1 \sim 6.3$, $m_2 \sim 2.9$, and $r_1 \sim 0.17$, $r_2 \sim 0.30$, so that $m_1 r_1 \sim 1.1$, and $m_2 r_2 \sim 0.9$, roughly as expected. The respectively predicted numbers of phenomena in the two sets are $n_1 \sim 45.2$, $n_2 \sim 45.3$; the total number of phenomena their Bernoulli distributions would predict is $n_1 + n_2 \sim 90.5$, which is essentially identical to the estimate of *n* highlighted in Figure 8.4, for the Bernoulli distribution based on a single homogeneous multiplicity applying to all phenomena, despite the factor of ~2.2 difference between multiplicities m_1 and m_2. The mean multiplicity for the two distributions, $(m_1 n_1 + m_2 n_2)/(n_1 + n_2) \sim 4.6$, also is not far from our originally calculated central value $m = 4.4$ in Figure 8.5.

The Fraction, *r*, of Phase Space Covered, and the Multiplicity, *m*, of 60 Cosmic Phenomena, Using Bernoulli-Statistics Estimates

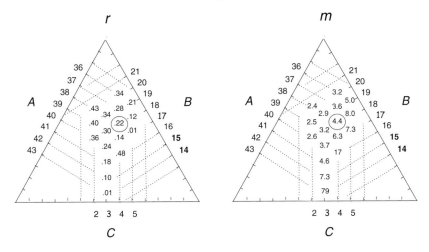

Fig. 8.5 The fraction of the phase space of observations, *r*, and the estimated multiplicity, *m*, of the 60 phenomena investigated by means of instrumentation available in 2019. As in Figure 8.4, the results are shown for $N = A + B + C = 60$ major phenomena, where *A* is the number of phenomena recognized through a single set of observations, *B* represents the number recognized by means of two independent sets, and *C* the number observed by three mutually independent means. Multiplicities, *m*, on the right, initially cluster fairly stably around the encircled value $m = 4.4$ marking the best fit for *A*, *B*, and *C* based on available observations, but then rather rapidly diverge downward in the figure, with declining fractions, AC/B^2. For the same reasons, a similar trend also is seen for increasing *C* values on the right. Low fractional values are implausible in a Bernoulli distribution unless the selection studied is based on the extraction of a very small ratio, *r*, from sets of phenomena occurring with extremely high multiplicity. This is confirmed by the rapidly diminishing values of *r* on the left, for each of the high values of *m* on the right. Throughout, the product *rm* is uniformly stable and of order $rm = 1$.

8.15   Comparison of Data Available in 1979 and 2019

In contrast to *n* and *m*, the fraction *r* of the phase space sampled should monotonically increase over time, as long as the cosmic search is actively pursued. And as the value of *r* approaches unity, *n* and *m* should converge to their final values. These should be independent of the year in which the populations *A*, *B*, *C*, or *D* may have been compiled, and therefore might be compared to those originally estimated in *Cosmic Discovery* for the year the 1979:

In 1979, I listed $N = 43$ different phenomena, and estimated $A = 35$, $B = 7$, $C = 1$. At the time, I assumed that $m \gg 1$, $r \ll 1$ and $mr \ll 1$, the last of which

TABLE 8.5

The Number of Phenomena Estimated in Different Years

Cosmic Phenomena	Original 1979	Revised 1979*	2019
Total discovered ($A + B + C$)	43	40	60
Singly recognized A	35	32	38
Doubly recognized B	7	7	18
Triply recognized C	1	1	4
Multiplicity, m	large	50	4.4
Fraction of phase space sampled, r	0.01	0.00885	0.219
Product, mr	—	0.44	0.96
Total number of cosmic phenomena, n	123	112	91
Recognized fraction of total	35%	36%	66%

* Revised as explained in the text.

seemed not unreasonable at the time, although $mr \sim 1$ would have been a far more credible approximation. With this, I made use of the approximation

$$\frac{A}{2B}(A + 2B) = \frac{nm(1-r)^m}{(m-1)}\left[1 + \frac{(m-1)r}{(1-r)}\right] \sim n \quad \text{for} \quad m \gg 1, rm \ll 1, \tag{8.11}$$

where I set $m/(m-1) \sim 1$ and $(1-r) \sim 1$ to obtain $n \sim 123$. The assumption $m \gg 1$ also made the value of $C \sim 0.93$ in equation (8.5) above, in reasonable approximation to unity. These values appear as entries in the column designated "Original 1979" in Table 8.5.

As already mentioned, three phenomena listed in 1979, the unidentified radio sources, the X-ray background, and the gamma-ray background, have now been superseded. This is because radio sources are by now much more satisfactorily identified, and the X-ray and gamma-ray backgrounds generally appear to resolve into discrete sources. So the revised number for that epoch may now be listed as $A = 32$, $B = 7$, $C = 1$, and $mr \sim 1$. With these revised data, and use of equations (8.7) to (8.9), instead of (8.11) we obtain a set of values entered in the Table 8.5 column "Revised 1979." Finally, column "2019" provides the most recent estimates for the same parameters.

With the counts A, B, and C, estimated in 1979 as I was writing *Cosmic Discovery*, and the denser sampling of the phase space available in 2019, the comparisons for the years 1979 and 2019 are listed in Table 8.5.

8.16   Verification

The number of major phenomena estimated four decades ago may have seemed far-fetched at the time, but their rough verification now may be explained rather simply, based on the knowledge gained in the meantime on

(i) the range of information the Universe is able to transmit, (ii) the nature of the physical processes at work in the Cosmos, and (iii) the way we defined major physical phenomena then, and as a consequence even as we conduct a comparable census today.

We can take up this third point first: We defined major phenomena as differing from all other such phenomena by a factor of a thousand in at least one observational trait. Looking at Figures 6.4, 6.8, and 6.9, respectively, for ranges of angular resolution, spectral resolution, and time resolution, and all three of these for energy ranges, we note that, for electromagnetic radiation, the available energy range covers 8 factors of a thousand; angular resolution covers 3; and spectral resolution 2.5; time resolution covers 5, not all of which are fully accessible. The product of these four yields a discovery space volume spanning 300 distinct four-dimensional cubes, and thus a potential for no more than ~300 distinct ways in which individual phenomena might exhibit a single feature in which they differ observationally by a factor of at least one thousand from all other phenomena.

As earlier mentioned, the total number of distinct major phenomena observationally discovered or rediscovered by electromagnetic means to date is roughly given by $A + 2B + 3C - 6 \sim 80$. The subtracted value coarsely takes into account discoveries, to date, attributable to observations by means of neutrinos, gravitational waves, or matter directly impacting Earth.

The ranges available for cosmic rays are incomparably smaller than those for electromagnetic radiation because magnetic fields make the locations from which these energetic particles arrive and their times of arrival uncertain, and because destructive collisions along their trajectories only add to uncertainty. Gravitational wave and neutrino processes could, in principle, add considerable volume to the phase space of all observations, but we have not yet had much observational contact with these, and so they have not contributed significantly to our tally.

Knowing what we do about neutrinos, however, we can be fairly confident that the total energy stored in primordial neutrinos is consistent with neutrino population densities that could have been present when the Universe initially became enriched in helium and other light elements. The only generation of neutrinos since then would have to have taken place in the interiors of stars or at the surfaces of black holes, and the total energy of generated neutrinos, as well as their inability to strongly interact with other radiations or matter, are unlikely to have produced highly coherent phenomena beyond the types of bursts we already observe in supernova explosions and might expect to see generated in hypernova explosions and other compact processes, such as the merger of massive neutron stars into black holes.

Gravitational wave generation may also span a fairly narrow range of potential phase space. Figure 4.4 shows that mergers of black holes and a variety of explosive processes can generate gravitational waves, and that the size of the bodies required to generate the waves leads to the production of extremely long-wavelength gravitational radiation, as well as extremely long wave trains. The first of these means that baselines needed to acquire high angular resolution will need to be extremely long; the second tells us that arrival times will be highly protracted. Location and timing of the energetic processes involved may then be most readily achieved through detection of auxiliary electromagnetic processes that may accompany gravitational wave generation. In practice, deep insight on gravitationally radiating processes is likely to be limited by the same phase space limitations that already restrict the number of major electromagnetic phenomena we are likely to encounter.

Lastly, there are the arguments put forward by Fukugita and Peebles (2004), which indicate that our current understanding of the Universe does not leave a great deal of room for energetic processes that might have escaped attention so far. Rather, we are apparently stuck with two major energy reservoirs, dark energy and potentially dark matter, which appear to lack the free energy needed to generate novel phenomena. The free energy the Universe does make available already appears to be efficiently accessed by means developed to date or likely to be made available within the next century or two, provided the pace of astronomical research continues at its current level.

An additional factor worth mentioning is that we do now see a larger number of significant phenomena than we detected four decades ago. Several phenomena are now clearly seen as constituting distinct processes crowded into the same boxes of the observational phase space, and clearly distinguishable from one another solely on the criterion introduced earlier, that major phenomena differ from one another by factors of a thousand in at least one observational trait.

This, in itself, suggests that the tally of major phenomena arrived at 40 years ago, by this factor-of-a-thousand distinction, was not far wrong. It also shows that we now recognize distinctions on much finer scales of resolution as yielding interesting insight of their own, and that the provisional classification scheme for phenomena introduced in earlier decades no longer provides a sufficiently fine sieve to serve a particularly useful purpose. At a time when our surveys are beginning to cover hundreds of millions of galaxies, and even larger numbers of stars, the age of studying individual smaller nuggets may have arrived, focusing attention on lesser anomalies rather than major trends.

These all are plausible arguments for why we now should not expect to discover a far larger number of major phenomena than we already recognize, and why the number of such phenomena predicted many decades ago, before we knew about all the bounds constraining the range of observational cosmic data that would ultimately become available, appear to have been roughly correct.

The same conclusion may also be partly due to our recognition that many phenomena earlier judged distinct now are known merely to be different perspectives of one and the same set of events. Thus, quasars, radio galaxies, and X-ray galaxies all are symptomatic of an active galactic nucleus. All of these also are related to the supermassive black holes that reside at the centers of many galaxies. I have listed those separately, above, because not all supermassive black holes are active; but a future count of cosmic phenomena might list all these phenomena as merely different aspects of supermassive black hole evolution. Such a future tally might well also list main sequence stars and giants as different evolutionary phases of ordinary stars whose lives could conclude in one of three separate ways, as white dwarfs, neutron stars, or stellar-mass black holes, depending only on their respective initial masses and partially also on initial chemical composition.

In this sense, the taxonomy we have undertaken to obtain some estimate of the number of cosmic phenomena or species is likely to follow the history of biological taxonomy. There too, groups that initially could have appeared to constitute different species were quickly found to represent no more than different phases of a single organism's evolution. Thus the life cycle of a monarch butterfly begins as an egg that turns into a larva, then transforms itself into a chrysalis, from which an adult butterfly eventually emerges. These could be considered to be different phenomena, but they merely represent the life cycle of one and the same species.

In astronomy, as in biology, we ultimately need to better understand the relationship between different species as well as their respective life cycles. In biology, we now know that different species are characterized by differences in genome, the sequencing of their DNA. Those are what largely drives the world of living matter and makes it understandable. In astrophysics, taxonomy can be useful in providing a sense for how complex the different constituents of our Universe may be. But we need to obtain a deeper understanding in terms of some grander motif, a small set of overarching physical laws that appear to be universal, mutually consistent, and thus may serve to *explain* the workings of the Universe in some rational schema. Despite great efforts, we have not yet gained that insight. Some crucial factor may be missing!

8.17   Unimodular Phenomena

We cannot rule out that certain discoveries will only be made if some highly specific observation is undertaken, and remain invisible to all other attempts at discovery. Dark matter and dark energy could fall into this category, in that neither might ever make its presence known except through the gravitational influence it exerts. None of the other discoveries made to date appear to clearly fall into this category. The astrophysics of all the other discoveries listed in Section 8.11 suggests that each of the phenomena revealed will ultimately be observed also through means other than those that led to its original discovery.

8.18   Predictions

Based on the binomial statistics this chapter has pursued, Table 8.5 predicts that astronomy will eventually discover a total of roughly 90 distinct major cosmic phenomena. With 60 of these already in hand, we should expect the rate of discovery to decline in the decades ahead. The slope of the curve of accumulated discoveries in Figure 8.2 should gradually flatten beyond the present 60th discovery shown. The discovery rates per quarter century should then also remain well below those shown in the lower curve of Figure 8.2 for the peak years 1965–95. And as we approach completion of the cosmic toolkit, the number of discoverable phenomena should also reach its final peak. Further major discoveries should dwindle to zero. As already discussed in Chapter 1, at current funding rates worldwide for astronomy, this completion could occur as early as within a century or two – roughly half the time elapsed since Galileo first pointed his innovated telescope at the heavens, and comparable to the years since Fraunhofer, Kirchhoff, and Bunsen first viewed the Sun's light through their prisms. With a current international workforce of tens of thousands, such a rapid ending would not be surprising. The next and final chapter will portray the reorganization of astronomy this accelerated growth has entailed.

Notes

1 The Number of Class A Phenomena Characterizing the Universe, M. Harwit, *Quarterly Journal of the Royal Astronomical Society* 16, 378–409, 1975

2 Cosmic Discovery – The Technical Influence on Astronomy, M. Harwit, in *Proceedings of the Alexander von Humboldt Foundation Bi-National Colloquium*, The Institute for Advanced Study, Princeton, NJ, August 1981, pp. 41–56

3 Physicists in Astronomy – Will You Join the Dance?, M. Harwit, *Physics Today* 34, 172–87, November 1981

4 *Cosmic Discovery: The Search, Scope and Heritage of Astronomy*, M. Harwit, Basic Books, 1981

5 The Relation between the Number of Species and the Number of Individuals in a Random Sample of an Animal Population, R. A. Fisher, A. S. Corbet, & C. B. Williams, *Journal of Animal Ecology* 12, 42–58, 1943

6 Ibid., Harwit, 1975

7 Ibid., Fisher, et al., 1943

8 Ibid., Harwit, 1975, p. 380

9   *In Search of the True Universe: The Tools, Shaping and Cost of Cosmological Thought*, M. Harwit, Cambridge University Press, 2013, Chapter 11

10 *Making the Invisible Visible: A History of the Spitzer Infrared Telescope Facility (1971–2003)*, R. M. Rottner, Monographs in Aerospace History No. 47, National Aeronautics and Space Administration, NASA SP-2017-4547

11 The Baryon Census in a Multiphase Intergalactic Medium: 30% of the Baryons May Still Be Missing, J. M. Shull, B. D. Smith, & C. W. Danforth, *Astrophysical Journal* 759, 23 (15 pp.), November 1, 2012

12 A Cosmology Calculator for the World Wide Web, E. L. Wright, *Publications of the Astronomical Society of the Pacific* 118, Issue 850, 1711–15, 2006, http://www.astro.ucla.edu/~wright/ACC.html

13 Seven-Year Wilkinson Microwave Anisotropy Probe (WMAP) Observations: Sky Maps, Systematic Errors, and Basic Results, N. Jarosik, et al., *Astrophysical Journal Supplement Series* 192, 14 (15 pp.), February 2011

14 Planck 2015 Results XIII. Cosmological Parameters, Planck Collaboration, P. A. R. Ade et al., *Astronomy & Astrophysics* 594, A13, June 17, 2016, *arXiv:1502.01589v3*

15 Ibid., Fisher, et al., 1943

16 An Account of the Discovery of Jupiter as a Radio Source, K. L. Franklin, *Astronomical Journal* 64, 37–39, 1959

17 Tycho Brahe

18 Discovery of X-Ray and Extreme Ultraviolet Emission from Comet Hyakutake C/1996 B2, C. M. Lisse, et al., *Science* 274, 205–9, 1996

19 David Fabricius

20 Galileo Galilei

21 Christian Huygens

22 Gemiano Montanari

23 Nicolas Louis de Lacaille

24 William Herschel

25 The Periods of the Variable Stars in the Cluster Messier 5, S. I. Bailey, *Astrophysical Journal* 10, 255–65, 1899

26 William Herschel

27 Far-Infrared Fine-Structure Line Emission from Galactic Nebulae, G. J. Melnick, *Cornell University PhD Thesis*, January 1981, pp. 93–103

28 ISO LWS Observations of Planetary Nebula Fine-Structure Lines, X.-W. Liu, et al., *Monthly Notices of the Royal Astronomical Society* 323, 343–61, 2001

29 Brandes & Benzenberg

30 Guiseppe Piazzi

31 Huggins & Miller

32 51.8 Micron [O III] Line Emission Observed in Four Galactic H II Regions, G. Melnick, G. E. Gull, & M. Harwit, *Astrophysical Journal* 227, L35–38, 1979

33 Johannes Hartmann

34 Carbon Monoxide Emission as a Precise Tracer of Molecular Gas in the Andromeda Galaxy, N. Neininger, et al., *Nature* 395, 871–73, 1989

35 A CO Survey on a Sample of Herschel Cold Clumps, O. Fehér, et al., *Astronomy & Astrophysics* 606, A102, 2017

36 Observations of the 63 Micron [O I] Emission Line in the Orion and Omega Nebulae, G. Melnick, G. E. Gull, & M. Harwit, *Astrophysical Journal* 227, L29–33, 1979

37 Hertzsprung & Russell

38 Viktor Hess

39 Direct Observations of Galactic Cosmic Rays, D. Müller, *European Physical Journal H* 37, 413–58, 2012

40 Walter Adams

41 E. E. Barnard

42 Slipher, Duncan, Hubble, etc.

43 Hey, Parsons, & Phillips

44 Extended Rotation Curves of High-Luminosity Spiral Galaxies V. NGC 1961, the Most Massive Spiral Known, V. C. Rubin, W. K. Ford, Jr., & M. S. Roberts, *Astrophysical Journal* 230, 35–39, 1979

45 Lemaître; Hubble

46 Mapping the Universe, M. J. Geller & J. P. Huchra, *Science* 246, 897–903, 1989

47 The Connection between Pisces-Perseus and the Local Supercluster, M. P. Haynes & R. Giovanelli, *Astrophysical Journal* 306, L55–59, 1986

48 Baade & Zwicky

49 Further Data Bearing on the Identification of the Crab Nebula with the Supernova of 1054 AD, Part I, J. J. L. Duyvendak, Part 2, N. U. Mayall & J. H. Oort, *Publications of the Astronomical Society of the Pacific* 54, 91–94 and 95–104, 1942

50 Broadband, Radio Spectro-Polarimetric Study of 100 Radiative-Mode and Jet-Mode AGN, S. P. O'Sullivan, et al., *Monthly Notices of the Royal Astronomical Society* 469, 4034–62, August 2017

51 Fluctuations in Cosmic Radiation at Radio-Frequencies, J. S. Hey, S. J. Parsons, & J. W. Phillips, *Nature* 158, 234, 1946

52 Identification of the Radio Sources in Cassiopeia, Cygnus A and Puppis A, W. Baade & R. Minkowski, *Astrophysical Journal* 119, 206–14, 1954

53 Zeeman Effect in Stellar Spectra, H. W. Babcock, *Astrophysical Journal* 105, 105–19, 1947

54 George Herbig; Guillermo Haro

55 Herschel GASPS Spectral Observations of T Tauri Stars in Taurus: Unraveling Far-Infrared Line Emission from Jets and Discs, M. Alonso-Martínez, et al., *Astronomy & Astrophysics* 603, A138, April 18, 2017

56 Radio Monitoring of Protoplanetary Discs, C. Ubach, et al., *Monthly Notices of the Royal Astronomical Society* 466, 4083–93, 2017

57 Extreme Radio Flares and Associated X-Ray Variability from Young Stellar Objects in the Orion Nebula Cluster, J. Forbich, et al., *Astrophysical Journal* 844, 109 (12 pp.), 2017

58 Evidence for X-Rays from Sources Outside the Solar System, R. Giacconi, et al., *Physical Review Letters* 9, 439–43, 1962

59 X-Ray Sources in the Galaxy, S. Bowyer, et al. *Nature* 201, 1307–8, 1964

60 Ultraluminous X-Ray bursts in Two Ultracompact Companions to Nearby Elliptical Galaxies, J. A. Irwin, et al., *Nature* 538, 356–58, 2016

61 An Extremely Luminous X-Ray Outburst at the Birth of a Supernova, A. M. Soderberg, et al., *Nature* 453, 470–74, 2008

62 Investigations of the Radio Source 3C273 by the Method of Lunar Occultations, C. Hazard, M. B. Mackey, & A. J. Shimmins, *Nature* 197, 1037–39, 1963

63 3C273, A Star-Like Object with Large Red-Shift, M. Schmidt, *Nature* 197, 1040, 1963

64 Molecular Outflows in Local ULIRGS: Energetics from Multitransition OH Analysis, E. Gonazález-Alfonso, et al., *Astrophysical Journal* 8367, 11 (41 pp.), 2017

65 A Measurement of Excess Antenna Temperature at 4080 Mc/s, A. A. Penzias & R. W. Wilson, *Astrophysical Journal* 142, 419–21, 1965

66 A Preliminary Measurement of the Cosmic Microwave Background Spectrum by the Cosmic Background Explorer (COBE) Satellite, J. C. Mather, et al., *Astrophysical Journal* 354, L37–40, 1990

67 Observations of Extremely Cool Stars, G. Neugebauer, D. E. Martz, & R. B. Leighton, *Astrophysical Journal* 142, 399–401, 1965

68 Discovery of a Shell around Alpha Lyrae, H. H. Aumann, et al., *Astrophysical Journal* 278, L23–27, 1984

69 Disappearance of Stellar Debris Disks around Main-Sequence Stars after 400 Million Years, H. J. Habing, et al., *Nature* 401, 456–58, 1999

70 Cosmic X-Ray Sources, Galactic and Extragalactic, E. T. Byram, T. Chubb, & H. Friedman, *Science* 152, 66–71, 1966

71 Observations of a Strong Unidentified Microwave Line and of Emission from the OH Molecule, H. Weaver, et al., *Nature* 208, 29–31, 1965

72 Observation of a Rapidly Pulsating Radio Source, A. Hewish, et al., *Nature* 217, 709–13, 1968

73 Magnetars, V. M. Kaspi & A. M. Beloborodov, *Annual Review of Astronomy & Astrophysics* 55, 261–301, 2017

74 Infrared Photometry of Extragalactic Sources, G. H. Rieke & F. J. Low, *Astrophysical Journal* 176, L95–100, 1972

75 Herschel Far-Infrared and Submillimeter Photometry for the Kingfish Sample of Nearby Galaxies, D. A. Dale, et al., *Astrophysical Journal*, 745, 95 (18 pp.), January 20, 2012

76 Appearance of Relativistically Expanding Radio Sources, M. J. Rees, *Nature* 211, 468–70, 1966

77 Quasars: Millisecond-of-Arc Structure Revealed by Very-Long-Baseline Interferometry, C. A. Knight, et al., *Science* 172, 52–54, 1971

78 Further Measurements of Magnetic Fields in Interstellar Clouds of Neutral Hydrogen, G. Verschuur, *Nature* 233, 140–42, 1969

79 Pulsar Rotation and Dispersion Measures and the Galactic Magnetic Field, R. N. Manchester, *Astrophysical Journal* 172, 43–52, 1972

80 Observations of Gamma-Ray Bursts of Cosmic Origin, R. W. Klebesadel, I. B. Strong, & R. A. Olson, *Astrophysical Journal* 182, L85–88, 1973

81 Cosmological Gamma-Ray Bursts and the Highest Energy Cosmic Rays, E. Waxman, *Physical Review Letters* 75, 386–89, 1995

82 A Superluminal Source in the Galaxy, I. F. Mirabel & L. F. Rodriguez, *Nature* 371, 46–48, 1994

83 The 164 and 13 Day Periods of SS 433: Confirmation of the Kinematic Model, B. Margon, S. A. Grandi, & R. A. Downes, *Astrophysical Journal* 241, 306–15, 1980

84 Extended Rotation Curves of High-Luminosity Spiral Galaxies V. NGC 1961, the Most Massive Spiral Known, V. C. Rubin, W. K. Ford, Jr., & M. S. Roberts, *Astronomical Journal* 230, 35–39, 1979

85 0957+561 A, B: Twin Quasistellar Objects or Gravitational Lens? D. Walsh, R. F. Carswell, & R. J. Weymann, *Nature* 279, 381–84, 1979

86 New VR Magnification Ratios of QSO 0957+561, L. F. Goicoechea, et al., *Astrophysical Journal* 619, 19–29, 2005

87 P. W. Porcas, et al., *Nature* 282, 385–86, 1979

88 Heavy-Element Enrichment in Low-Density Regions of the Intergalactic Medium, L. L. Cowie & A. Songaila, *Nature* 394, 44–46, 1998

89 A Uniform Metallicity in the Outskirts of Massive, Nearby Galaxy Clusters, O. Urban, et al., *Monthly Notices of the Royal Astronomical Society* 470, 4583–99, 2017

90 A New Test of General Relativity: Gravitational Radiation and the Binary Pulsar PSR 1913+16, J. H. Taylor & J. M. Weisberg, *Astrophysical Journal* 253, 908–20, 1982

91 Ibid., Aumann, et al., 1984

92 Ibid., Habing, et al., 1999

93 The Remarkable Infrared Galaxy Arp 220 = IC 4553, B. T. Soifer, et al., *Astrophysical Journal* 283, L1–4, 1984

94 Strong Radio Sources in Bright Spiral Galaxies, J. J. Condon, *Astrophysical Journal* 242, 894–902, 1980

95 NICMOS Imaging of the Nuclei of Arp 220, N. Z. Scoville, et al., *Astrophysical Journal* 492, L107, 1998

96 The Hubble Deep Field: Observations, Data Reduction, and Galaxy Photometry, R. E. Williams, et al., *Astronomical Journal* 112, 1335–89, image plates: 1734–52, 1996

97 Observation of a Neutrino Burst from the Supernova SN 1987A, K. Hirata, et al., *Physical Review Letters* 58, 1490–93, 1987

98 Observation of a Neutrino Burst Coincident with Supernova 1987 A in the Large Magellanic Cloud, R. M. Bionta, et al., *Physical Review Letters* 58, 1494–96, 1987

99 The Connection between Pisces-Perseus and the Local Supercluster, M. P. Haynes & R. Giovanelli, *Astrophysical Journal* 306, L55–59, 1986

100 Mapping the Universe, M. J. Geller & J. P. Huchra, *Science* 246, 897–903, 1989

101 A Planetary System around the Millisecond Pulsar PSR1257+12, A. Wolszczan & D. A. Frail, *Nature* 355, 145–47, 1992

102 A Jupiter-Mass Companion to a Solar-Type Star, M. Mayor & D. Queloz, *Nature* 378, 355–59, 1995

103 A Review of the Homestake Solar Neutrino Experiment, R. Davis, *Progress in Particle and Nuclear Physics* 32, 13–32, 1994

104 Real-Time, Directional Measurements of ^{8}B $\sim$ 10 MeV Solar Neutrinos in the Kamiokande II Detector, K. S. Hirata, et al., *Physical Review D* 44, 2241–60, 1991

105 Infrared Spectrum of the Cool Brown Dwarf Gl 229B, B. R. Oppenheimer, S. R. Kulkarni, K. Matthews, & T. Nakajima, *Science* 270, 1478–79, 1995

106 A Jupiter-Mass Companion to a Solar-Type Star, M. Mayor & D. Queloz, *Nature* 378, 355–59, 1995

107 The Chemical Compositions of Very Metal-Poor Stars HD 122563 and HD 140283: A View from the Infrared, M. Afşar, et al., *Astrophysical Journal* 819, 103 (11 pp.), March 10, 2016

108 Metal Enrichment and Ionization Balance in the Lyman α Forest at z = 3, A. Songaila & L. L. Cowie, *Astrophysical Journal* 112, 335–51, 1996; *R-Process Nucleosynthesis* in Supernovae, J. J. Cowan & F.-K. Thielemann, *Physics Today* 47–53, October 2004

109 Latest Results from the Pierre Auger Observatory, I. Lhenry-Yvon for the Pierre Auger Collaboration, *EPJ Web of Conferences* 121, 03003, 2016, DOI: 10.1051/epjconf/201612103003

110 Observational Evidence from Supernovae for an Accelerating Universe and a Cosmological Constant, A. G. Riess, et al., *Astronomical Journal* 116, 1009–38, 1998

111 A Survey of z > 5.8 Quasars in the Sloan Digital Sky Survey. I. Discovery of Three New Quasars and the Spatial Density of Luminous Quasars at $z \sim$ 6, Xiaohui Fan, et al., *Astronomical Journal* 122, 2833–49, 2001

112 Stellar Orbits around the Galactic Center Black Hole, A. M. Ghez, et al., *Astrophysical Journal* 620, 744–57, 2005

113 A Geometric Determination of the Distance to the Galactic Center, F. Eisenhauer, et al., *Astrophysical Journal* 597, L121–24, 2003

114 A Bright Millisecond Radio Burst of Extragalactic Origin, D. R. Lorimer, et al., *Science* 318, 777–80, 2007

115 The Host Galaxy of a Fast Radio Burst, E. F. Keane, et al., *Nature* 530, 453–56, 2016

116 A Repeating Fast Radio Burst, L. G. Spitler, et al., *Nature* 531, 202–5, 2016

117 A Nearby Massive Galaxy Cluster behind the Milky Way, R. C. Kraan-Korteweg, et al., *Nature* 379, 519, 1996

118 On the Origin of the Local Group's Peculiar Velocity, D. D. Kocevski & H. Ebeling, *Astrophysical Journal* 645, 1043–53, 2006

119 A Measurement of Large-Scale Peculiar Velocities of Clusters of Galaxies: Results and Cosmological Implications, A. Kashlinsky, et al., *Astrophysical Journal* 686, L49–52, 2008

120 Cosmological Parameters from the First Results of Boomerang, A. E. Lange, et al., *Physical Review D* 63, 042001, January 19, 2001

121 Nine-Year Wilkinson Microwave Anisotropy Probe (WMAP) Observations: Final Maps and Results, C. L. Bennett, et al., *Astrophysical Journal Supplement Series* 208, 20 (54 pp.), 2013

122 Ibid., Abbott, et al., 2016

123 Fluctuations in Cosmic Radiation at Radio Frequencies, J. S. Hey, S. J. Parsons, & J. W. Phillips, *Nature* 158, 234, 1946

9

The Human Aspect of
the Cosmic Search

9.1 A Turn to a New Future

To put the human aspects of modern astronomy into perspective, we need to note the way the field was traditionally budgeted and conducted, the rapid transformations it is currently undergoing, how those changes have affected career paths, and where these may ultimately lead. The inspiring goals of the field, some appearing well within reach, the enthusiasm and optimism of young astronomers, and some of the more sobering questions arising, all are at play.

The major unresolved questions will hinge on how far we may continue along the road within view before arriving at an abyss that may block our ever journeying further. What this may mean for humanity will then quite literally lie in the balance because potential aspirations and alternatives have never been seriously discussed.

* * *

By far the most consequential influence on astronomical progress in the post-World War II era was the book-length report *Science – The Endless Frontier*, the brain child of its lead author Vannevar Bush.

In the pre-war years, Bush had been Dean of Engineering at MIT. As World War II loomed President Franklin Delano Roosevelt appointed Bush head of a powerful, newly created Office of Scientific Research and Development, whose major wartime accomplishments included the establishment of the Manhattan Project to build the atomic bomb and the MIT Radiation Laboratories to advance wartime radar. The Office also had oversight of the production of penicillin, sulfa drugs, and improved vaccines, as well as the insecticide DDT, all of which lowered US military death rates during the war.[a]

As the waning months of the war came into focus in November 1944, Bush solicited President Roosevelt's instructions to produce a scientific roadmap designed to guide the nation's post-war efforts. An unpredictable means of funding US science and engineering, during World War I and the interwar years, had taught Bush that this was no way for a country to compete technologically on an international level.[2] With Roosevelt's consent and the help of some of the finest scientific, technical, and legal minds in the United States, Bush completed this task in just seven months. By then Roosevelt had died, but on July 5, 1945, Bush handed his visionary document to Roosevelt's successor, President Harry S Truman, who approved.

Bush then spent the next five years persuading the US Congress to enact the report's major recommendations.

The name of this oft-cited report, *Science – The Endless Frontier*, could easily be misconstrued as implying that science would forever remain endless, its frontiers receding to ever-greater distances as rapidly as scientists could advance on them. Even as we solved one problem, others would take its place.

It would be wrong to believe that this was what the document's title was intended to convey.

Bush's immediate concern was to impress on the US Congress that the nation could maintain its wartime military and industrial prominence only as long as it dedicated itself to a continuing campaign of technological innovation, which would require a federal involvement as guarantor of seed investments.

Looking back at those post-war years, today, Bush's concerns about the centrality of science appear to have focused mainly on medical advances and industrial and military prowess. Understanding the Universe, through advances in astronomy and cosmology, would most likely not have struck Bush as scientific topics qualifying for his purposes.

Where the dividing lines should be drawn might not even have been obvious in the immediate post-war era; but by the turn of the millennium around the year 2000, it had become striking that virtually all major post-war astronomical discoveries had come about through technological advances initiated for military, industrial, or medical purposes. Here, the flow of funds was reversed from the direction Bush appeared to have had in mind. In astronomy, the fruits of military and industrial research were being invested in science, rather than the

---

[a] For an incisive analysis of the support of science in the United States in the years following World War II, see the review by Daniel J. Kevles, "Principles and Politics in Federal R & D Policy, 1945–1990 – An Appreciation of the Bush Report."[1]

fruits of science giving rise to technical advances that would ultimately raise standards of living or guarantee national security.

By 1970, the trend in military spending on unspecified scientific activities sufficiently alarmed US Senate Majority Leader Michael Joseph (Mike) Mansfield for him to introduce an amendment to the Military Procurement Authorization Act of Fiscal Year 1970 specifically forbidding the Defense Department the use of appropriated funds to carry out research of any kind unless it was directly related "to a specific military function."[3]

This slowed down the technological transfer from the military, but the prohibition was not always rigorously enforced, and military support for basic astronomical projects could occasionally still be found.

Shared interests continued to prevail where they suited the aims of the former partners best. Military and industrial advances in computing and in artificial intelligence found quick application in astronomy, particularly as the cost of computing and storage of information sharply dropped and the costs of accumulating and storing large banks of data became increasingly affordable. But astronomers found direct expenditures on their investigations rapidly rising as the full cost of developing detectors for sensing novel classes of messengers – high-energy cosmic rays, neutrinos, and gravitons, none of which had obvious industrial, medical, or military applications – now had to be borne by astrophysics alone.

9.2   Planning Astronomy

By the 1960s, astronomy had become sufficiently expensive and variegated that decadal planning sessions were held to determine the community's major priorities and to recommend successive 10-year programs of communally prioritized missions to the funding agencies. These would complement the many smaller projects that individual scientists or groups would continue to propose to the funding agencies for less expensive, often more speculative and challenging investigations.

Over the half century to follow, the decadal reviews became progressively more influential in determining the largest, most expensive ventures. Major undertakings that once had been directly or indirectly supported by the US military now were being funded by the National Science Foundation, NSF; the National Aeronautics and Space Administration, NASA; the US Department of Energy, DoE; through international collaborations; or through the support of wealthy individuals or non-profit organizations. Astronomers also needed to reorganize their working style as the cost of astronomical missions escalated. The amount of useful information extracted from each new venture now had to be maximized. This meant carrying out large surveys of the sky and storing their huge amounts of data in massive new data banks; and it would need to be done at minimal cost. These were useful new ways of conducting astronomy, even if the style they emphasized was foreign to an earlier generation of instrumentalists used to following up each new result through further investigations to come up with novel insights. Those follow-on investigations would now have to await some further major mission dedicated to some other major purpose. That wait would likely amount to decades.

This era of large-scale, standardized surveys came into being around the year 2000, with large teams of astronomers, other scientists and engineers, sometimes exceeding a thousand members, dedicating themselves to the construction and servicing of a single astronomical facility. How well this particular approach to major astronomical investments will ultimately work still remains to be fully assessed, but a number of difficulties have begun to trouble astronomical organizations charged with implementing national and international plans.

9.3   An International Endeavor and the Perception of US Leadership

For half a century after World War II, the United States held an almost indisputable lead in virtually all astronomical and cosmological investigations. With the turn of the millennium, some of the most ambitious astronomical projects world-wide began to be initiated in Europe, China, South Africa, Australia, and South America, as increasing wealth in these regions permitted the allocation of funds to curiosity-driven research. Yet, coordination with these newer efforts has not been easy for the United States.

Matt Mountain, President of the Association of Universities for Research in Astronomy, AURA, and Adam Cohen, President and chief executive of Associated Universities, Inc., AUI, both located in Washington, DC, have recently collaborated on a white paper, *US Astronomers Face Hard Decisions*.[4] Pointing to escalating costs of the major astronomical projects currently being undertaken or planned for the immediate future, their investigation warns that "building billion-dollar facilities in the 2020s and beyond will be impossible with the current model for funding and collaboration."

Mountain and Cohen argue that in the United States national plans for astronomy continue to be developed in "semi-isolation." Although the United States often partners with other nations on joint investigations, the sources of funding to which the decadal reviews are addressed confine themselves to US governmental structures. And neither the US Congress nor the White House Bureau of the Budget, though cognizant of such collaborations, appear

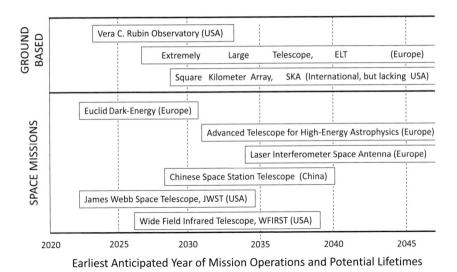

Fig. 9.1 Major ground-based telescopes and astronomical space missions currently being planned world-wide. This figure is based on a recently updated white paper prepared by Matt Mountain, President of the Associated Universities for Research in Astronomy, AURA, and Adam Cohen, President of Associated Universities Inc., AUI, in the United States. Their paper's primary aim was to exhibit the political reality that countries other than the United States often take lead responsibility for major missions, and that US participation and influence in international planning for astronomy is more limited than widely believed. In 2020 NASA renamed WFIRST the "Nancy Grace Roman Space Telescope" in honor of NASA's first Chief of Astronomy.

to realize that the United States no longer is the engine driving many of these international projects.

To illustrate these new realities, Mountain and Cohen compiled information on some of the leading astronomical missions currently under consideration, here reproduced in a somewhat edited format in Figure 9.1. Their sobering assessment is that "International astronomy projects are underway or planned, irrespective of the US Decadal Survey. US astronomers must decide in 2020 what competitive capabilities they need in the 2030s and beyond."

In terms of actual purchasing power, Mountain and Cohen assert, little has changed in the effective budget of NASA since the 1980s. Inflation has steadily led to boosted mission expenditures. This has given the false impression that the United States may be investing more in its space efforts today than at any time in the nation's history. Scaled to the buying power of a dollar in fiscal year 2014, however, NASA's budget has remained essentially flat at roughly $20B per year.

An added concern of Mountain and Cohen is that the National Science Foundation, responsible for providing much of the funding of ground-based observatories, has witnessed a steadily rising fraction of its astronomy budget devoted to keeping the major national observatories operational. Given that the NSF's budget has remained flat since 2010, this has led to diminishing expenditures on research grants, the lifeblood of scientific innovation and the primary means for supporting the work of PhD candidates in astronomy at US universities.

All of these considerations are significant and will require both political will to cooperate on an international scale and to periodically adjust for unpredictable fluctuations in the economies of partnering nations. The European Space Agency supported by a host of member nations has shown itself adept at dealing with such vagaries, painful as these often may be. If, as Mountain and Cohen propose, the United States were willing to coordinate its astronomical efforts more closely with its partners abroad, the Congress as well as the Administration would need to find ways of remaining flexible under comparably uncertain conditions. It would likely be the price of regaining the international leadership Mountain and Cohen fear the United States may be losing, but would bring with it both the opportunities and responsibilities of planning astronomical efforts on a far broader multinational front.

9.4   Astronomy's Evolving Social Structure

The complexities of international collaboration are not the sole problems to be overcome. For young researchers astronomical research currently imposes awkward career paths, often involving a succession of post-doctoral positions before a nascent scientist may expect to gain a secure research position.

Is this the best way to organize the conduct of science?

If one believes that those who ultimately gain permanent positions through hardship and persistence can best be entrusted with scientific work, it could be. But such almost monastic dedication may not be the best criterion for selecting gifted individuals to positions of responsibility.

The United States spends considerable sums of public moneys on work performed along currently pursued lines, but these may not be the best ways to meet the nation's responsibilities, both to the public paying for the research and to the young scientists who sacrifice long-term security because they are extraordinarily committed to their work.

Fortunately, a number of studies supported by the American Institute of Physics have begun to shed light on the processes involved, and may ultimately suggest improvements. I will return to these below, but it may first be worth reviewing how the current system of training astronomers and space scientists was originally shaped by urgencies of the Cold War in the late 1950s.

9.5   A Transformed Scientific Community

By the mid-1950s, astrophysical research, everywhere, had been con-ducted for almost a century along lines that Kirchhoff and Bunsen would have recognized as quite customary. In mid-nineteenth-century Heidelberg, senior professors directed scientific investigations partly through personal dedication to the work, partly by supervising lower-paid full-time assistants, and partly by advising students, some of whom might eventually succeed them, becoming university professors themselves.

In the United States, all this changed almost overnight with the Soviet Union's launch of its Sputnik satellite in 1957. This was a success of immense proportions, an intricate feat the United States had never attempted. Space science had suddenly revealed itself as firmly tied to national security!

Where rocketry and space flight had previously been considered largely a military priority, President Eisenhower recognized that the United States had not pursued space science with sufficient imagination and vigor. This needed to be bolstered, and could not be done simply with military and industrial interests in mind. He directed that a new agency, the National Aeronautics and Space Administration, NASA, be established to secure the broad scientific basis on which the study of space and the world beyond could be dispassionately pursued: By 1958 this new agency was in place.

With the advent of NASA, larger teams of astronomers began to form, as astrophysicists, engineers, and eventually computer experts joined forces to undertake comprehensive surveys, often conducted with fully dedicated telescopes in space. These projects were expensive, required expertise that no single individual could offer, and could only be reliably conducted by well-organized teams. Scientists had to learn to work as members of teams, rather than as individuals, or better yet to do both.

Chapter 7 hinted at the progressive structural changes in astrophysics in the last third of the twentieth century and the early years of the new millennium. These introduced a novel style of work, eventually pursued by teams often assembling more than a hundred, occasionally more than a thousand scientists dedicated to building and operating a single, highly complex, costly observatory.

The LIGO/Virgo project to directly detect gravitational waves may starkly illustrate this. By the time the first direct detection of gravitational waves was announced in the *Physical Review Letters* in early 2016, the co-authors of that scientific paper numbered around 1500, and came from 133 different institutions around the globe, roughly one-third of them from the United States, the other two-thirds from 16 other countries.

These structural changes in the astrophysical community, so strongly entrenched today, largely altered the character of novel discoveries. Often they were anchored firmly to Galison's logic tradition, as exhibited by efforts using the Hubble Space Telescope to establish a coherent cosmic history of galaxy formation and evolution. Only seldom did these large-scale projects yield single golden events, as in the identification of the first outburst of neutrinos observed in the explosion of the Magellanic Cloud supernova in 1987, or the first powerful burst of gravitational waves in late 2015.

The lives of the individuals conducting these searches similarly were recast. Launch and operation of missions in space, and construction of complex ground-based observatories, required experts in data extracting and storage systems to handle the veritable tsunamis of scientific data now flooding the field.

Much of this work came to be carried out by young post-doctoral workers willing to travel wherever they were most needed – a nomadic way of life that could be sustained only into one's mid-thirties, even if one was willing to postpone the start of traditional family life.

What was likely to happen next is only just being gleaned. Young astronomers engaged in some of the most esoteric research often have to wait years before obtaining secure research positions. And, after dedicating themselves to research for more than a decade, they may ultimately find their hopes for a permanent position dashed. The skills and flexibility these young people generally have acquired often do lead to well-paid careers in industry or government. But their hard-won PhD degrees in astronomy may have been for naught, though their expertise in dealing with complex problems can prove to be a highly valued skill.

The astronomical and astrophysical communities dependent on the work of these young scientists are only now coming to grips with these societal shifts.

9.6   Studies Initiated by the American Institute of Physics

Most candidates obtain a doctoral degree in astronomy or astrophysics at age 26 to 29. Their PhD does not guarantee a permanent research position. Most wishing to continue pursuit of interesting problems first need to apply for a post-doctoral position on an established research team. This provides a two-to-three year opportunity to demonstrate initiative, capability, drive, and possibly also a disposition to participate in team work. Those who pass many of these tests may then be offered a second or even a third post-doctoral position, often on other large teams, where a quite different set of scientific problems and distinct styles of work may be pursued. By then the candidate may be 35 years old, highly skilled and looking for a more stable, long-term position.

The American Institute of Physics, AIP, a federation of physical science societies including the American Astronomical Society, AAS, and the American Physical Society, APS, has been studying the employment problems raised by these transformations in research. A particularly informative longitudinal study, "Women's and Men's Career Choices in Astronomy and Astrophysics," has been pursued for well over a decade by Rachel Ivie, Susan White, and Raymond Y. Chu of the AIP. It follows the career paths of graduate students who were enrolled in astronomy or astrophysical studies in US universities during 2006–7.[5]

Several factors appear to have directed the outcomes of these careers. A particular emphasis of the study was a determination of circumstances that enabled candidates to remain in astronomy, rather than dropping out to pursue careers elsewhere. Some of these factors appeared to be gender-dependent, and determined by whether the candidates were single or married. Others were more universal. Two indispensable circumstances helping a candidate to be retained in the field, long-term, appeared to be the candidate having completed at least one post-doctoral post, and finding a beneficial advisor in the profession. Occasionally, candidates reported that finding a new advisor helped.

On the first of these two points, the report is unequivocal. "A respondent who had completed a postdoc was 2.49 times more likely to be working in physics and astronomy than a respondent who had not completed a postdoc."

On the second, the implication of the word "advisor" in the context of graduate and post-doctoral work appears to be central. The authors of the study do not further define the word, but for graduate students and many young PhDs the principal advisor is likely to be the current or a former research supervisor, the person who has hired the post-doc, provided the opportunity of participating on a significant research project, and pays the post-doc's salary. The advisor may also be the only person who could write an informed letter of recommendation promoting a candidate's chances for retention in the field – a desirable outcome for everyone who has an investment in the candidate's efforts.

In many ways, education at the PhD level is a parenting system. German research scientists have long acknowledged this by referring to their former thesis advisor as their "Doktor Vater" – more recently perhaps also as their "Doktor Mutter." The same parent–offspring nurturing of nascent scientists at the highest levels also emerges in Harriet Zuckerman's genealogical study, *Scientific Elite: Nobel Laureates in the United States*.[6]

At first glance, a young researcher's advisor appears to play an altruistic role. This type of relationship has by now been studied in considerable depth in a range of different contexts and appears to follow a prescription referred to as Hamilton's rule.[7; 8] In the mid-1960s, William Donald Hamilton, at the time a 28-year-old evolutionary biologist, asked why any organism would carry out an altruistic act on behalf of its immediate kin? Why would a parent make sacrifices for the benefit of an offspring or other family member?

Hamilton suggested that such altruistic acts would make sense if the cost c to the parent, led to benefits b to the offspring, and the offspring was willing to repay a reasonable fraction r of those benefits to the parent, so that

$$br \geq c. \tag{9.1}$$

In this arrangement a parent could support an offspring at considerable cost, though expecting the offspring to return the favor, so the parent could equitably benefit.

Seen in this light, altruism and parental nurturing appear to just make good business sense and could be reasonably expected to also govern the interaction of a research advisor with a PhD candidate or a young post-doctoral worker.

To the young researcher the benefits b provided by participating in a significant research project may be high, and if the fractional returns of those benefits r to the research supervisor are high as well, generally meaning that r approaches unity, $r \sim 1$, the supervisor will find the costs c incurred in hiring the post-doc to be well spent.

This will bode well for the prospects that the younger researcher will be retained in the field. Any satisfied supervisor for whom the cost of supporting a post-doc is handsomely repaid by the post-doc is likely to promote the career of this young researcher whose work has significantly advanced the aims and findings of the supervisor's project.

The post-doc, however, may see this in a beneficial way as well. If the cost to the post-doc c' leads to benefits b' to the research project in return for which the supervisor provides a favorable recommendation r' that further promotes the young researcher's career, everybody will have benefitted: $b'r' > c'$.

Hamilton's principle is thus reciprocal. It works to everyone's benefits. It may also find support in the main findings of Ivie, White, and Chu, who report that their "respondents – the young researchers – who had changed advisors and who gave their advisors less desirable ratings were more likely to have [also] reported already leaving the field." And, "A respondent who had changed advisors was 2.33 times more likely not to be working in physics and astronomy than respondents who had not changed advisors." The relations between young researchers and their supervisors are usually mutual; either both find the relationship satisfactory or both find it disappointing.

Further studies may not support Hamilton's rule as applicable; but in terms of the research conducted on the careers of young researchers, to date, it seems to be roughly borne out.

9.7   The Choice of a Research Field

A similar business relationship also appears to more generally prevail.

Academic scientists are rewarded for individual achievements. Universities hire scientists who have shown themselves exceptionally and demonstrably proficient at what they do. The most prestigious scientific prizes tend to be awarded to individuals, rather than to groups. Such factors provide incentives for establishing oneself as an independent thinker, a scientific leader clearly set apart from others, working on projects of one's own.

The decision on whether to work on a popular subject, possibly as a member of a large team, or to devote oneself to a potentially more rewarding project nobody else happens to consider important at the time, involves weighing the benefits B of working alone against the costs C of undertaking such solitary work. The cost may be that the field of choice is too far removed from the center of current emphasis to which most established scientists flock and where those scientists recommend that research be funded.

For established professionals, cooperation can also be an advantage if it provides the sole means for gaining access to an expensive scientific facility – a high-speed computer, or a major new observatory to study the Universe. The scientist may then need to weigh not only the benefits B and costs C of working alone but also the corresponding costs c and benefits b of working in a more popular field, as part of a more influential scientific community.

Usually, joining the crowd makes life rather easier because the benefits of working on projects selected by the research community normally are predictably high, whereas the returns R from an untested field are indeterminate, as are the potential benefits B, even if the costs of investing in an unproven field may deliberately be kept low, or perhaps because such costs normally *have* to be kept low to receive funding.

This is why the search for gravitational waves was so unsuccessful in the 1970s. Investments, i.e., costs C, had to be kept low in such an untrodden field, and so the benefits B also were insignificant, even with an expectation that the returns R could be maximal.

Investing in the direct detection of gravitational waves became feasible only when the tiny decline of a binary pulsar's orbital period observed by Hulse, Taylor, and Weisberg confirmed the predictions of general relativity, so that confidence was high that these waves could someday be directly detected.[9; 10] In the long-term, the behavior of the scientific community may come close to establishing equilibrium between cooperative thinking by established groups and the peculiar preferences of gifted individuals willing to shoulder considerable personal risk to work on less popular projects of their own.

At least this may be our hope. If nothing else, we should at least strive toward such a state of equilibrium in astrophysics. The discipline will always need some measure of balance between large expensive projects for which team work is essential, and lone researchers attempting to scout out new directions, most of which may lead nowhere, but some that will propel astrophysics into new directions.

9.8   Future Projects

Regrettably, the division of labor just described appears to be difficult to maintain. A recent study titled "Large Teams Develop and Small Teams Disrupt Science and Technology" has focused on this relationship.[11] Here, the word *disrupt* is used in a positive sense to indicate that long-term progress requires deeper analysis of the broad advances large teams are able to bring about – unquestioned acceptance of which could lead to missteps.

From an analysis of a staggering 65 million papers, patents, and software products published between 1954 and 2014 the study's authors conclude, "Both small and large teams are essential to a flourishing ecology of science and technology. Taken together, the increasing dominance of large teams, a flurry of scholarship on their perceived benefits, and our findings, call for new investigations into the vital role that individuals and small groups have in advancing science and technology."

As this paper's investigations further show, the funding for these two types of research disturbingly paints a "unified portrait of underfunded solo investigators and small teams who disrupt science and technology by generating new directions on the basis of deeper and wider information search. These results suggest the need for government, industry and non-profit funders of science and technology to investigate the critical role that small teams appear to have in expanding the frontiers of knowledge, even as large teams rapidly develop them."

This message, perhaps not unexpectedly, echoes the warnings of Mountain and Cohen in their analysis of NSF astronomy funding, and the continuing reductions of small grants for research distributed to individual astronomers or small groups, for just the kind of disruptive analyses a balanced program of astronomy needs if it is to remain vibrant.

9.9   The Dollar Value of Curiosity-Driven Astronomy

Like many other types of estimates, means to establish the dollar value of curiosity-driven astronomy are not immediately obvious. To determine whether the national and international budgets allotted to these efforts are compatible with their perceived value, a number of diverse perspectives may be examined. First, however, we need to understand how the expenditures of this research are allocated.

In the early part of the twenty-first century a threshold investment of one or two billion dollars constructing a major new observatory, whether on Earth or in space, was not unusual. These expenditures represented the initial outlay required to keep an army of a thousand technical experts, mainly scientists and engineers, dedicated to the effort for a typical 10-year span leading to completion. This included support personnel and infrastructure amounting to an expenditure of the order of one to two hundred thousand dollars per technical expert per year at full-cost accounting. These experts concentrated on diverse essentials for the mission, designing, assembling, and exhaustively testing the assembled facility before it could reliably enter use.

Even during the construction phase of observatories it is not the cost of materials – glass for large mirrors, highly purified solid-state materials for detectors, aluminum for light-weight structures – that adds significantly to expense. Rather, the main costs include the design and construction of instruments, the assembly of optimal observing sequences, and data reduction and archiving programs designed and cross-checked by skilled specialists. Skill and persistence are these teams' most valuable attributes.

With this in mind, we may turn to the individuals and organizations investing in curiosity-driven astronomy, to reach some understanding of the dollar value they themselves assign their work. Four different groups come to mind.

The first of these comprises the post-docs who are engaged in curiosity-driven work and wish to continue on this career path. When they no longer can be retained for lack of funding, they may reluctantly enter the general workforce as computer programmers, specialists in arcane activities, or other positions in industry or government where they usually earn considerably more, and could all along have earned considerably more, than they did as post-docs.

An investigation, *Astronomy Degree Recipients: One Year After Degree*, published by Patrick Mouldy and Jack Pold of the American Institute of Physics, used data from a survey of graduate students who completed their degrees in 2014, 2015, and 2016. They found that PhD recipients who had remained in academic or governmental scientific research during that first year after receiving their degree typically earned about $60,000 that year, whereas those working in the private sector had garnered salaries roughly $45,000 higher.[12]

This dollar difference indicates the price individual scientists are willing to pay for remaining in curiosity-driven astronomy, even with no guarantees of longer-term employment as astronomers. Granted that this dollar difference may just be the price they are personally willing to pay for the thrill of participating in the enterprise, it nevertheless is a measure not to be discounted.

A second indicator comes from wealthy individuals who offer to fund the construction of a major observatory, sometimes in today's price range of a hundred million dollars. Often these individuals' wealth has been self-made, indicating that the donor knows what a dollar can buy and is willing to provide the funding, convinced that the information the observatories will help us gain will be worth a substantial donation. The twin Keck telescopes on Mauna Kea in Hawaii are examples of such high investments that have yielded an extraordinary wealth of curiosity-driven data. Granted that wealthy individuals often enable construction of observatories that may also serve as monuments to their generosity, their choice of how this munificence should be directed still speaks volumes.

A third impression of the worth of curiosity-driven astronomy comes from members of the public volunteering to scour complex astronomical data in search of anomalies that would be difficult to find without such dedicated help. Though performing this service without pay, these individuals consider their contributions worthwhile, as do the professionals who benefit from this otherwise unaffordable support.

These projects span a wide range of astronomical observations ranging from studies of coronal mass ejection in the course of a total Solar eclipse, to a variety of studies of extragalactic sources. Results from the Solar project were published by a list of 250–300 authors from 93 different institutions that included volunteers from university departments in the biological and physical sciences, municipal high schools, local astronomical societies, and even a couple of elementary schools.[13]

Many extragalactic projects have been organized by a consortium named *Zooniverse*, which has attracted over a million volunteers. Their many investigations have, by now, led to more than 200 peer-reviewed publications.[14]

The "Crowdsourcing and Citizen Science Act" of 2017 cites a *Sense of Congress* that "granting Federal science agencies the direct, explicit authority to use crowdsourcing and citizen science will encourage its appropriate use to advance Federal science agency missions and stimulate and facilitate broader public participation in the innovative process, yielding numerous benefits to the Federal Government and citizens who participate ..."[15]

A fourth source of support for astronomy, of course, is governmental. In a democracy it is the people's representatives who, after due deliberation, allocate most of the research funding requested. Fortunately governmental agencies nowadays join science writers and journalists to explain scientific advances in terms the public can understand. Citizens at any level then can convey their approval or disapproval to their representatives, so they may better judge the levels at which curiosity-driven astronomy is publicly appreciated.

9.10   An Economy of the Commons

Ideally, the dollar value of curiosity-driven astronomy might also be dispassionately estimated in a manner resembling that of the encyclopaedic enterprise *Wikipedia*. Articles for this website covering the most broadly ranging subjects are submitted at no cost by volunteers world-wide. Financial support to sustain *Wikipedia* comes entirely from voluntary contributions. The annual report from *Wikipedia's* parent organization, *Wikimedia*, for the financial year July 1, 2016 to June 30, 2017, cited: 15 billion visits world-wide to its website each month; 5 million new articles contributed by volunteers in the course of the year; and 6.1 million donations amounting to more than 87 million US dollars, which made the organization's operations accessible in more than 50 countries. Among these millions of donations, only a total of as few as 20 exceeded $50,000. The average contribution was roughly $15, indicating the broad popularity of the site among ordinary citizens contributing at whatever level they could afford.[16]

The economics at play here are similar to those traditionally encountered in *the economics of the commons* most notably introduced by the work of Elinor Ostrom and her colleagues in the 1990s. In a landmark presentation of 1994, Ostrom defined other efforts in which a common resource – here the availability of funding for curiosity-driven astronomy – has to be shared by a larger population than available resources are able to satisfy. She wrote:[17]

> A common-pool resource, such as a lake or ocean, an irrigation system, a fishing ground, a forest, the internet, or the stratosphere, is a natural or man-made resource from which it is difficult to exclude or limit users once the resource is provided by nature or produced by humans.

Clearly *Wikipedia* fits this mold. It is a man-made resource whose contributors are almost exclusively also its beneficiaries and thus are fully cognizant of its value.

Curiosity-driven astronomy similarly is a public resource, though the discoveries it enables often are too specialized to be directly evaluated by the public or by individual volunteers, but instead are supported by a mix of governmental funding, individuals willing to work at lower salaries than they could obtain in industry, or wealthy donors who happen to be astronomy enthusiasts.

9.11   A Policy for Assembling the Foreseeably Required Toolkit

An inventory of available carriers of information reveals the interplay between physical processes at work in the Universe and the instruments that could successfully observe them. Chapter 6 showed that the range of capabilities defining these instruments can be graphically displayed in a multidimensional space, highlighted regions of which designate observational capacities already in hand. Adjacent intervening and encircling blank domains within this space correspondingly call out capabilities not yet within our grasp. They hint at the existence of a *discovery space*, calling attention to instrumentation promising to yield further discoveries if messengers can be found to reliably transmit information in these domains.

Lying beyond these graphically displayed regions, however, a more forbidding range emerges, in which the Universe either generates no messengers, or else where all the messengers the Cosmos does generate are systematically disfigured or destroyed along any and all trajectories they would need to traverse to reach Earth.

The number of years it may take us to assemble and implement the complete set of instruments required to fully tap all the channels of information the Universe then does provide, and the entire set of messengers each channel can transmit, should give us a useful estimate of the minimum time and expenditures required for compiling a full inventory of all the major phemomena the Universe reliably exhibits.

In earlier chapters we estimated that, at current levels of investment in astronomy, we should have most of the requisite instrumentation in hand within a couple of centuries. The most economic astronomical program to pursue then will be determined by the cost of building the observatories required to capture all the information the Universe faithfully transmits. Continued astronomical searches may then need to also ferret out whatever information content the messengers we observe will still provide. At some point, however, we could find ourselves satisfied that the information gathered suffices, and that further observations will progressively yield diminishing returns.

If the astronomical community agrees that such a sequential approach is viable, it should be possible to figure out the most economical way to pursue these ends.

9.12   Multi-Messenger Science

The prospects of *multi-messenger observations* combining the advantages of distinct carriers of information have been gaining increasing attention.

It may no longer make sense to expect each individual subdiscipline, such as radio astronomy, or cosmic-ray physics, or molecular spectroscopy, to provide panoramic insight on complex astrophysical processes, when progress may more readily and more reliably be provided by complementing means.

Even today, long-term astrophysical plans are being developed along instrumental lines, with emphasis on sustaining a wide range of areas of expertise in each individual discipline, as all these disciplines vie to solve new problems previously beyond their respective technological reach. In the years following World War II, this was a healthy way to continually advance the state of the art and maintain communal expertise and corporate memory. By now, however, it may not be the most effective way to move forward, and a different balance of efforts could prove more effective.

Once the combination of tools in a complete toolkit is in hand, some designed to discern seldom occurring exceedingly short events lasting no longer than a millisecond, others highlighting gradual evolution over millennia, the pace and scale of observational astronomy may also substantially decline even if it continues to persist. Most likely, this work will by then be highly automated and far less costly to conduct.

9.13   A Curiosity-Driven Fragile Enterprise and Its Costs

We still do not know whether the current trend of dedicating new observatories to relatively narrowly defined but massively pursued ends is the most productive way for astronomy to move forward. Neither the real costs nor the resulting benefits of any new course of investigation can ever be certain in advance, because our knowledge of the world remains fragmentary. The information at hand is often isolated, its relation to as-yet-unfathomed factors unknown. Histories of previous successes and failures can be a guide, though none may match precisely the conditions current decisions face. The only guide on how to proceed today often amounts to no more than a collection of vaguely resembling previous histories illustrating the consequences of earlier decisions.

How Much Astronomical Effort Will Ultimately Be Deemed Enough?

The search for habitable life on other planets, and perhaps just as significantly the search for intelligent life reluctant to broadcast its existence, might lead to a desire to obtain high-resolution images of a remote planet's surface with telescopes deployed within our own Solar System. If so, where should we draw the line between affordable and unaffordably expensive observatories dedicated to just such an effort?

A potentially plausible proposal might be a desire to see whether industrial activities on a remote planet were sufficiently advanced that any exploratory mission we could launch could encounter a highly sophisticated hostile defense. During the Cold War, the United States overflew the Soviet Union on many occasions to search for new factories, missile launch facilities, and other large-scale facilities to make an inventory of likely Soviet military capabilities. If a similar program was to provide a first impression of such capabilities in observations carried out locally from the Solar System, what would we need to aim for in angular resolving power?

A reasonably sized factory or military facility would likely exhibit a footprint around a hundred meters in diameter for some of the nearest planets; at a distance of $\sim 3pc = 10^{17}$ meters, this would require an angular resolution of $\sim 10^{-15}$ radians or $\sim 10^{-10}$ arcseconds.

Comparably heroic spectroscopic efforts would probably be required to ascertain the genetic structure of a planet's living organism to determine whether life on the planet was based on DNA coding similar to that found on Earth.

Should such extreme investigations be considered part of a scientific program of investigations, or will we then be approaching the point where science begins to turn into engineering – where, in the first of the cited examples we are mainly interested in, whether exploration of a planet would have to be combined with military considerations and potentially defensive action, while in the biological case we might be concerned with encountering and potentially countering hostile bacterial or viral attacks.

In both these instances we will no longer be merely studying the Universe, or more specifically the planetary systems around us, but will instead be focused on how we might engineer a defensive system that could protect us against hostile environments. Our efforts by then will have shifted from a dispassionate quest to increase our understanding of the Universe and life within it, to an increased concern with engineering the Universe to make it hospitable.

This task would almost certainly be difficult, though not necessarily beyond other heroic human ventures of the past. Where we are likely to find ourselves so far removed from normal practice, the sole clear counsel to any undertaking will have to be reliance on the Hippocratic oath: "First, do no harm." Lacking greater wisdom, this oath has, in somewhat evolving versions over more than two thousand years, constituted the safest approach at our disposal for dealing with dire medical problems as well as arising societal challenges.

9.14   The Interplay between Curiosity-Driven Astronomy and Practical Applications

Any publicly funded or privately endowed effort owes supporters an account of its activities, achievements, and moneys spent. Curiosity-driven science is no different in this respect from any other undertaking.

Astronomy, of course, is not driven solely by curiosity. Its curiosity-driven core often reveals opportunities for improving life on Earth, and the costs as well as potential benefits of these more practical efforts may then be placed on a balance to determine whether a potential effort is worth further pursuit. An example may show how the interplay between curiosity-driven projects and practical applications tends to play out.

The WISE and NEOWISE Missions

Early this twenty-first century, NASA launched a low-cost mission, the Wide-field Infrared Survey Explorer, WISE, which Edward L. (Ned) Wright of the University of California at Los Angeles had designed to catalog all the most readily detected near-infrared sources in the Universe.

Once this mission was completed and discontinued, this highly successful observatory, by then idling in space but still operable, was resurrected to further detect and catalog the subset of the WISE mission's sources exhibiting measurable motions across the celestial sphere. Only Solar System planets, comets, and asteroids were sufficiently nearby to display observably high velocities.

Asteroids are aggregates of rock that readily absorb sunlight and reradiate this energy at near-infrared wavelengths – properties of crucial interest because each asteroid follows a well-defined orbit about the Sun, as does our planet Earth. Improved measures of the asteroidal orbits were crucial to determining whether any of them might ultimately intersect Earth's trajectory, potentially wreaking a cataclysmic collision. The WISE mission was thus reactivated, renamed NEOWISE – for Near-Earth Object Wide-field Infrared Survey Explorer – and put to use to catalog the orbits of all Solar System asteroids in sizes ranging down to a few hundred meters, and sometimes less. To date, it has revealed about 9000 asteroids, more than 200 of which are designated Near-Earth Asteroids, NEA, meaning that they might someday strike Earth.[18] Impacts by even the smallest among these, barely 100 meters in diameter, would generate explosions surpassing, by orders of magnitude, those of the most powerful hydrogen bombs ever detonated. If such an asteroid likely to impact Earth were found, attempts would be undertaken to deflect it into a more benign trajectory.

In 2017 NASA began planning just such a mission in which a refrigerator-sized spacecraft accelerated to a speed of 6 kilometers per second would strike a small asteroid, Didymos B, orbiting a larger companion, Didymos A. The resulting change in the smaller asteroid's orbit would then be observed from Earth to determine how well the impact had transferred momentum to the asteroid(s).[19]

While many practical concerns of this type deal with avoidance of serious threats, others share more positive aspects.

9.15   Asteroids Arriving from Interstellar Space

A more serious threat than collisions of Solar System asteroids impacting Earth may come from asteroids arriving randomly from interstellar space.

A surprising visitor in 2017 was just such an asteroid. It passed through the inner Solar System at speeds of the order of 100 km per second, and was approximately 1 km long with a diameter of order 0.1 km. Darryl Seligman and Gregory Laughlin of Yale University in New Haven, Connecticut, analyzed the orbital motion of this cigar-shaped exo-asteroid now named Oumuamua.[20; 21] Their paper suggested that we should be able to sample considerable numbers of asteroids that have escaped their parent star's circumstellar environs in the course of Galactic history, and pass through the Solar System from time to time. This conjecture received almost immediate support when, just two years later, on August 30, 2019, Gennady Borisov, a Russian amateur astronomer using a 65 cm telescope he had constructed himself, discovered another Solar System intruder. The International Astronomical Union has now named this comet-like object 2I/Borisov. The designation 2I indicates its being the second such intruder discovered. Additional objects arriving from interstellar space are planned to be sequentially numbered and similarly named after their discoverers.[22]

Some estimates suggest that the number density of asteroids ejected from planetary systems around other stars in the Galaxy could be so high that at least one would be passing through the Solar System at any given time along a trajectory that passes through a sphere of radius $\sim$10 AU about the Sun, a radius not much larger than the orbital radius of the planet Uranus. A recent assessment suggests that the actual number of these interstellar interlopers could be as high as one asteroid or comet arriving within $\sim$1 AU from the Sun, every few years, a region encompassed by Earth's orbit about the Sun.

9.16   Trade-Off between Exploring Remote Exoplanets and a Solar-System-Wide Interception Observatory

Payloads launched at speeds of about 100 km per second from dedicated stations strategically deployed across a spider web of stations continuously orbiting the Sun at radial distances of order 10 AU, or $\sim 10^9$ km from the Sun, could rapidly intercept extrasolar planets, asteroids, or comets arriving from interstellar space – detecting them sufficiently early in their trajectories to intercept and study them.

On landing on these interlopers, the probes could check their surface layers for soil bacteria, likely precursors for more complex forms of life. A program to erect such a spider web of nodes across the Solar System could begin operations within a few decades, its greatest technical challenge being the need to build at least small-scale rockets capable of reaching speeds of order 100 km per second required for intercepting asteroids expected to be swept up by the Solar System as it wanders through ambient interstellar space at a velocity of order 30 km per second.

Such a research program would undoubtedly be far less expensive than any exploratory voyages to even the nearest exoplanetary systems no more than a few parsecs away. And a payload aboard a spacecraft journeying to any of these nearest systems, even at comparably high speeds of order 100 km per second, would not reach its target for another thirty thousand years.

We now estimate that interstellar space may be populated by roughly as many planets ejected from other planetary systems as are found in those systems themselves. With patience, we might in this way wait for some of these interstellar planets to come to us to be explored. If exoplanetary systems are as rich in asteroids as the Solar System, the number density of asteroids in interstellar space could be as much as several hundred times higher even than the number of interstellar planets.

The collection of soil samples and searches for chemical or mineralogical differences from, or similarities to properties of Solar System asteroids would be particularly interesting to determine. They could deliver insights on the formation of asteroids elsewhere in the Galaxy. Searches for microbial organisms that could potentially have spawned in those other planetary systems would perhaps also provide insight on how, when, and where living matter may first have emerged.

9.17   Natural and Engineered Universes

The past four decades have witnessed a silent scientific transformation. Seldom specifically called out, it neatly divides the Universe in two: One natural, the other engineered. Until quite recently, the astronomers' Universe was purely part of nature. In this respect, it fundamentally differed from the engineered universes that physicists or geneticists had begun making their own. Even today, astronomy largely remains a purely observational science in virtually all its activities, unable to disturb the Cosmos it studies – although, as we will see below, this may soon change.

In contrast, physics has taken a rather different course. Pure observations of naturally occurring processes have mostly been overtaken by investigations conducted on systems nature does not provide. Physicists now work with engineered systems sometimes operating at temperatures measured in nano-degrees Kelvin, 10^{-9} K, one part in a billion of a degree above absolute zero. Temperatures as low as this are not encountered in nature, where all matter and all radiations are continually bathed by, and in contact with, a roughly 3 K *cosmic microwave background radiation* and, most likely also, an equally pervasive theoretically predicted ~1.9 K *cosmic neutrino/antineutrino background*.

Many researchers in physics, and also in genetics, no longer see a sharp dividing line separating their deliberately engineered universes from the natural Universe. Partly responsible for this are useful perspectives the engineered systems cast back on how nature operates or, more important perhaps, how nature might be re-engineered to serve new purposes.

For our understanding of the Universe, a particularly striking piece of genetic engineering has been a recent realization that the series of four amino acids directing and sustaining all life on Earth is not the sole basis on which life could arise. Detailed investigations led by Steven A. Benner, originator of the *Foundation of Applied Molecular Evolution* in Alachua, Florida, show that the series of four amino acids directing the evolution of all terrestrial species is not the sole basis on which life could arise. An alternative alphabet based on an eight-letter DNA, instead of the four-letter DNA on which terrestrial life is based, could readily sustain biological processes as well. This alphabet maintains DNA's double helix structure, and is able to faithfully transcribe this engineered DNA into complementing RNA, a key step for forming proteins and sustaining life. As the inventors of the scheme conclude, this "expands the scope of the structures that we might encounter as we search for life in the cosmos."[23] The implication of this thrust is that life elsewhere in the Universe might not just be a variant of life here on Earth as often tacitly assumed. A number of fundamentally distinct forms of life might naturally exist elsewhere, or at least be engineered to exist there.

These structural changes in the ways science is conducted reflect intensifying efforts to investigate and perhaps construct systems of all kinds having potential practical applications with no immediate counterparts in nature.

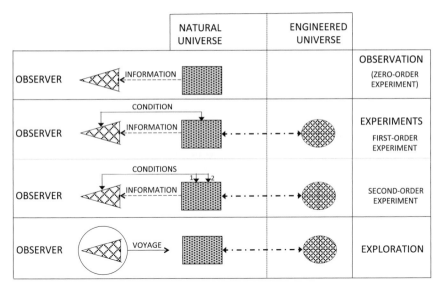

Fig. 9.2 Scientific investigations have, for centuries, taken three different forms: Passive observation; experimentation in which a system is subjected to one or more calibrated stimuli or *conditions*, such as changes in temperature, pressure, or magnetic environment, to which the system's response is recorded; or voyages of exploration conducting observations and experiments in otherwise inaccessible domains. Recent decades have seen such methods of inquiry extended also to deliberately engineered systems not found in nature. This has permitted checking whether recognized laws of nature extend also to these engineered structures, or need to be modified; whether such altered laws then also apply to nature in ways not previously realized; and whether the engineered systems may serve practical purposes that nature fails to provide. Dash-dotted lines joining systems in the engineered universe to those in the natural Universe are meant to indicate that experimental and exploratory techniques formerly dedicated mainly to studies of naturally occurring systems can similarly be applied to the study of engineered systems. Beyond the confines of the Solar System, the natural Universe can currently be studied solely by observational means; we lack the means to forestall astronomical events or affect cosmic evolution.

As Figure 9.2 emphasizes, scientific investigations conducted today take on three distinct forms:

- Passive observations of whatever naturally occurring phenomena we may encounter;

- Experiments on phenomena of progressively increasing complexity, to discern how deliberately applied stimuli may display the physical foundations on which a system's appearance and behavior are based; and

- Exploration, taking scientists out of their laboratories and observatories on voyages to witness events otherwise inaccessible.

Among these three approaches, we may first need to differentiate between the natural and the applied sciences, even if the borders separating them have all but vanished.

Astronomers try to understand the Universe and its evolution through a mix of observations, experiments, calculations, simulations, and the inferences we may draw from these. High-energy particle accelerators, laboratory spectroscopy, chemical experiments, and basic theoretical considerations all contribute significant perspectives as well. But remove the capstone of direct astronomical observations, and all these supplementary efforts offer little. Without direct observational evidence we might still know how physical and chemical processes work here on Earth, but whether or not the same laws of nature apply to the Cosmos would remain uncertain. Theoretical predictions need to be verified.

The study of solids or fluids at extremely low temperatures could simulate quantum processes which, someday, may be found to mirror processes active in supermassive *black holes* or exhibit processes operating in *neutron stars* composed almost entirely of neutrons – particles normally accounting for more than half the mass of all large atoms. But our understanding of the Universe today, first and foremost, crucially depends on information astronomical observatories directly provide. The supplementary information the engineered universes yield may, or may not, prove useful. The dot-dashed lines in Figure 9.2 are meant to suggest that the engineered universes can shed light on naturally occurring processes, though only once these correspondences have been verified and are firmly established. In astronomy and cosmology, such links are not yet clear, but ultimately may emerge.

It is this distinction, between a largely observational science and the experimental sciences, which will ultimately restrict astronomy's scope without simultaneously also limiting the scope of the experimental sciences. This discerns why our observational efforts to study the Universe may ultimately come to an end, even as the experimental sciences continue to flourish.

9.18   Engineering the Universe?

Before science can begin its investigations, its tacit purposes need to be defined: Scientists are society's emissaries tasked with investigating the world we inhabit and reporting on our findings in language the general public can readily grasp.

As in any language, this requires a vocabulary that, over time, is honed and pared down to essentials. Here and there, the language may differ from one field of science to another; but of necessity all of these fields must conform to sets of widely shared expectations. Explicit language is one of the most essential returns scientists can provide our fellow citizens so they may find our efforts useful in shaping our community's future.

Emphasis on some of the distinct ways in which we attempt to advance science may help:

Like all other forms of life, humans have consistently survived through adaption to new norms of fitness. Modern biologists and geneticists deliberately search for ways to succeed in this struggle by changing the universe into which we were born, for instance by permanently eradicating smallpox, or by altering the genes of crops to make them more resistant to ecological threats. The worlds that physicists are creating for better communication, improved materials, and reliable transportation may similarly help our race to survive more readily. And if astronomers can succeed in identifying asteroids whose catastrophic impacts on Earth may be prevented, or find ways to halt and reverse climate change, such efforts should certainly not be dismissed, even if many aspects of these attempts to engineer our environs might have seemed unimaginable in Charles Darwin's times. The distinctions between an engineered habitat and the natural universe are thus losing the clarity they once had. To date, the astronomical universe has remained largely immune to human engineering – though we cannot predict how long it may remain that way. If Earth were at some epoch to become uninhabitable, humans might wish to find some exoplanet where life could continue.

9.19   The Alternatives Offered by Exploratory Voyages

Although limits on the information the Universe transmits may signal an approaching end to our quest to understand the Cosmos, it could equally well also define requirements for a renewed start. If we cannot observe processes of interest from a distance, we may need to switch to an entirely different, far more unwieldy and costly program of *exploration*, in which we might send instruments and possibly emissaries on voyages across the Universe to gather any additional information on remote locales. The enormous increase of effort, cost, and persistence that this would necessarily entail would need to be assessed to define the scope and prospects of such efforts.

Exploratory journeys within the Solar System have by now provided information that observations alone did not reveal. But systematic exploration on a Galactic or cosmic scale would certainly be far more costly and incomparably more time consuming than the observational means we have traditionally pursued.

Extensive exploration appears largely beyond present technological reach. Our Universe is large; we cannot easily travel anywhere we wish to investigate phenomena of interest. The time required to traverse cosmic realms at the speed of light, the maximum speed the Universe permits, often would be of the order of millions if not billions of years.

Today, proposals are being considered for propelling tiny instrument-laden spacecraft to alpha Centauri, the nearest analogue to our Solar System. Even these modest steps will require intergenerational plans, coherence, steadfast funding and endurance spanning many decades and potentially centuries. To date, human experience on such long-term projects has been limited to just a few ventures, such as the erection of cathedrals in the Middle Ages. Whether competing social priorities would prevent such dedicated efforts from being similarly applied to cosmic searches is unclear.

9.20   The Search for Extraterrestrial Intelligence

Though appearing fanciful, right now, traversal to other planetary systems may lie within the range of achievable possibilities if conditions demanded it. But it would not be easy, and it would most likely not correspond to science-fiction recitals of colonization by cryogenically preserved humans taken out of refrigeration just in time to land in a new world to take it over.

Many curiosity-driven astronomers have hoped to find insight on life on exoplanets with a Search for Extraterrestrial Intelligence, SETI, which could at least reveal where intelligent life has emerged beyond Earth or the Solar System. Detecting advanced civilizations on at least some of these inhabited planets would seem to rank among some of the greatest prizes curiosity-driven science might deliver.

9.21   How Would Implantation of Life on Some Barren Planet Be Engineered?

If we were more interested in engineering the Universe so humans might propagate themselves we would probably search for habitable worlds exhibiting a total absence of living matter.

Why?

If humans seriously wished to colonize space, the last place we might wish to consider would be a planet where life had already taken a hold. Inevitably, we are part of a larger biological community, and the denizens of extrasolar planets undoubtedly would have to be too. Explorers landing on such planets could well release enormous pandemics, in which the microbes we necessarily brought with us would decimate the natives, while their own microbes would attack the arriving explorers.

Given such uncertainties, attempting to colonize a barren habitable planet might be preferable. Even this would however require careful planning. A best way might be to simulate the historical order in which life on Earth evolved, though preferably on a far shorter time scale than the several billion years life has taken to evolve here to date. Such an accelerated development might, or might not, be possible. To do this, we would need first to better understand the biological history of our own planet – even if we continued to fail in our quest to find how life originated in the first place.

We still are far too uncertain about how life evolved on Earth. That Earth's atmosphere originally was devoid, or almost devoid, of oxygen appears to be well documented by now; just how different species evolved as concentrations of atmospheric oxygen increased remains uncertain. As Andrew H. Knoll and Erik A. Sperling at Harvard have pointed out, "In the annals of Earth history, few associations have proven more iconic or durable than that between animals and oxygen."[24]

Our planet Earth formed roughly 4.5 billion years ago. The fossil record indicates that single-celled organisms date back at least about 3.6 billion years.[25] Preserved microfossils document an early prevalence of photosynthetic bacteria and, later, eukaryotic microorganisms defined by a nucleus and genetic material enclosed within a membrane. Life existed long before large amounts of oxygen found their way into the atmosphere, perhaps as recently as half a billion years ago.

As the concentration of oxygen increased, sponges, a primitive sister group to animals, could have begun to thrive at oceanic-surface-layer oxygen levels of the order of 1% of present-day abundances, permitting survival of precursors to some of the earliest animals documented by fossil records.

All of this complicates the search for a planet that might favor colonization. We do not even know the full complement of other organisms on which human life depends here on Earth. An exoplanet habitable by people might need implantation of all of these complementing forms of terrestrial life as well to permit a sustainable ecology to thrive, much of which would need to be spawned at the very least by imported microorganisms, sperm and ova, and other seeds of life to keep transportation and importation costs low. Accompanying robots would have to see these first generation organisms through to maturity and succession, tend to the education of the young human contingent, and carry out a vast variety of emerging tasks to assure continuing progress.

In any enterprise of this magnitude the implications of Charles Darwin's theory of survival of the fittest would need to be broadened. Survival on a new habitable planet would have to be based on an inclusive ecological strategy. A significant portion of the species we recognize today would likely need to thrive if any of its components were to survive. Humans cannot tackle such a migration on their own. We thrive or wither jointly: bacteria active within our digestive system; pollinators such as honey bees and the plants they visit; insects, animals, fish, and plants. Life on another planet thus would likely have to be planned in stages. We might need to first implant suitable soil bacteria or comparable oceanic microorganisms, and then gradually introduce successively higher life forms, dependent for their survival on those precursors. But we should not simply doubt that transformation of a planet to accommodate humans might ultimately be possible.

* * *

These eventualities may perhaps never arise, or be reversed, or abandoned in favor of other societal priorities – as comparably drastic changes in human aspirations brought about by wars and pandemics in our own times have often shown.

Catastrophes have their own way of channeling history.

As scientists our task is to shed light on potential paths forward, so humanity may find more informed ways of deciding what to try next. Curiosity-driven astronomy may help us envision that future.

Notes

1 Principles and Politics in Federal R & D Policy, 1945–1990 – An Appreciation of the Bush Report, D. J. Kevles, in a retrospective foreword to *Vannevar Bush: Science – The Endless Frontier, A Report to the President on a Program for Postwar Scientific Research*, reprinted by the National Science Foundation on the 40th Anniversary, 1950–1990, pp. ix–xxxiii

2 *The Physicists: The History of a Scientific Community in Modern America*, D. J. Kevles, Harvard University Press, Cambridge, MA, 1987, pp. 148, 190–99, and 268

3 *Military Procurement Authorization Act for FY 1970, Public Law 91–121, Section 203*

4 US Astronomers Face Hard Decisions, M. Mountain & A. Cohen, based on an article "Billion-Dollar Telescopes Could End up beyond the Reach of US Astronomers" by the same authors published in *Nature* 560, 427–29, 2018

5 Women's and Men's Career Choices in Astronomy and Astrophysics, R. Ivie, S. White, & R. Y. Chu, *Physical Review Physics Education Research* 12, 020109, 2016

6 *Scientific Elite: Nobel Laureates in the United States*, H. Zuckerman, The Free Press, New York, 1977

7 The Genetical Evolution of Social Behaviour, I. & II., W. D. Hamilton, *Journal of Theoretical Biology* 7, 1–16; 17–52, 1964

8 A Generalization of Hamilton's Rule for the Evolution of Microbial Cooperation, J. Smith, D. Van Dyken, & P. C. Zee, *Science* 328, 1700–03, 2010

9 Discovery of a Pulsar in a Binary System, R. A. Hulse & J. H. Taylor, *Astrophysical Journal* 195, L51–53, 1975

10 A New Test of General Relativity: Gravitational Radiation and the Binary Pulsar PSR 1913+16, J. H. Taylor & J. M. Weisberg, *Astrophysical Journal* 253, 908–20, 1982

11 Large Teams Develop and Small Teams Disrupt Science and Technology, Lingfei Wu, Dashun Wang, & J. A. Evans, *Nature* 566, 378–82, 2019

12 Astronomy Degree Recipients: One Year after Degree, P. Mouldy & J. Pold, *Statistical Research Center of the American Institute of Physics*, August 2019, p. 10

13 Acceleration of Coronal Mass Ejection Plasma in the Low Corona as Measured by the Citizen CATE Experiment, M. J. Penn et al., *Publications of the Astronomical Society of the Pacific* 132,014201, 134 pp., January 2020

14 People-Powered Discovery, L. A. Shanley, *Science* 367(6747), book review, p.153, 2020

15 Public Law 114-329 114th Congress – January 6, 2017. Short Title: American Innovation and Competitiveness Act, 42 USC 1861: An Act to invest in innovation through research and development, and to improve the competitiveness of the United States. Section 402. Crowdsourcing and Citizen Science Page 130 STAT. 3019

16 Wikimedia Annual report for 2016–2017: https://annual.wikimedia.org/2017/

17 Coping with the Tragedies of the Commons, Elinor Ostrom – a talk given in 1994, https://pdfs .semanticscholar.org/7c6e/92906bcf0e590e6541eaa41ad0cd92e13671.tif

18 NEOWISE Reactivation Mission Year Two: Asteroid Diameters and Albedos, C. R. Nugent, et al., *Astronomical Journal* 152, 63 (12 pp.), 2016

19 NASA's First Asteroid Deflection Mission Enters Next Design Phase, June 30, 2017, https://www.nasa.gov/ feature/nasa-s-first-asteroid-deflection-mission-enters-next-design-phase

20 A Brief Visit from a Red and Extremely Elongated Interstellar Asteroid, K. J. Meech, et al., *Nature* 552, 378–81, 2017

21 The Feasibility and Benefits of In Situ Exploration of Oumuamua-Like Objects, D. Seligman & G. Laughlin, *Astronomical Journal* 155, 217, May 2018

22 Alien Comets May Be Common, Analysis of 2I/Borisov Suggests, W. Wayt Gibbs, *Science* 366(6465), 558, November 1, 2019

23 Hachimoji DNA and RNA: A Genetic System with Eight Building Blocks, S. Hoshika, et al. *Science* 363, 884–87, 2019

24 Oxygen and Animals in Earth History, A. H. Knoll & E. A. Sperling, *Proceedings of the National Academy of Sciences* 111, 3907–8, 2014

25 Drifting on Pieces of Earth's Cracked Shell, N. Angier, *New York Times International Edition*, December 19, 2018, p. 7

Appendix

Symbols, Glossary, Units and Their Ranges

Symbols

~	Denotes that a particular numerical value is approximate.
″	Denotes an angle measured in seconds of arc.
′	Denotes an angle measured in minutes of arc.
°	Denotes an angle measured in degrees.
α	Greek letter "alpha." See also **alpha particle**.
Å	See **Angstrom unit**.
β	Greek letter "beta." See also **beta particle** and **electron**, e^-.
c	The speed of light in vacuum, 2.998×10^{10} cm/s.
γ	Greek letter "gamma." See also **gamma rays**.
δ	Greek letter "delta."
e	The electron charge, 4.803×10^{-10} electrostatic units.
e^-	Electron. See also **electron**, e^-.
e^+	Positron. See also **positron**, e^+.
G	The gravitational constant, 6.674×10^{-8} cm^3 g^{-1}s^{-2}.
h	Plank's constant, 6.626×10^{-27} erg s.
$\hbar$	Planck's constant divided by 2π, $\hbar = h/2\pi = 1.055 \times 10^{-27}$ erg s.
k	Boltzmann's constant, 1.381×10^{-16} erg/K.
K	Degree Kelvin. See also **Kelvin temperature, K**.
$L_\odot$	The Solar luminosity, 4×10^{33} erg/s.
λ	Greek letter "lambda." See also **wavelength**, λ.
Λ	Greek letter "Lambda." See also **cosmological constant**.

$M_\odot$ The Solar mass, 2×10^{33} g.

μ Greek letter "mu" for "micro," one part in a million.

N   Neutron. See also **neutron**.

ν Greek letter "nu." See also both **neutrino**, ν, and **frequency**, ν.

$\bar\nu$ Designation for an antineutrino. See also **antineutrino**, $\bar\nu$.

P   Proton. See also **proton**.

σ Greek letter "sigma." See also **cross section for an interaction**, σ.

Glossary

absolute magnitude: See **magnitude of a star or galaxy**.

adiabatic expansion: A gas expanding without input or loss of heat is said to expand adiabatically. An expansion of the Universe without infusion or extraction of heat is similarly said to be adiabatic.

alpha particle, α: The highly energetic nucleus of a helium atom consisting of two protons and two neutrons.

Angstrom unit, or angstrom, Å: A measure of length equaling 10^{-8} cm.

annihilation of matter: The destruction of matter on encountering antimatter, with an accompanying liberation of energy and the formation of pairs of photons or particles and their antiparticles. See also **antimatter**.

antimatter: Matter consisting of antiparticles.

antineutrino, $\bar\nu$: Antiparticle of the neutrino, ν. See **neutrino**, ν.

antiparticle: Matter consists of atoms that contain neutrons, protons, and electrons. Corresponding to each of these three particles there exists an antiparticle with identical mass, but opposite charge, if any. Neutrinos, which are neutral particles, are distinguished from antineutrinos by the direction of their spin. Particles and antiparticles annihilate on encounter. See also **annihilation of matter**.

arcsecond, ″: An angle that subtends 1/3600 of a degree.

asteroid: A small planet, often barely large enough to be detected with a telescope. The smallest observed asteroids have diameters of 50 to 100 meters.

atomic mass unit, amu: A mass equivalent to one-twelfth the rest-mass of an isolated atom of carbon ^{12}C in its ground electronic and nuclear states.

AXAF: Advanced X-ray Astrophysics Facility, a powerful X-ray astronomical space observatory renamed **Chandra** after launch in July 1999.

beta particle: An electron or positron emitted from a nucleus, usually at high energy.

billion: 10^9.

binary star: Two gravitationally bound stars orbiting a common center of mass.

bit: A contraction for "binary digit," usually designated by symbols 0 or 1; a unit of information.

black hole: A highly compact massive object, whose gravitational attraction is so strong that no matter or radiation can escape. See also **Schwarzschild radius**.

blue shift: A shift of light to higher frequencies. See also **redshift**.

Boltzmann constant, k: A constant of nature enabling the conversion of an equilibrium temperature T into units of energy. For an ideal gas the product of pressure P and the volume V occupied by a particle, that is, an atom or molecule of the gas, provides a measure of the mean energy of the individual particles, $PV = kT$, with $k = 1.381 \times 10^{-16}$ erg/K.

brown dwarf: A body of mass lower than that of stars. Stars are sufficiently massive to convert hydrogen into helium in their interior. Brown dwarfs are incapable of doing this. But even low-mass brown dwarfs can derive energy from fusing deuterium nuclei, an activity that planets are not sufficiently massive to exhibit. The distinctions between the least massive brown dwarfs and the most massive planets are not always well defined.

carrier of information: Any particle or wave that transmits information from a source.

cascade: A succession of processes, each of which triggers the next.

caustic: Concentrated image formed by a parallel beam of radiation through reflection, refraction, or other deflection by a surface or body altering the course of incident rays. Regions of catastrophic deflections generally are referred to as *caustics*.

Cepheid variable: A bright yellow variable star that pulses regularly with a period as short as 2 days for some Cepheids and as long as 40 days for others. Cepheids are sufficiently luminous to be detected in nearby galaxies. Because their periods are directly related to their luminosities, Cepheids can serve as distance indicators.

cluster of galaxies: A grouping of galaxies that may contain as many as several thousand individual or interacting galaxies. A small cluster usually is called a group of galaxies.

column density: The total number of particles contained within a column of unit cross section, often taken to be 1 cm^2, and a length defined by two locations, one of which usually is the location of the observer's telescope, the other a point of particular interest, e.g., the most remote edge of the Milky Way plane.

comet: A Solar System body that disintegrates on approaching the Sun, leaving a trail of debris and often a long tail of ionized gas pointing away from the Sun.

cosmic background radiation: See **microwave background radiation.**

cosmic maser: A monochromatic source of radiation whose high surface brightness is produced through the same process active in man-made masers and lasers. See also **maser.**

cosmic ray: A highly energetic particle that travels at a speed close to the speed of light. Electrons, positrons, protons, and nuclei of atoms have all been found in the rain of cosmic-ray particles that continually impacts Earth's upper atmosphere from as-yet-unspecified cosmic realms.

cosmic variance: A unique characteristic of the Universe for which no particular explanation exists.

cosmological constant, Λ: A uniform vacuum energy density pervading the Universe.

Cosmos: The Universe; all that we can survey.

cross section for an interaction, σ: A significant interaction of two particles takes place only if the encounter is sufficiently close. The particles act as though they had a cross section for interaction whose diameter exceeds the center-to-center distance between the two particles whenever they interact.

curvature of space: In many cosmological models, light does not propagate along straight lines, but rather along curved trajectories. The curvature of these trajectories is the curvature of the space.

dark energy: This is a hypothetical energy permeating the Universe and tending to accelerate cosmic expansion. Little is known about its physical origins. It might take the form of a cosmological constant Λ, an energy content of the vacuum that constantly regenerates itself as the vacuum expands. But it could equally well represent other energy distributions, which might vary with time and place.

dark matter: This hypothetical form of matter appears to dominate the gravitational forces exerted on stars and interstellar matter within galaxies. It also dominates the gravitational interactions of galaxies within clusters. Its nature is not understood.

diffraction: The spreading of a light beam around a body that blocks part of the beam. Also the spreading of different wavelengths of light along different directions.

disk: A thin, circular aggregate of gas, dust, or stars, gravitationally attracted to and circling about a massive central body.

Doppler shift: The systematic shift of an entire spectrum of radiation toward longer wavelengths – lower frequencies – when the source of radiation rapidly recedes from the observer, and toward shorter wavelengths – higher frequencies – when the source approaches.

dust: Fine grains of solid matter. In interstellar space dust appears aggregated in dark irregular clouds.

dwarf star: A low-luminosity main sequence star.

eclipsing binary: A pair of mutually orbiting stars in which one star passes in front of the other and blocks its light.

electromagnetic radiation: A class of radiation comprising radio waves, infrared rays, visible light, ultraviolet radiation, X-rays, and gamma rays, differing from each other only in their wavelengths, which also determine the frequency and energy of each quantum of radiation.

electromagnetic theory: A mathematical depiction of electric and magnetic processes.

electromagnetic waves: See **electromagnetic radiation**.

electron, e^-**:** A negatively charged particle often found orbiting an atomic nucleus, but able to travel through space by itself if removed from the atom.

electronvolt, eV: A unit of energy. The energy carried by a quantum of yellow light is about 2 eV. An X-ray photon has an energy of several thousand eV. See Table A.2.

erg: A unit of energy. See Table A.2. A mass, m, of one gram moving with a velocity, v, of 1 centimeter per second has an energy $mv^2/2 = 0.5$ erg.

eruptive variable: A variable star that suddenly changes its output of light, usually from a normally low to a far higher eruption level. See also **nova**.

eV: See **electronvolt** and Table A.2.

event horizon: A surface separating events that can be observed from others that disappear where the recession velocity of the events involved exceeds the speed of light.

evolved star: A star that has already converted its available store of hydrogen into helium and no longer is found on the main sequence. See also **main sequence**.

extreme ultraviolet: Ultraviolet rays of very short wavelength and X-rays of very long wavelength define the upper and lower bounds of this wavelength range.

Faraday rotation: A plane polarized wave passing through an ionized gas along the direction defined by a local magnetic field experiences a rotation of its direction of polarization. This rotation is largest at long radio wavelengths and decreases monotonically at shorter wavelengths.

flare star: A star whose luminosity can increase by a substantial factor in a matter of minutes, with a subsequent, slower decline to normal levels. In some rare stars the luminosity increases by more than a factor of 100. In most it increases only by a factor of 2 to 5.

flat space: Cosmic space is said to be flat if the Universe is infinite and its geometry obeys Euclid's postulates, in particular, that parallel straight lines, no matter how far they are extended, never intersect. In Einstein's general relativity this happens only when the mass–energy density of the Universe has a critical value, ρ_{crit}, related to the Hubble constant, H, and the gravitational constant, G, by $\rho_{crit} = 3H^2/8\pi G$. See also **Hubble constant, H.**

flux: The energy passing through unit area each second within the measured range of frequencies.

frequency, ν: The number of crests of a wave, for example, an electromagnetic wave, passing an observer during an interval lasting one second. See also **hertz (Hz).**

galactic plane: The disk-shaped aggregation of stars and gas in the central plane of a spiral galaxy.

galaxy: A galaxy is an isolated grouping of 10^9 to 10^{11} stars and associated interstellar matter, mutually bound by gravity. The **Galaxy** – where the word **galaxy** is capitalized – is the particular galaxy in which the Solar System resides. It contains $\sim 10^{11}$ stars, measures a hundred thousand light years across, and contains some stars older than 10^{10} years. The Galaxy is sometimes called the Milky Way because, seen from the Solar System, which lies $\sim 25,000$ light years from the Galaxy's center, it looks like a white diffuse band stretching across the night sky.

gamma ray, γ: An electromagnetic wave whose wavelength is less than 10^{-10} cm and whose energy is higher than 10^5 eV. See Table A.2.

gamma-ray burst: Periodic burst of gamma rays reaching Earth from different parts of the sky, at unpredictable times, often lasting no longer than a few seconds. Most arrive from distant galaxies and appear to be generated in hypernova explosions or by the merger of two neutron stars. See also **hypernova.**

gauss: A unit of magnetic field strength. Measured on Earth, magnetic field strengths are around half a gauss but vary with latitude. In magnetic stars the field strength can exceed 10^4 gauss. In interstellar space the field strength tends to be less than 10^{-5} gauss.

GeV: See Table A.2.

GHz, gigahertz: 10^9 Hz. See Table A.2. See also **Hz.**

giant star: Any highly luminous star. See also **red giant.**

globular cluster: A closely packed spherical aggregate of 10^5 to 10^7 stars bound by gravity. The Galaxy contains several hundred of these clusters, all formed about 10^{10} years ago.

Gpc, gigaparsec: See Table A.1.

gram, g: A unit of mass. A cubic centimeter of water, approximately a thimbleful, has a mass of one gram. See Table A.3 and Figure A.2 for the relations between different units of mass.

gravitational radiation: A form of radiation liberated by massive, accelerating bodies. In late 2015, gravitational waves were first directly observed. Previously, their existence was only inferred indirectly from gradual changes in the orbits of binary neutron stars. The quantum of gravitational radiation is sometimes called a graviton.

gravitational waves: See **gravitational radiation**.

Heisenberg uncertainty principle: See **uncertainty principle**.

hertz, Hz: A measure of frequency named after the nineteenth-century German physicist Heinrich Hertz. 1 Hz equals 1 cycle per second. See Table A.2.

Hertzsprung–Russell diagram: A drawing that plots the brightness of a group of stars as a function of temperature or spectral appearance.

homogeneity: The property of having a uniform consistency throughout.

Hubble constant, H: The expansion rate of the Universe, generally expressed in kilometers per second, per megaparsec, km s^{-1} Mpc^{-1}. See Table A.1.

hypernova: An extremely powerful supernova.

Hz: See **hertz, Hz**.

ideal gas: A gas idealized as consisting of non-interacting point-like particles in random motion. The concept is useful in analyzing the behavior of highly dilute gases.

information: A quantitative measure of data content. See also **bit**.

infrared galaxy: A galaxy that emits most of its energy at infrared wavelengths.

infrared radiation: Radiation of wavelength in the 0.7 micrometer to 1 millimeter range. See also Table A.1.

infrared star: A star that emits most of its energy as infrared radiation.

intensity of radiation: The energy content of a beam of radiation; a measure of brightness.

interferometer: Apparatus used for interferometry. See also **interferometry**.

interferometry: The use of interference between superposed beams of electromagnetic radiation to measure the angular or spectral structure of a radiating source.

interplanetary space: The region between the planets orbiting the Sun.

interstellar cloud: A cloud of gas or dust in the space between stars.

ion: An atom charged through addition or (more often) removal of electrons orbiting the nucleus.

ionized hydrogen region: A domain of interstellar space containing hydrogen ionized through removal of electrons from their parent nuclei. It largely contains freely moving electrons, protons, and smaller concentrations of other ions and atoms.

isotope: The number of protons in the nucleus of an atom defines the atom's chemical properties – as an element. Atoms of a given element may differ in the number of neutrons their nuclei contain. Such atoms are said to be different isotopes of the element.

isotropy: Independence of orientation; having identical characteristics along any direction.

Kelvin temperature, K: A temperature measured on a scale with units identical to those of the conventional Celsius (centigrade) scale, but with its zero point at absolute zero, where the energy of all matter attains an absolute minimum. The zero point on the Kelvin scale corresponds to -273.15 on the Celsius scale.

keV: See Table A.2.

kHz: See Table A.2.

limb: The edge of the projected disk of the Sun or Moon, or stars, or planets, as viewed by an observer.

logarithm: The logarithm of a number N, measured to base 10, is n, if $10^n = N$. We write $\log_{10} N = n$.

logarithmic scale: A scale on which numbers that differ by a constant ratio, say a factor of 10, are plotted a constant distance apart. See, for example, this glossary's Figure A.2.

luminosity: The energy emitted in unit time by a radiant source.

magnetar: A rapidly rotating highly magnetized neutron star – a highly magnetized pulsar.

magnetic field: A field that exerts a force on moving charges and magnetized bodies. See also **gauss**.

magnetic variable: A star having an extraordinarily high magnetic dipole field, often exhibiting a magnetic field thousands of times higher than the Sun's.

magnitude of a star or galaxy: The brightness of a star or a galaxy. The magnitude of a star or galaxy, as seen from Earth, is called its apparent magnitude. Seen from a standard distance, chosen as 10 parsecs – about 30 light years – the magnitude of the star or galaxy is called its absolute magnitude. Two stars or galaxies that differ in luminosity by one magnitude differ by a factor of ~2.5 in brightness. Five magnitudes amount to a brightness difference of 100. Magnitudes measured in the visual wavelength band are called visual magnitudes.

main sequence: On a plot of luminosity of stars as a function of color or temperature, more than 90% of all stars fall along a band that stretches from bright, hot, blue stars to faint, cool, red stars. This band marks the main sequence. See also **Hertzsprung–Russell diagram.**

main sequence star: A star that falls anywhere along the main sequence in a Hertzsprung–Russell diagram. See also **main sequence; Hertzsprung–Russell diagram.**

maser: An intense source of electromagnetic radiation emitted in an avalanche of photons all of which have identical wavelength, polarization, and direction of travel.

megahertz, MHz: A million hertz = 10^6 cycles per second. See Table A.2.

meteor: A grain of interplanetary matter that burns and disintegrates as it enters the upper atmosphere at high speed and produces a streak of light along its trajectory. Colloquially, meteors are often called shooting stars.

meteorite: A sizeable meteor, much of which survives passage through the atmosphere to hit Earth as a solid chunk of matter. See also **meteor.**

MHz: See **megahertz.**

messenger: Any carrier of information arriving from the Universe can act as a messenger. Radio waves, infrared radiation, visible light, X-rays, and gamma radiation arriving from the direction of a celestial source can act as messengers providing information about that source. The helium isotope ^{4}He, found in abundance in the Universe today, also is a messenger providing evidence that the early Universe was sufficiently hot to convert hydrogen into helium. Meteorites arriving at Earth from interplanetary space provide evidence on mineralogical processes elsewhere in the Solar System and potentially even beyond. All these can act as messengers though correctly interpreting the messages they convey may be difficult.

Michelson interferometer: An optical device that splits the photons in an incident light beam into two, one group traveling along one optical path, the other usually along an orthogonal path. When the two beams are then optically recombined, the combined beam will appear identical to the incoming beam only if the two traversed paths are equally long. If one path is longer than the other, the beam will exhibit diminished intensities at wavelengths at which the phases of the combined beams differ. The interferometer thus can indicate whether the relative lengths of the two optical paths change, for example through periodic stretching by a gravitational wave.

micrometer: 1 micrometer = 10^{-4} cm = 10^{-6} m. See also Table A.1.

microwave background radiation: Isotropically arriving microwave radiation from the Universe. The intensity of this radiation is roughly equivalent to that which would be found inside a cavity whose walls were kept at a temperature of 3 degrees Kelvin.

microwave radiation: Electromagnetic radiation in the wavelength range from approximately 1 mm out to beyond 10 cm.

moon: A planet-like body gravitationally bound to, and orbiting, a larger parent planet. Earth has only one moon; Jupiter has well over a dozen. See also **planet**.

neutrino, ν: A subatomic particle with zero charge, a spin of 1/2, a rest-mass close to zero, and a speed usually close to the speed of light. The neutrino's antiparticle is the antineutrino. See also **antineutrino**.

neutron, N: A neutrally charged particle with a mass somewhat in excess of a hydrogen atom. In all atomic nuclei, except for ordinary hydrogen, it is found bound to protons and other neutrons. When isolated from a nucleus, a neutron decays with a mean life of 885 seconds, giving rise to an electron, a proton, and an antineutrino.

neutron star: A collapsed, compact star whose core consists largely of neutrons.

noise: Spurious signal registered by a detector.

nova: A star that erupts in a matter of hours or days to increase in luminosity by a factor of $\sim 10^5$, returning to its original luminosity, roughly comparable to the Sun's, in a matter of weeks. Novae are binaries consisting of a cool red giant and a compact, hotter companion, which tidally strips the giant's outer layers, gathering this material until a critical surface mass triggers a thermonuclear explosion. Such cycles periodically repeat.

nucleon: Generic name for protons and neutrons in atomic nuclei.

observation: A passive form of study in which the observer has no way of, or refrains from, stimulating the system under investigation.

occultation: The extinction of light from a celestial source by a body that passes between the source and the observer.

old star: A star that has been evolving for more than $\sim 10^8$ years. A young star, in contrast, may have been formed less than 10^7 to 10^8 years ago.

orbital velocity: The speed with which an orbiting body moves along its trajectory.

orthogonal dimensions: Two or more lines that are perpendicular to each other are said to be orthogonal. Lengths measured along such lines specify the dimensions of a physical system.

parallax: The change of apparent source position in the sky as an observer moves from one location to another. In judging the distances of stars, the motion of Earth is taken to be equivalent to one radial distance between Earth and the Sun, 1.5×10^{13} cm, along a base line perpendicular to the line of sight to the star.

parameter: A trait that can be quantified and whose value describes the physical state of a system.

parsec, pc: The distance of a star at which its parallax is one second of arc. This distance is 3×10^{18} cm, roughly corresponding to three light years, or 30 trillion kilometers.

phase: A stage of development of an evolving system.

phase of matter: Atomic and molecular matter can exist in a number of different states, called phases. Major distinctions in phase are the ionized, gaseous, liquid, or solid phases. But other phases, such as different crystalline states of a solid, can also exist.

phenomenon: A class of objects or patterns of events that drastically differs from all other classes.

photometry: A low-spectral-resolution brightness measurement that generally encompasses a bandwidth comparable to the mean frequency of the observed radiation.

photon: A quantum of radiation.

Planck's constant, h: A constant whose value is $\sim 6.6 \times 10^{-27}$ erg s. When multiplied by the frequency v of a photon, Planck's constant gives the photon's energy, $E = hv$.

planet: The Sun is orbited by eight major bodies called "planets," six of which in turn are orbited by moons, and at least three of which exhibit rings. In order of increasing distance from the Sun, the planets are Mercury, Venus, Earth, Mars, Jupiter, Saturn, Uranus, Neptune.

planetary nebula: A cloud of gas ejected from an evolved star and ionized by the ultraviolet radiation emitted by the star's white dwarf remnant.

planetary system: A grouping of planets gravitationally bound to a star about which they orbit. The Solar System is a planetary system, but by now a sizeable fraction of other stars is known to also exhibit planetary systems.

plasma: An ionized gas. See also **ionized hydrogen region**.

positron, e^+: Antiparticle of the electron and identical to it in all ways except that it carries a positive, instead of a negative, electrical charge.

proton, P: Nucleus of a hydrogen atom.

pulsar: A source of sharp radio or gamma-ray pulses emitted at regular intervals. In some pulsars the intervals can be as short as milliseconds; in others as long as several seconds. Most pulsars are known to be rotating, highly magnetized neutron stars that are remnants of supernova explosions.

pulsating variable: A star whose radius and brightness vary regularly or semi-regularly.

quasar: A compact source of radiation occupying the nucleus of a galaxy. Quasars harbor giant black holes with masses ranging up to $\geq 10^9 M_{\odot}$. They often are highly redshifted and exhibit irregular variations in brightness.

radar: A technique that transmits radio pulses to distant objects and measures that distance by the delay in arrival of the returning reflected pulse.

radio star: A star that emits radio waves, generally a rare star that becomes a radio emitter when its luminosity suddenly increases.

radio wave: An electromagnetic wave whose wavelength exceeds a millimeter.

recession velocity: Speed of recession along an observer's line of sight.

red giant: A luminous red star whose red color indicates a cool surface temperature, and whose high luminosity implies an extremely large surface area.

redshift: A move of an entire spectrum toward longer wavelengths, lower frequencies. A galaxy from which the observed arriving wavelength emitted by a remote atom is λ_0, whereas the same spectral line emitted by atoms locally is λ_1, and is said to have a redshift $z = [(\lambda_0 - \lambda_1)/\lambda_1]$. See also **blueshift**.

relativistic: An adjective indicating that a body's motion is governed by the laws of relativity.

relativistic particles: Subatomic particles that move at velocities close to the speed of light.

resolving power: Spectral resolving power, R_s, is the ratio of the wavelength, λ, at which an observation is carried out, to the wavelength difference, $\Delta\lambda$, which can just be resolved:

$$R_s = \lambda/\Delta\lambda.$$

If an angle $\Delta\theta$ radians can just barely be resolved, the angular resolving power – also called the spatial resolving power – is

$$R_\theta = 1/\Delta\theta.$$

rest-mass: A body's mass at rest is at a minimum; at high velocities its mass increases significantly.

satellite: There are two kinds of satellites: natural and artificial. Natural satellites of planets are moons. Artificial satellites are devices placed in orbit around a planet or moon, generally for scientific, communications, or military purposes. See also **moon**.

scalar quantities: A scalar quantity can be described by a single magnitude, such as temperature, mass, density, or speed, none of which specify a spatial direction.

scalar particles, fields or perturbations: A scalar particle is a hypothetical particle having no directional properties. Thus, it has no spin. Scalar fields are similarly hypothetical and are the fields generated by scalar particles. In principle, these fields can perturb the density distribution in the space to produce scalar perturbations.

Schwarzschild radius: The radius of a sphere constituting the surface of a black hole through which no outward directed radiation or matter can escape. See also **black hole**.

sensitivity: The capacity of an instrument to detect weak signals.

SETI: A project to Search for Extraterrestrial Intelligence, primarily by analysis of radio signals that could be reaching us from nearby stars with planetary systems.

shock: A sudden impact that changes the state of a system.

Solar mass, $M_\odot$**:** The mass of the Sun. A unit of mass equaling the mass of the Sun. See Table A.3.

Solar System: The Sun and the system of planets, moons, asteroids, comets, and all other matter orbiting the Sun. See also **planet, moon, asteroid, comet**.

spatial resolution: See **resolving power**.

spectral energy distribution: The distribution of radiated energy across the range of wavelengths or frequencies at which a star or a galaxy radiates.

spectral line: A narrow, dark or bright feature in a spectral display of light due to an excess or a lack of radiation at one particular color or wavelength.

spectral resolution: See **resolving power**.

spectral type: The spectral type of a star, determined by its spectrum, largely depends on the star's surface temperature. Surface chemistry can also be a factor. Stars whose spectra exhibit similar features are said to be of the same spectral type.

spectrometer: Apparatus used in spectroscopy. See also **spectroscopy**.

spectroscopic binary: A binary in which the presence of the two stars is discerned by virtue of two sets of superposed spectra, one spectrum corresponding to each star. The two spectra exhibit opposite Doppler shifts that vary with time, as each star orbits a common center of mass, alternately approaching and receding from the observer.

spectroscopy: The separation of light into its wavelength or color components.

spectrum: The display of the different color components in light, or wavelength components in other types of radiation, the intensity of each component being separately displayed.

speed of light, c**:** The speed of light is about 3×10^{10} centimeters per second or, equivalently, 300,000 kilometers per second.

spin: Every fundamental particle is characterized by a spin angular momentum. For electrons, protons, neutrons, and neutrinos that spin has a value $(1/2)\hbar$. For light quanta the spin is $\hbar$. For gravitational waves the spin is believed to be $2\hbar$, where $\hbar$ is Planck's constant, h, divided by 2π. Some subnuclear particles also have zero spin.

spiral galaxy: A galaxy that exhibits stars, gas, and dust arranged in lanes or segments of lanes that stretch outward from the galaxy's center in a spiral pattern. Barred spirals are galaxies in which spiral arms appear at the ends of an elongated bar-shaped aggregate of stars at the galaxy's center.

stability: The ability to withstand small disturbances and return to equilibrium.

star: A gravitationally bound compact mass containing between 10^{32} and 10^{35} grams of matter. It can keep shining as long as nuclear or gravitational energy keeps being released by activity in the star's highly compressed central regions. See also **binary star, brown dwarf, evolved star, flare star, giant star, main sequence star, old star, red giant, variable star, young star.**

statistical reasoning: Reasoning based on the probabilities of randomly occurring events in a system of known structure.

superluminal source: A source whose components are expanding at a rate appearing to be faster than the speed of light.

superluminal velocities: Speeds that appear to be greater than the speed of light.

supermassive black holes: Black holes in the nuclei of galaxies whose masses may reach $\geq 10^{10} M_\odot$.

supernova: A star whose luminosity can increase by a factor of $\sim 10^8$ over a period of hours or days as the star explodes. The most luminous supernovae, termed hypernovae, are the most luminous individual stars known. The supernova bright phase declines over a period of months. See also **hypernova.**

surface brightness: The energy emanating from each unit of area in a second.

thermonuclear conversion: The conversion of hydrogen into helium, or more generally of any element into another, at temperatures sufficiently high for nuclear reactions to take place. The nuclear reactions in a star's interior liberate energy that ultimately is emitted at the star's surface as starlight.

time dilatation: The slowing down of the pace at which a clock will run. Time dilation occurs in bodies that move at velocities close to the speed of light and in bodies placed in the close proximity of massive objects.

ultraviolet radiation: Electromagnetic radiation in a range of wavelengths shorter than those of violet light and ranging in wavelength from ~ 4000 Å down to 100 Å. The human eye does not see (sense) ultraviolet radiation.

uncertainty principle: The uncertainty principle due to Werner Heisenberg states that certain complementary pairs of properties of matter, or radiation, cannot be simultaneously measured with arbitrary accuracy. Thus, the frequency and time of arrival of a particle or photon cannot both be determined beyond a certain level of precision.

Universe: The entire world in which we live; all we can survey. See also **Cosmos**.

variable star: See **Cepheid variable, eruptive variable, flare star, pulsating variable**.

visual binary: A system in which two mutually orbiting stars are sufficiently far apart to be spatially resolved when viewed through a telescope.

watt, W: A unit of power equal to 10^7 erg per second.

wavelength, λ: The distance between successive crests of a wave.

white dwarf: A compact star that has gravitationally contracted to its present size, after exhausting all its sources of thermonuclear energy. Its size is comparable to that of Earth. Its mass is comparable to the Sun's.

X-ray: A photon whose energy is 10^3 to 10^5 times that of visible light.

X-ray background radiation: Diffuse X-radiation from the sky, largely due to X-ray emission from supermassive black holes in galaxy nuclei. Its wavelengths range from $\sim$1000 Å down to 0.1 Å.

X-ray galaxy: A galaxy that emits an appreciable fraction of its energy in the X-ray wavelength range.

X-ray star: A star that emits an appreciable fraction of its energy in the X-ray wavelength range.

young star: See **old star**.

Units and Their Ranges

Very large or very small numbers are expressed in powers of ten. Thus 10^6 stands for 1,000,000, that is, a million, where the exponent 6 indicates the number of zeroes following the number 1. Similarly 10^{-6}, standing for 0.000,001, represents one-millionth, or one part in a million; and 7×10^{-6} is seven parts per million.

TABLE A.1

Relation between Units of Length

Unit	Length in Centimeters	Length in Meters
Angstrom unit, Å	10^{-8}	10^{-10}
Micrometer, μm	10^{-4}	10^{-6}
Millimeter, mm	10^{-1}	10^{-3}
Centimeter, cm	1	10^{-2}
Meter, m	10^2	1
Kilometer, km	10^5	10^3
Light year, ly	9×10^{17}	9×10^{15}
Parsec, pc	3×10^{18}	3×10^{16}
Kiloparsec, kpc	3×10^{21}	3×10^{19}
Megaparsec, Mpc	3×10^{24}	3×10^{22}
Gigaparsec, Gpc	3×10^{27}	3×10^{25}

TABLE A.2

Relation between Units of Photon Energy, Wavelength, and Frequency

Unit	Energy (erg)	Photon Wavelength (cm)	Photon Frequency (Hz)
1 electronvolt (eV)	1.6×10^{-12}	1.2×10^{-4}	2.5×10^{14}
1 keV = 10^3 eV	1.6×10^{-9}	1.2×10^{-7}	2.5×10^{17}
1 MeV = 10^6 eV	1.6×10^{-6}	1.2×10^{-10}	2.5×10^{20}
1 GeV = 10^9 eV	1.6×10^{-3}	1.2×10^{-13}	2.5×10^{23}
1 TeV = 10^{12} eV	1.6	1.2×10^{-16}	2.5×10^{26}
1 PeV = 10^{15} eV	1.6×10^3	1.2×10^{-19}	2.5×10^{29}

(Wavelength, λ) × (Frequency, ν) = (Speed of Light, c) : $\lambda\nu = c = 3 \times 10^{10}$ cm/s
(Energy, E) = (Planck's constant, h) × (Frequency ν)

TABLE A.3

Relation between Units of Mass

Unit	Mass in Grams
Microgram, μg	10^{-6}
Milligram, mg	10^{-3}
Gram, g	1
Kilogram, kg	10^3
Metric ton	10^6
Solar Mass, $M_\odot$	2×10^{33}

TABLE A.4

Some Global Cosmological Parameters[a]

Parameter	Value
Contemporary Hubble constant, H_0	70 ± 3 km s^{-1} Mpc^{-1}
Age of the Universe, T	13.8 ± 0.02 Gyr
Redshift at matter–radiation density equality, z_{eq}	3371 ± 23
Recombination redshift, z_{rec}	1090 ± 0.3
Re-ionization redshift, z_{re}	8.8 ± 1.2
Helium ^{4}He mass fraction	0.250 ± 0.003
Number of neutrino species, N_{eff}	3.0 ± 0.3
Summed neutrino mass of all species, Σm_ν	$< 0.3.5 \times 10^{-34}$ g
Scalar fluctuation spectral index, n_s	0.967 ± 0.004
Galaxy density fluctuation amplitude, σ_8	0.82 ± 0.01

[a] Based on data from the WMAP (2011) and Planck (2015) Collaborations (see Chapter 1 notes 46 and 47).

TEMPERATURES IN THE UNIVERSE

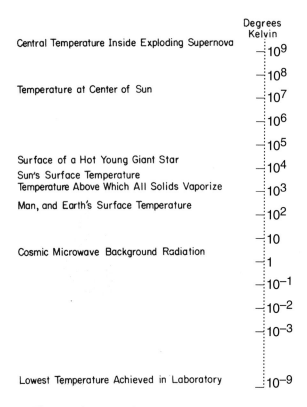

Fig. A.1 The range of temperatures encountered in the Cosmos

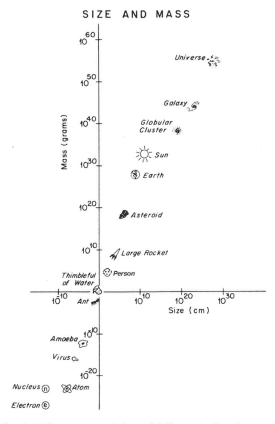

Fig. A.2 The masses and sizes of different bodies shown on a logarithmic scale, expressed in the powers of 10 by which they exceed that of a cubic centimeter, roughly a thimbleful, of water. The range stretches from the electron, whose mass is $\sim 10^{-27}$ g and whose radius is $\sim 10^{-13}$ cm, to the visible part of the Universe with mass $\sim 10^{55}$ g and size $\sim 10^{28}$ cm.

Index